Numerical Solution of Nonlinear Boundary Value Problems with Applications

Milan Kubíček
Department of Chemical Engineering
Prague Institute of Chemical Technology

Vladimír Hlaváček
Department of Chemical Engineering
State University of New York at Buffalo

Dover Publications, Inc.
Mineola, New York

Bibliographical Note

This Dover edition, first published in 2008, is an unabridged
republication of the work originally published by Prentice-Hall,
Inc., Englewood Cliffs, New Jersey, in 1983.

Library of Congress Cataloging-in-Publication Data

Kubíček, Milan.
    Numerical solution of nonlinear boundary value problems
with applications / Milan Kubíček and Vladimír Hlaváček.
        p. cm.
    Includes bibliographical references and index.
    ISBN-13: 978-0-486-46300-1
    ISBN-10: 0-486-46300-1
    1. Originally published: Englewood Cliffs, N.J. : Prentice-Hall,
1983. 2. Nonlinear boundary value problems. 3. Numerical calculations. 4. Engineering mathematics—Formulae. I. Hlaváček,
Vladimír, Ing. II. Title.

TA347.B69K8 2008
515'.355—dc22

                                                    2007037259

Manufactured in the United States of America
Dover Publications, Inc., 31 East 2nd Street, Mineola, N.Y. 11501

Contents

Preface

During the past decade there has been a remarkable growth of interest in problems associated with systems of linear and nonlinear ordinary differential equations with split boundary conditions. Throughout engineering and applied science, we are confronted with nonlinear two-point boundary value problems that cannot be solved by analytical methods. With this interest in finding solutions to particular nonlinear boundary value problems has come an increasing need for techniques capable of rendering relevant profiles. Although considerable progress has been made in developing new and powerful procedures, notably in the fields of fluid and celestial mechanics, and chemical and control engineering, much remains to be done. It is apparent that although physical models of boundary value type are evident in many branches of modern engineering and applied science, the application of methods has remained largely within the sphere of chemical engineering. On the other hand, in this text we do not overlook the importance of physical systems that lie outside the realm of chemical engineering: for example, orbital mechanics, theory of elasticity, and mathematical biology.

This book is concerned with the development, analysis, and practical application of various numerical techniques that can be adapted successfully for the solution of nonlinear boundary value problems. One cannot expect a particular technique to be superior to others for all problems. We have tried to present an account of what has been accomplished in the field. Accordingly, it seemed appropriate to shape this text to those interested in numerical analysis as a working tool for physicists and engineers. Our emphasis is on description and straightforward application of numerical techniques without presenting in detail the underlying theory. The theory selected reflects our own interest and experience with the application of diverse numerical algorithms. We believe that the techniques described in this book will provide investigators with tools that will permit them to solve difficult problems in modern engineering, applied science, and other fields.

It is assumed that readers are acquainted with numerical analysis to the extent that it is taught in the usual engineering courses. They also must have some experience with applied analysis and programming.

Many people encouraged us in the writing of this text. We gratefully acknowledge the assistance of N. R. Amundson, R. Aris, H. Hofmann, M. Holodniok, M. Marek, M. G. Slinko, and J. Villadsen.

MILAN KUBÍČEK
Prague, Czechoslovakia

VLADIMÍR HLAVÁČEK
Buffalo, New York

Numerical Solution of Nonlinear Boundary Value Problems with Applications

Occurrence and Solution of Nonlinear Boundary Value Problems in Engineering and Physics

1

1.1 Occurrence of Nonlinear Problems for Ordinary Differential Equations

There are a large number of problems in engineering and physics that can be described through the use of nonlinear ordinary differential equations. When the (boundary) conditions, which together with the differential equations describe the behavior of a particular physical system, are determined at various points, the resulting problem is referred to as a boundary value problem.

The boundary conditions may be classified according to various criteria: (1) linear–nonlinear boundary conditions, (2) separated conditions–mixed conditions, (3) two-point–multipoint problems, and so on.

A great number of nonlinear boundary value problems are represented by equations of diffusional type. Here the boundary conditions result after specification of dependent variables (or fluxes) at the boundary of the system. These conditions are usually of the separated type; that is, at a given value of the independent variable, the values of the dependent variables, derivatives, or a combination of both are prescribed. The boundary conditions for diffusion problems may be of nonlinear type, especially for radiation problems. If there is an external relation between boundary conditions (e.g., recycle problems in chemical engineering), the boundary conditions are of the mixed type; that is, for a given value of the independent variable, a combination of dependent variables (or derivatives) with different arguments results. Sometimes the boundary conditions may be given in integral form; for example, if the total amount of heat transferred is specified.

In problems of mechanics, multipoint boundary value problems occur; for example, for a multibody system, the velocity at different points may be specified.

A number of nonlinear boundary value problems result after the formulation of a model for a particular physical situation. Examples include diffusion occurring in the presence of an exothermic chemical reaction [1], heat conduction associated with radiation effects [2], deformation of shells [3], and so on†. For these examples, the nonlinear equations represent the true physical situation. However, there are a number of nonlinear boundary value problems which result after certain mathematical transformations. To illustrate this family of equations, boundary layer problems will be presented [4]. Although the flow of a viscous fluid is described by rather complicated nonlinear partial differential equations (Navier–Stokes equations), certain transformations make it possible to convert them to nonlinear ordinary differential equations (boundary value problems). The new dependent variables include some dependent and independent variables occurring in the original problem. Whereas solution of the original Navier–Stokes equations represents a difficult numerical problem, the transformed equations (i.e., the boundary layer equations) are more conducive to numerical treatment. Similar transformations may be used to convert the nonlinear parabolic equations to a nonlinear boundary value problem. Sometimes this transformation is referred to in the literature as the Boltzmann transformation [5]. Finally, the nonlinear boundary value problem for ordinary differential equations results after proper discretization of nonlinear elliptic partial differential equations with two independent variables [6, 7]. A specific group of problems is created by the family of optimization problems which are required to establish the "optimum profiles." The equations formulated by use of the Euler–Jacobi variational equations, as well as by the Pontrjagin's maximum principle, belong to this family [8].

1.2 Existence of a Solution

Generally speaking, for a nonlinear boundary value problem it is difficult to prove rigorously the existence of a solution. However, engineers and physicists are more interested in finding in a numerical way a region of parameters where the given nonlinear boundary value problem does not exhibit a solution. Fortunately, for a great number of correctly formulated nonlinear boundary value problems there exists at least one solution. Nevertheless, there are physical problems which for particular values of governing parameters do not possess a solution. For instance, the diffusion equation incorporating a strongly exothermic reaction of zero order need not always exhibit a solution [1]. Another example is represented by the boundary layer equations describing spiral flow in a porous pipe [9].

---

†Bracketed arabic numbers throughout refer to references at chapter end.

1.3 Problems of a Multiplicity of Solutions

There are a number of nonlinear boundary value problems that may exhibit more than one solution. It is a difficult mathematical problem to investigate the domain in which multiple solutions may occur. From the physical point of view, a strong exothermic or autocatalytic reaction, radiation effect, or other feedback mechanism is responsible for multiple steady states of a particular physical model. Table 1-1 surveys some physical models represented by nonlinear boundary value problems for ordinary differential equations that may exhibit multiple solutions.

1.4 Nonlinear Phenomena

Nonlinearities occurring in boundary value problems are caused by a number of different physical effects. In chemical engineering problems, the following nonlinear phenomena are frequently encountered:

1. Chemical reactions
2. Adsorption phenomena
3. Volume change resulting from the mole change accompanying a chemical reaction
4. Radiation effects and problems connected with nonlinear heat transfer
5. Dependence of the rate, equilibrium, and transport coefficients on concentration and temperature
6. Dissipation of energy
7. Flow of non-Newtonian fluids
8. Gravitation and Coulomb forces

Nonlinearities caused by a chemical reaction may be divided into two major groups: (1) concentration dependences and (2) temperature dependences. The first- and zero-order reaction-rate expressions are the linear relations that occur in transport equations; all other reaction-rate expressions are of the nonlinear type. For instance, an esterification reaction occurring in a liquid phase is represented by a second-order reversible reaction (i.e., the nonlinearity is of quadratic type). For catalytic reactions the reaction-rate expression is of the Langmuir-Hinshelwood type [17] (rational function) or of integer power form. In the realm of bioengineering the reaction-rate expressions for an enzymatic reaction, which are formally equivalent to the Langmuir–Hinshelwood expressions, are referred to as Michaelis–Menten kinetics [18]. The temperature dependences are always nonlinear; for example, the reaction-rate or adsorption constants are exponentials:

$$k = k_0 \exp\left(\pm \frac{E}{RT}\right)$$

Here minus is for the reaction-rate expression, while plus must be used for

Problem	Equations	Number of Solutions
Diffusion and exothermic zero-order reaction in a slab [1]	$y'' = -\delta e^y$ $y'(0) = 0, \quad y(1) = 0$	$0 = \delta_0 < \delta < \delta^*$: two solutions $\delta > \delta^*$: no solution
Diffusion and exothermic first-order reaction in a slab [10]	$y'' = \phi^2 y \exp\left[\dfrac{\gamma\beta(1-y)}{1+\beta(1-y)}\right]$ $y'(0) = 0, \quad y(1) = 1$	$\gamma\beta > \dfrac{4\gamma}{\gamma-4}, \quad \phi_1 < \phi$ $< \phi_2$: three solutions Outside this region: one solution
Diffusion and exothermic first-order reaction in a sphere [11, 12]	$y'' + \dfrac{2}{x}y' = \phi^2 y \times$ $\exp\left[\dfrac{\gamma\beta(1-y)}{1+\beta(1-y)}\right]$ $y'(0) = 0, \quad y(1) = 1$	Up to 15 solutions have been established
Diffusion, convection, and isothermic reaction with an adsorption kinetic term [13]	$\dfrac{1}{\text{Pe}}y'' - y' - \text{Da}\left(\dfrac{1+B}{1+By}\right)^2 y \times$ $\dfrac{y+C}{1+C} = 0$ $y'(0) = \text{Pe}[y(0) - 1],$ $\quad y'(1) = 0$	For certain values of the governing parameters, three solutions exist for $\text{Da}_1 < \text{Da} < \text{Da}_2$ Outside this region: one solution
Diffusion, convection, and exothermic reaction occurring in a tubular reactor [14]	$\dfrac{1}{\text{Pe}}y'' - y' + \text{Da}(1-y) \times$ $\exp\left(\dfrac{\theta}{1+\theta/\gamma}\right) = 0$ $\dfrac{1}{\text{Pe}}\theta'' - \theta' - \beta(\theta - \theta_c)$ $+ B\,\text{Da}(1-y) \times$ $\exp\left(\dfrac{\theta}{1+\theta/\gamma}\right) = 0$ $y'(1) = \theta'(1) = 0$ $y(0) = \dfrac{1}{\text{Pe}}y'(0),$ $\theta(0) = \dfrac{1}{\text{Pe}}\theta'(0)$	Up to five solutions have been established
Equilibrium of suspended charged drops [15]	$y'' + \dfrac{1}{x}y' = \dfrac{\beta}{y^2}$ $y'(0) = 0, \quad y(1) = 1$	For $\beta < 0.42$: one solution For $0.42 < \beta < 0.78$: two or three solutions
Flow between two rotating disks [16]	$F'' = \sqrt{\text{Re}}\,HF$ $+ \text{Re}(F^2 - G^2 + k)$ $G'' = 2\,\text{Re}\,FG + \sqrt{\text{Re}}\,G'H$ $H' = -2\sqrt{\text{Re}}\,F$ $F(0) = F(1) = H(0) = H(1)$ $\quad = 0$ $G(0) = 1, \quad G(1) = s$	For greater values of Re, more than ten solutions have been found

adsorption effects [17]. Of course, this exponential dependence is the main source of numerical difficulties accompanying the particular physical problem.

Another source of nonlinearities are the adsorption processes since the majority of adsorption isotherms are represented by rational functions. Evidently, as a result, diffusion problems in which adsorption phenomena must be considered represent a nonlinear boundary value problem. One should notice that the expression describing the rate of growth of microorganisms (Monod expression) is formally identical to the Langmuir adsorption isotherm [18].

A chemical reaction may be accompanied by expansion or contraction phenomena. For instance, diffusion, convection, and second-order reactions are described by a simple differential equation,

$$\frac{dy}{dx} = L \frac{d^2y}{dx^2} + ky^2$$

while for the volume-change case the equation is more complicated:

$$\frac{dy}{dx} = L \left[\frac{1}{1 + E(1 - y)} \frac{d^2y}{dx^2} + \frac{E}{[1 + E(1 - y)]^2} \left(\frac{dy}{dx}\right)^2 \right] + ky^2$$

Here L and E are the parameters describing dispersion and volume change, respectively [19]. We may note that both reaction and adsorption give rise to a nonlinear "source term," while the volume-change effect results in nonlinear derivatives or nonlinear coefficients.

Among the most complicated nonlinear problems are those connected with radiation. The heat flux caused by radiation may be written

$$q = \sigma(T^4 - T_0^4)$$

This relation can be incorporated into transport equations in different ways. For instance, for tube radiation through a jacket, the nonlinear radiation term is a part of the differential equation [20], while for radiation by means of the inlet or outlet surface of the tube, the radiation effect is a part of the boundary conditions. There are very complicated physical situations in which the radiation effect is accompanied by a chemical reaction or by a velocity distribution (radiation boundary layer) [21]. For some extreme situations Newton's law for convective transfer does not describe properly the typical features of the process: for instance, convective heat transfer to boiling helium is described by a cubic law [22]:

$$q = \alpha(T - T_0)^3$$

If the concentration or temperature dependences of rate, equilibrium, or transport coefficients are considered in the physical model, nonlinearities in derivatives and coefficients result (see, e.g., [23]). For small variations of concentration or temperature for a particular physical process, average values of coefficients may be used; however, for problems with extreme variations of the order of 1000°C (e.g., supersonic flow of real gases around blunt objects), the temperature dependences of coefficients must be considered [24]. The average values of coefficients cannot be used if jump variations can be expected; for example, as a result of higher temperatures, recrystallization may occur, which can cause an essential change in thermal conductivity. However, the most practical problem is the temperature and concentration dependence of viscosity [25].

Fluids in which the vector of the shear stress τ and the gradient of velocity dy/dx is not linearly proportional are usually referred to as non-Newtonian fluids. A number of semiempirical nonlinear relations used in the literature are presented in Table 1-2 [26, 27]. Problems that are linear for Newtonian fluids may be strongly nonlinear for non-Newtonian fluids.

If energy dissipation is considered in hydrodynamic problems, then in Fourier–Kirchhoff convective heat-transfer equations, a new term occurs which is proportional to the second power of the gradient of velocity. Of course, this term must be considered for those problems in which extreme gradients result, which, however, can also be the case for relatively slow flow conditions.

Finally, the gravitational or Coulomb forces should be mentioned.

TABLE 1-2

SOME SEMIEMPIRICAL RELATIONS BETWEEN VECTOR
OF SHEAR STRESS τ AND GRADIENT OF VELOCITY

τ	Model
$-m \left\| \dfrac{dv}{dx} \right\|^{u-1} \dfrac{dv}{dx}$	Ostwald–de Waele
$A \operatorname{arcsinh} \left(-\dfrac{1}{B} \dfrac{dv}{dx} \right)$	Eyring
$-(\varphi_0 + \varphi_1 \mid \tau \mid^{\alpha-1})^{-1} \dfrac{dv}{dx}$	Ellis
$a \left[1 - \exp \left(-\alpha \dfrac{dv}{dx} \right) \right]$	Taganov

1.5 Types of Nonlinear Boundary Value Problems in Chemical Engineering

1. Calculation of chemical engineering equipment
   (a) Calculation of temperature profiles in countercurrent heat exchangers if the heat-transfer coefficients are dependent on temperature.
   (b) Calculation of concentration and temperature profiles in countercurrently operated packed mass exchangers (rectification columns, absorbers) [28].

(c) Calculation of concentration profiles in extraction columns [29].

(d) Calculation of concentration and temperature profiles in tubular reactors with countercurrent cooling.

(e) Calculation of concentration and temperature profiles in tubular recycle reactors [30].

(f) Calculation of concentration and temperature profiles in tubular reactors if the inlet concentration and the amount of heat transferred are specified.

(g) Calculation of concentration and temperature profiles for a tubular reactor with an external heat exchanger [31].

(h) Calculation of concentration and temperature for a tubular reactor with an internal heat exchanger [31].

(i) Calculation of concentration and temperature profiles for complex configuration "reactor-heat exchangers": for instance, various types of ammonia reactors (TVA, Casal, Quench, NEC, Fauser-Montecatini, OSAG, Haldor-Topsoe, etc.) [32].

(j) Design of catalytic convertors with a short bed [33].

2. Chemical reaction engineering problems

(a) Mass transfer and effectiveness factor evaluation for an isothermal catalytic reaction occurring on a single pellet having nonlinear kinetics [34].

(b) Heat and mass transfer and effectiveness factor evaluation for a nonisothermal catalytic reaction occurring on a single porous pellet [10, 11].

(c) Calculation of the characteristics of a laminar flame [35].

(d) Calculation of the critical dimensions of an explosive sample [1].

(e) Absorption and chemical reactions with nonlinear kinetics (or nonisothermal) occurring in a liquid film [36].

(f) Heat transfer in a nonporous catalytic particle.

(g) Transport phenomena and substrate consumption in a microbial film [37].

(h) Calculation of optimum temperature and pressure profiles in tubular reactors [38].

(i) Calculation of optimum "catalyst profiles" in tubular reactors [38].

(j) Calculation of boundary layer problems with a chemical reaction [39].

3. Problems of heat and mass transfer

(a) Heat and mass transfer in a plate, cylinder, and sphere if thermal conductivity is dependent on temperature [40] and diffusivity is dependent on concentration [41].

(b) Radiation problems (e.g., radiation of a sphere or a fin) [20].

(c) Solution of combined conduction–radiation problems in an optically thick medium [42].

(d) Heat and mass transfer in boundary layer problems (natural convection from a vertical wall [43], transpiration cooling [44], massive blowing [45], radiation in the boundary layer at subsonic and supersonic velocities [46], thermal boundary layer on rotating bodies [4], heat transfer in a compressible boundary layer [4], etc.).

1.6 Types of Nonlinear Boundary Value Problems in Physics

1. Problems of hydrodynamics and hydrostatics
   (a) Flow of non-Newtonian fluids on a vertical plate (or Newtonian fluids with variable properties).
   (b) Calculation of the shape of velocity profiles in a non-Newtonian fluid flowing between two rotating cylinders.
   (c) Calculation of boundary layer problems for both Newtonian and non-Newtonian fluids (flow on a flat plate [4], flow on a plate with surface curvature [47], flow through a diverging channel [4], flow caused by rotating disks [4], multidimensional flow [44], flow through a porous structure [9, 48], flow of a compressible fluid [4], problems of a supersonic boundary layer [4], problems of a non-steady-state boundary layer [49]), magnetohydrodynamic hypersonic flow [50].
   (d) Effect of fluid motion on a free surface shape [51].
2. Problems of electrodynamics and electrostatics
   (a) Equilibrium of suspended electrically charged drops [15].
   (b) Breakdown of dielectrics [52].
   (c) Ionic boundary layers [53].
   (d) Problems of calculation of photoionization chambers [54].
   (e) Calculation of semiconductor-device current characteristics [55, 56].
3. Problems of magnetohydrodynamics and plasma theory
   (a) Problems leading to the magnetohydrodynamic boundary layer (Hartmann flow) [57].
   (b) Diffusion of magnetic field into plasma [58].
   (c) Problems of plasma radiation (Troesch equation) [59, 60].
   (d) Problems of radiative magnetohydrodynamic channel flow [61].
   (e) Calculation of the effect of a cold wall on a hot plasma [62].
   (f) Calculation of the flow of weakly ionized gases [63].
4. Problems of classical mechanics: theory of elasticity
   (a) Calculation of N-body trajectories [64, 68].
   (b) Nonlinear oscillations.
   (c) Deformation of thin shells [66].
   (d) Finite bending of thin-walled tubes [65].
   (e) Stress analysis of solid propellant grains [67].

5. Problems of astrophysics
   (a) Relativistic boundary layer.
   (b) Chandrasekhar's theory of gravitational equilibrium of stellar structures [69].

REFERENCES

[1] FRANK-KAMENETSKII, D. A.: *Diffusion and Heat Transfer in Chemical Kinetics*, 2nd ed., Plenum Press, New York, 1969 (translated by J. P. Appleton).

[2] CHEN, J. L. P., AND CHURCHILL, S. W.: *Combust. Flame 18*, 27 (1972).

[3] ARCHER, R. R.: *J. Math. Phys. 41*, 165 (1962).

[4] LOITSIANSKI, L. G.: *Laminare Grenzschichten*, Akademie Verlag, Berlin, 1967.

[5] DRAKE, R. L., AND ELLINGTON, M. B.: *J. Comp. Phys. 6*, 200 (1970).

[6] KELSEY, S. J.: The application of quasilinearization to nonlinear systems, *73rd National Meeting of AICHE*, Minneapolis, Minn., 1972.

[7] JONES, D. J.: *Comp. Phys. Commun. 4*, 165 (1972).

[8] LEE, E. S.: *Quasilinearization and Invariant Imbedding*, Academic Press, New York, 1968.

[9] TERRILL, R. M., AND THOMAS, P. W.: *Phys. Fluids 16*, 356 (1973).

[10] MAREK, M., AND HLAVÁČEK, V.: *Sci. Papers Inst. Chem. Technol., Prague*, K3, 19 (1969).

[11] WEISZ, P. B., AND HICKS, J. S.: *Chem. Eng. Sci. 17*, 265 (1962).

[12] COPELOWICZ, I., AND ARIS, R.: *Chem. Eng. Sci. 25*, 885 (1970).

[13] ROBERTS, G. W., AND SATTERFIELD, C. N.: *IEC Fundam. 5*, 317 (1966).

[14] HLAVÁČEK, V., HOFMANN, H., AND KUBÍČEK, M.: *Chem. Eng. Sci. 26*, 1629 (1971).

[15] ACKERBERG, R. C.: *Proc. Soc. Lond., Ser. A 312*, 129 (1969).

[16] HOLODNIOK, M., KUBÍČEK, M., AND HLAVÁČEK, V.: *J. Fluid. Mech., 81*, 689 (1977), *108*, 227 (1981).

[17] HOUGEN, O. A., AND WATSON, K. M.: *Chemical Process Principles*, J. Wiley, New York, 1945.

[18] AIBA, S., HUMPHREY, A. E., AND MILLS, N. F.: *Biochemical Engineering*, Academic Press, New York, 1965.

[19] DOUGLAS, J. M., AND BISCHOFF, K. B.: *IEC Proc. Des. Dev. 3*, 130 (1964).

[20] COBBLE, M. H.: *J. Frankl. Inst. 227*, 206 (1964).

[21] CHENG, E. H., ÖZISIK, M. N.: *Appl. Sci. Res. 28*, 185 (1973).

[22] LYON, D. N.: *Adv. Cryog. Eng. 10*, 371 (1965).

[23] WEEKMAN, V. W., JR.: *J. Catal. 5*, 44 (1966).

[24] BADE, W. L.: *Phys. Fluids 5*, 150 (1962).

[25] CHENG, S. C., BIRTA, L. G., AND KUMAR, A.: *Simulation 100* (September 1972).

[26] BIRD, R. B., STEWART, W. E., AND LIGHTFOOT, B. N.: *Transport Phenomena*, J. Wiley, New York, 1960.

[27] SCHLICHTLING, H.: *Boundary Layer Theory*, McGraw-Hill, New York, 1960.

[28] STEWART, G., AND BEVERIDGE, G. S. G.: *IFAC Symp. Digital Simul. Continuous Processes*, Györ, Hungary, 1971.

[29] MACKLENBURGH, J. C., AND HARTLAND, S.: *Can. J. Chem. Eng.* 47, 453 (1969).
[30] REILLY, M. J., AND SCHMITZ, R. A.: *AICHE J.* 12, 153 (1966).
[31] CAHA, J., HLAVÁČEK, V., AND KUBÍČEK, M.: *Chem. Ing. Tech.* 45, 1308 (1973).
[32] KJAER, J.: *Calculation of Ammonia Convertors on an Electronic Digital Computer*, Akademisk Forlag, Copenhagen, 1963.
[33] HLAVÁČEK, V., AND HOFMANN, H.: *Chem. Eng. Sci.* 25, 173 (1970).
[34] EMIG, G., AND HLAVÁČEK, V.: Diffusion und Reaktion in porösen Kontakten, *Fortschr. Chem. Forsch.* 13, 3/4 (1970).
[35] KIRKBY, L. L., AND SCHMITZ R.: *Combust. Flame 10*, 205 (1966).
[36] ONDA, K., et al.: *Chem. Eng. Sci.* 25, 753, 761, 1023 (1970).
[37] ATKINSON, B., AND DAOUD, I. S.: *Trans. ICHE 46*, T19 (1968).
[38] GUNN, D. J., AND THOMAS, W. J.: *Chem. Eng. Sci.* 20, 89 (1965).
[39] LEVINSKY, S. S., AND BRAINERD, J. J.: *IAS Paper 63–63* (1963).
[40] COHEN, D. S., AND SHAIR, F. H.: *Int. J. Heat Mass Transfer 13*, 1375 (1970).
[41] RUTHEVEN, D. M.: *J. Catal.* 25, 259 (1972).
[42] HEINISCH, R. P., AND VISKANTA, R.: *AIAA J.* 6, 1409 (1968).
[43] POTTER, W. A., AND THRONE, J. L.: *Simulation 129* (March 1970).
[44] JENG, D. R., AND WILLIAMS, D. W.: *AIAA J.* 11, 1560 (1973).
[45] LIN, M. T., AND NACHTSHEIM, P. R.: *AIAA J.* 11, 1584 (1973).
[46] HOSHIZAKI, H., AND WILSON, K. H.: *AIAA J.* 3, 1614 (1965).
[47] MURPHY, J. S.: *AIAA J.* 3, 2043 (1965).
[48] TERRILL, R. M., AND THOMAS, P. W.: *Appl. Sci. Res.* 21, 37 (1969).
[49] ROY, S.: *AIAA J.* 11, 1581 (1973).
[50] SMITH, M. C., AND CHING-SHENG WU: *AIAA J.* 2, 963 (1964).
[51] CHIN, J. H., AND GALLAGHER, L. W.: *AIAA J.* 2, 2215 (1964).
[52] COPPLE, J.: *J. Inst. Electr. Eng. (Lond.)* 85, 56 (1939).
[53] CHUNG, P. M.: *AIAA J.* 3, 817 (1965).
[54] WARDLAW, A. B., AND COHEN, I. M.: *Phys. Fluids 16*, 637 (1973).
[55] SEIDMAN, T. I., AND CHOO, S. C.: *Solid State Electron.* 15, 1229 (1972).
[56] DE MARI, A.: *Solid State Electron.* 11, 33 (1968).
[57] BRANDT, A., AND GILLIS, J.: *J. Comput. Phys.* 3, 523 (1969).
[58] FALK, T. J., AND TURCOTTE, D. C.: *Phys. Fluids 5*, 1288 (1962).
[59] TROESCH, B. A.: *Intrinsic Difficulties in the Numerical Solution of a Boundary Value Problem.* Rept. NN-142 (1960), TRW, Inc., Redondo Beach, Calif.
[60] KUBÍČEK, M., AND HLAVÁČEK, V.: *J. Comp. Phys.* 17, 95 (1975).
[61] HELLIWELL, J. B.: *J. Eng. Math.* 7, 347 (1973).
[62] CHU, M. S.: *Phys. Fluids 16*, 1441 (1973).
[63] CHUNG, P. M., HOLT, J. F., AND LIN, S. W.: *AIAA J.* 6, 2372 (1968).
[64] LANCASTER, J. E.: *AIAA Paper No. 70–1060* (August 1970).
[65] PERRONE, N., AND KAO, R.: *J. Appl. Mech., Trans. ASME 38*, Ser. E, 371 (1971).
[66] THURSTON, G. A.: *J. Appl. Mech.* 28, 557 (1961).
[67] PISTER, K. S., AND EVANS, R. J.: *AIAA J.* 4, 1914 (1966).
[68] LONG, S. L.: *AIAA J.* 3, 1937 (1965).
[69] CHANDRASEKHAR, S.: *Introduction to the Study of Stellar Structure*, University of Chicago Press, Chicago, 1939.

Initial and Boundary Value Problems for Ordinary Differential Equations, Solution of Algebraic Equations **2**

In this chapter initial and boundary value problems are formulated and differences between both formulations are discussed. For initial value problems, numerical methods are presented since numerical solution of initial value problems is important to get a solution of boundary value problems. In this chapter problems of numerical solution of linear and nonlinear algebraic equations are discussed since numerical techniques for the solution of boundary value problems are often requir⌄d to solve algebraic equations. The unicity and existence of a solution for a boundary value problem are illustrated by means of examples. Since this chapter presents supplementary material, we are not going to go into details; a reader who wishes detailed information is referred to the Bibliography at the end of the chapter.

2.1 Numerical Solution of Initial Value Problems

The set of first-order differential equations

$$\frac{dy_1}{dx} = f_1(x, y_1, \ldots, y_n)$$
$$\cdot$$
$$\cdot \qquad\qquad\qquad\qquad\qquad (2.1)$$
$$\cdot$$
$$\frac{dy_n}{dx} = f_n(x, y_1, \ldots, y_n)$$

represents an initial value problem if n necessary conditions are specified at a single point a; for example,

$$y_1(a) = y_{1,0}, \ldots, y_n(a) = y_{n,0} \tag{2.2}$$

Equations (2.1) and (2.2) may be rewritten in vector form ($' = d/dx$):

$$y' = f(x, y) \tag{2.1'}$$

$$y(a) = y_0 \tag{2.2'}$$

A detailed analysis of the existence and unicity of solutions to (2.1) and (2.2) can be found in textbooks on differential equations.

Formulation of the problem in the form (2.1) and (2.2) is rather general since for an arbitrary differential equation of higher order, which can be explicitly solved with respect to the highest derivative, that is,

$$y^{(k)} = F(x, y, y', y'', \ldots, y^{(k-1)}) \tag{2.3}$$

the form given by (2.1) can easily be obtained ($y_1 \equiv y$):

$$y_1' = y_2, \qquad y_2' = y_3, \ldots, y_{k-1}' = y_k, \qquad y_k' = F(x, y_1, y_2, \ldots, y_k) \tag{2.4}$$

The initial conditions [the given conditions are $y(a), y'(a), \ldots, y^{(k-1)}(a)$] can be rewritten in an analogous way.

In the last few decades a number of powerful algorithms for solution of initial value problems have been developed. Frequently, we try to replace the solution of a boundary value problem by a solution of certain initial value problems (see Sections 3.2–3.5, 4.6, 5.1, 5.5, 5.6). For this reason a short summary of reliable methods for the solution of initial value problems is presented next.

2.1.1 A Short Summary of Commonly Used Methods

Frequently, three groups of methods are used: (1) one-step methods, (2) multistep methods, and (3) extrapolation methods.

A solution $y(x)$ is required for a set of mesh points $x_0 = a, x_1, x_2, \ldots$. Usually, an equidistant mesh is considered (i.e., $x_{i+1} - x_i = h$). The values $y_1, y_2, \ldots$ that are to be calculated approximate the solution at the mesh points [i.e., $y(x_1), y(x_2), \ldots$]. Evidently, $y_0 = y(a)$ is given as an initial condition. The difference

$$e_i = y_i - y(x_i) \tag{2.5}$$

will be referred to as the error of approximation. We assume that the values y_i are not subjected to round-off error. For a given value of $x = b$ and a constant integration step h, the accuracy obtained, and also $e_N[N = (b - a)/h]$, depend on the value of step h. For a particular method the order of approximation p may be defined as

$$e_N = O(h^p), \qquad N = \frac{b - a}{h} \tag{2.6}$$

An important part of a numerical algorithm for the solution of initial value problems is an estimation of the error e_N and thus determination of the step size h that is necessary to reach the preassigned accuracy of the solution. A priori estimates of the error for a particular method are usually complicated and pessimistic. In practical calculations, however, a posteriori estimates combined with automatic step-size adjustment are employed. A posteriori estimate of an error may be accomplished by the Richardson extrapolation or deferred approach to the limit. For methods of pth order, the error of approximation can be written in the form

$$e = Ch^p + O(h^{p+1}) \tag{2.7}$$

Let us investigate the asymptotic behavior of the error e for $h \to 0+$. It can be shown that the term $O(h^{p+1})$ decreases rapidly and can be neglected. Based on the results Q_1 and Q_2 (approximations of the solution $y(b)$ for two different step sizes h [h_1 and h_2, respectively ($h_1 \neq h_2$)], an essentially improved value of $Q = y(b)$ can be obtained. For step size h_1, we have

$$Q_1 \doteq Q + Ch_1^p \tag{2.8a}$$

and for h_2,

$$Q_2 \doteq Q + Ch_2^p \tag{2.8b}$$

Solving (2.8a) and (2.8b) for two unknowns Q and C, we get

$$Q \doteq \frac{(h_1/h_2)^p Q_2 - Q_1}{(h_1/h_2)^p - 1} \tag{2.9}$$

Usually, $h_1/h_2 = 2$ is chosen. For this ratio the following extrapolation formulas result:

$$Q \doteq 2Q_2 - Q_1 \qquad p = 1 \tag{2.10a}$$

$$Q \doteq \tfrac{4}{3}Q_2 - \tfrac{1}{3}Q_1 \qquad p = 2 \tag{2.10b}$$

$$Q \doteq \tfrac{16}{15}Q_2 - \tfrac{1}{15}Q_1 \qquad p = 4 \tag{2.10c}$$

The value of Q is by an order of magnitude better than the original results. The differences $Q_1 - Q$ and $Q_2 - Q$ represent a good approximation of the error of Q_1 and Q_2, respectively.

2.1.2 One-Step Methods: Runge–Kutta Methods

The one-step methods make use of already calculated approximation $\bar{y} \doteq y(\bar{x})$ and calculate an approximation of the solution at the point $\bar{x} + b$ [i.e., $\hat{y} \doteq y(\bar{x} + b)$]. These types of methods are represented by the Runge–Kutta methods.

For a v-step Runge–Kutta process, we may write

$$g_i = f(\bar{x} + c_i h, \bar{y} + h \sum_{j=1}^{v} a_{ij} g_j) \qquad i = 1, 2, \ldots, v \tag{2.11}$$

and

$$\hat{y} = \bar{y} + h \sum_{i=1}^{v} b_i g_i \tag{2.12}$$

Recently, Fehlberg[†] has developed methods by which the local testing error (which is used for automatic step-size adjustment) may be estimated by means of the relation

$$\text{TE} = \hat{y} - \bar{\hat{y}} \tag{2.13}$$

Here for $\bar{\hat{y}}$ can be written

$$\bar{\hat{y}} = \bar{y} + h \sum_{i=1}^{\nu} b'_i g_i \tag{2.14}$$

The coefficients a_{ij}, b_i, b'_i, and c_i are numerical constants that characterize the particular procedure. Such a procedure can be determined by an array of the coefficients:

$$
\begin{array}{c|cccc}
c_1 & a_{11} & a_{12} & \ldots & a_{1\nu} \\
c_2 & a_{21} & a_{22} & \ldots & a_{2\nu} \\
\cdot & \cdot \\
\cdot & \cdot \\
\cdot & \cdot \\
c_\nu & a_{\nu 1} & a_{\nu 2} & \ldots & a_{\nu\nu} \\
\hline
 & b_1 & b_2 & & b_\nu \\
 & b'_1 & b'_2 & & b_\nu
\end{array}
\tag{2.15}
$$

The procedures for which $a_{ij} = 0$ (for $i < j$) are referred to as semi-implicit; if in addition also $a_{ii} = 0$, the procedure is called explicit. Usually, explicit procedures are used. Cases that do not belong in either of these categories are called implicit.

The coefficients of the common procedures are listed in Table 2-1. The for-

TABLE 2-1
SOME RUNGE–KUTTA METHODS[a]

0	0		
	1		

Euler method
$e = O(h)$

0	0	
1	1	
	0.5	0.5

Improved Euler method
$e = O(h^2)$

0	0	
0.5	0.5	
	0	1

Modified Euler method
$e = O(h^2)$

0	0		
0.5	0.5		
1	−1	2	
	$\frac{1}{6}$	$\frac{2}{3}$	$\frac{1}{6}$

Kutta method
$e = O(h^3)$

0	0			
0.5	0.5			
0.5	0	0.5		
1	0	0	1	
	$\frac{1}{6}$	$\frac{1}{3}$	$\frac{1}{3}$	$\frac{1}{6}$

Standard Runge–Kutta method
$e = O(h^4)$

[a]Blanks in the array of coefficients represent zero.

[†]E. Fehlberg: *Computing 4*, 93 (1969) and *6*, 61 (1970).

TABLE 2-1 Continued

0	0					
0.25	0.25					
0.25	0.125	0.125				
0.5	0	-0.5	1			
0.75	0.1875	0	0	0.5625		
1	$-\dfrac{3}{7}$	$\dfrac{2}{7}$	$\dfrac{12}{7}$	$-\dfrac{12}{7}$	$\dfrac{8}{7}$	
	$\dfrac{7}{90}$	0	$\dfrac{32}{90}$	$\dfrac{12}{90}$	$\dfrac{32}{90}$	$\dfrac{7}{90}$

Butcher method
$e = O(h^5)$

0.5	0.5
	1

Implicit method
$e = O(h^2)$

$0.5 - \dfrac{\sqrt{3}}{6}$	0.25	$0.25 - \dfrac{\sqrt{3}}{6}$
$0.5 + \dfrac{\sqrt{3}}{6}$	$0.25 + \dfrac{\sqrt{3}}{6}$	0.25
	0.5	0.5

Implicit method
$e = O(h^4)$

0	0				
$\dfrac{1}{3}$	$\dfrac{1}{3}$				
$\dfrac{1}{3}$	$\dfrac{1}{6}$	$\dfrac{1}{6}$			
0.5	0.125	0	0.375		
1	0.5	0	-1.5	2	
	$\dfrac{1}{6}$	0	0	$\dfrac{2}{3}$	$\dfrac{1}{6}$
	0.1	0	0.3	$\dfrac{6}{15}$	$\dfrac{1}{5}$

Merson method
$e = O(h^4)$

0	0		
0.5	0.5		
1	$\dfrac{1}{256}$	$\dfrac{255}{256}$	
	$\dfrac{1}{256}$	$\dfrac{255}{256}$	
	$\dfrac{1}{512}$	$\dfrac{255}{256}$	$\dfrac{1}{512}$

Fehlberg method [FEHL 1(2)]
$e = O(h^2)$

0	0			
$\dfrac{1}{4}$	$\dfrac{1}{4}$			
$\dfrac{27}{40}$	$-\dfrac{189}{800}$	$\dfrac{729}{800}$		
1	$\dfrac{214}{891}$	$\dfrac{1}{33}$	$\dfrac{650}{891}$	
	$\dfrac{214}{891}$	$\dfrac{1}{33}$	$\dfrac{650}{891}$	
	$\dfrac{533}{2106}$	0	$\dfrac{800}{1053}$	$-\dfrac{1}{78}$

Fehlberg method [FEHL 2(3)]
$e = O(h^3)$

TABLE 2-1 Continued

0	0				
$\frac{2}{7}$	$\frac{2}{7}$				
$\frac{7}{15}$	$\frac{77}{900}$	$\frac{343}{900}$			
$\frac{35}{36}$	$\frac{805}{1444}$	$-\frac{77{,}175}{54{,}872}$	$\frac{97{,}125}{54{,}872}$		
1	$\frac{79}{490}$	0	$\frac{2175}{3626}$	$\frac{2166}{9065}$	
	$\frac{79}{490}$	0	$\frac{2175}{3626}$	$\frac{2166}{9065}$	
	$\frac{229}{1470}$	0	$\frac{1125}{1813}$	$\frac{13{,}718}{81{,}585}$	$\frac{1}{18}$

Fehlberg method [FEHL 3(4)]
$e = O(h^4)$

mula for $\hat{y}$ was not developed if the row with $b'_1, b'_2, \ldots, b'_v$ is not presented. Figure 2-1 displays a procedure for an automatic step-size adjustment (based on evaluation of TE).

The parameter $p \in (0, 1)$ is chosen as a safety factor in order that the predicted step size would not be immediately unsuccessful. The value $p = 0.8$ works well. It seems reasonable to provide the algorithm with a lower and upper bound for the step size. This provision is necessary in order to suppress the effect of an inappropriate guess of ϵ.

The second group of the one-step methods are the procedures based on the Taylor expansion. For the sake of simplicity one scalar equation will be considered. The solution $y(\bar{x} + h)$ can be obtained in the form

$$y(\bar{x} + h) = y(\bar{x}) + hy'(\bar{x}) + \frac{h^2}{2!}y''(\bar{x}) + \frac{h^3}{3!}y'''(\bar{x}) + \cdots \qquad (2.16)$$

For a sufficiently differentiable function, we may write

$$y'' = \frac{d}{dx}f(x, y) = f_x(x, y) + f_y(x, y)y' = f_x(x, y) + f_y(x, y)f(x, y)$$

and in an analogous way,

$$y''' = f_{xx} + 2ff_{xy} + f_{yy}f^2 + f_x f_y + ff_y^2$$

and so on. A shortcoming of this procedure is the necessity of performing the differentiation operations analytically. On the other hand, the methods are very simple.

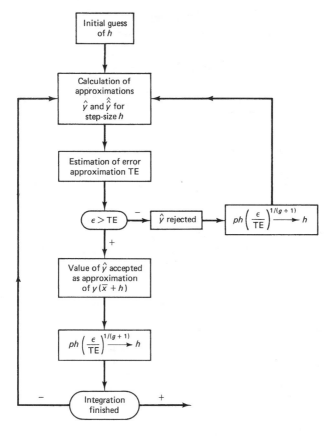

Figure 2-1 Flowchart of a Runge–Kutta method with an automatic step-size adjustment.

2.1.3 Linear Multistep Methods

A general linear multistep method may be written in the form

$$\alpha_k y_{n+k} + \alpha_{k-1} y_{n+k-1} + \cdots + \alpha_0 y_n$$
$$= h\{\beta_k f_{n+k} + \beta_{k-1} f_{n+k-1} + \cdots + \beta_0 f_n\} \qquad (2.17)$$

Here $\alpha_k \neq 0$, $\alpha_0^2 + \beta_0^2 > 0$, y_i denotes the approximation $y(x_i)$, $h = x_{i+1} - x_i$ is a step of an equidistant mesh, and $f_i = f(x_i, y_i)$ is the right-hand side at the point x_i [in a general case y and f are vectors; see (2.1)]. The coefficients for the explicit Adams–Bashforth and implicit Adams–Moulton methods are reported in Tables 2-2 and 2-3.

Let us discuss some properties of multistep methods. A shortcoming, in

TABLE 2-2
COEFFICIENTS OF ADAMS–BASHFORTH FORMULAS

$$y_{n+1} = y_n + h \sum_{i=0}^{q} \beta_{qi} f_{n-i} \quad \text{for several } q$$

i	0	1	2	3	4	5	e
β_{0i}	1						$O(h)$
$2\beta_{1i}$	3	−1					$O(h^2)$
$12\beta_{2i}$	23	−16	5				$O(h^3)$
$24\beta_{3i}$	55	−59	37	−9			$O(h^4)$
$720\beta_{4i}$	1901	−2774	2616	−1274	251		$O(h^5)$
$1440\beta_{5i}$	4227	−7673	9482	−6798	2627	−425	$O(h^6)$

TABLE 2-3
COEFFICIENTS OF ADAMS–MOULTON FORMULAS

$$y_n = y_{n-1} + h \sum_{i=0}^{q} \beta_{qi}^* f_{n-i} \quad \text{for several } q$$

i	0	1	2	3	4	5	e
β_{0i}^*	1						$O(h)$
$2\beta_{1i}^*$	1	1					$O(h^2)$
$12\beta_{2i}^*$	5	8	−1				$O(h^3)$
$24\beta_{3i}^*$	9	19	−5	1			$O(h^4)$
$720\beta_{4i}^*$	251	646	−264	106	−19		$O(h^5)$
$1440\beta_{5i}^*$	475	1427	−798	482	−173	27	$O(h^6)$

comparison with the one-step methods, is the necessity of multiple starting values. These methods are not of the self-starting type, and the starting values must be calculated by a different method—for example, by a one-step method. The advantage of the multistep methods is in the fact that for a method of the same order, the number of evaluations of the right-hand sides is lower than for a one-step method. On the other hand, the automatic step-size adjustment may be a difficult problem.

Explicit multistep methods are of the recurrent type and iterations are not necessary. The implicit methods posses lower error and exhibit better stability; however, iterations are necessary to calculate the new value of y. Frequently, for a sufficiently short h, the method of successive substitutions converges, but usually the Newton method is adopted.

Evidently, the Newton method requires evaluation of the Jacobian matrix of the right-hand sides. As an initial approximation the result of an explicit method is usually used. An appropriate combination of both these algorithms (e.g., of Adams–Bashforth–Moulton formulas of the same order) is the PECE (predictor–

corrector) algorithm. In this combination one or two iterations are performed in the implicit method.

Recently, a number of programs toward solution of initial value problems have appeared. Because of space limitations we cannot go into dteails. Roughly speaking, the PECE methods and Runge–Kutta–Fehlberg formulas are the most popular. The reader who wishes to be acquainted with recent developments in this area is referred to the survey papers by Hull et al.[†] and Shampine at al.[‡], where nonstiff initial value problems are discussed. Stiff initial value problems are dealt with in the following section.

2.1.4 Numerical Solution of Stiff Initial Value Problems

Often in practical problems differential equations with widely separated eigenvalues—called stiff ordinary differential equations—appear. Consider first a set of linear ordinary differential equations

$$\frac{dy}{dx} = Ay \tag{2.18}$$

For widely separated eigenvalues of the matrix A, the differential equations (2.18) are called stiff. For example, for

$$A = \begin{bmatrix} -500.5 & 499.5 \\ 499.5 & -500.5 \end{bmatrix} \tag{2.19}$$

and for initial conditons $y_1(0) = 2$, $y_2(0) = 1$, the solution of (2.18) is

$$y_1(x) = 1.5e^{-x} + 0.5e^{-1000x}$$
$$y_2(x) = 1.5e^{-x} - 0.5e^{-1000x} \tag{2.20}$$

The eigenvalues of (2.19) are $\lambda_1 = -1000$ and $\lambda_2 = -1$. A number of algorithms for numerical integration require that both $|h\lambda_1|$ and $|h\lambda_2|$ be bounded by a small number ϵ, for which usually $\epsilon \in \langle 1, 10 \rangle$. For instance, for the Euler method it is necessary that $|-1000h| < 2$. This condition imposes a severe restriction on the maximum length of the integration step h.

Dahlquist[§] defined the A-stability of an integration method for the simple differential equation

$$\frac{dy}{dx} = \alpha y \tag{2.21}$$

where α is an arbitrary complex parameter, Re $\alpha < 0$. The method is said to be A-stable if for a sequence y_n [$= y(x_n)$, $x_n = x_0 + nh$, y_0 is arbitrary] and for any

---

[†]T. E. Hull et al, *SIAM J. N Numer. Anal. 9*, 603 (1972).

[‡]L. F. Shampine, H. A. Watts, and S. Davenport, "Solving Non-Stiff Ordinary Differential Equations," *The State of the Art*, SAND75–0182 Rept., Sandia Labs., Albuquerque, N.M. 1975.

[§]G. G. Dahlquist, *BIT 3*, 27 (1963).

arbitrary $h > 0$,

$$\lim_{n \to \infty} y_n = 0$$

After using a particular method for (2.21), a recurrent relation results:

$$y_{n+1} = P(\alpha h)y_n \tag{2.22}$$

Here P is an algebraic expression, usually a rational function of αh. The A-stability assumes that

$$|P(\alpha h)| < 1, \qquad \text{Re } \alpha < 0, \quad h > 0 \tag{2.23}$$

In addition, the method is L-stable if

$$\lim_{h \to \infty} |P(\alpha h)| = 0$$

A detailed discussion of stiff procedures may be found in the paper by Seinfeld, Lapidus, and Hwang[†] and in the book by Lapidus and Seinfeld (1971).

An important class of implicit Runge–Kutta methods has been developed by Rosenbrock,[‡] Calahan[§], and Caillaud and Padmanabhan[‖]. These methods are called semi-implicit since the calculation can be performed in a recurrent way.

Consider an autonomous system

$$y' = f(y) \tag{2.24}$$

with the Jacobian matrix

$$.\,\Gamma(y) = \left\{ \frac{\partial f}{\partial y} \right\} \tag{2.25}$$

For a third-order method we may write

$$\begin{aligned}
k_1 &= h[I - ha_1\Gamma(y_n)]^{-1}f(y_n) \\
k_2 &= h[I - ha_2\Gamma(y_n + c_1k_1)]^{-1}f(y_n + b_1k_1) \\
y_{n+1} &= y_n + w_1k_1 + w_2k_2
\end{aligned} \tag{2.26}$$

Parameters of particular methods are presented in Table 2-4 (all methods are A-stable).

Liniger and Willoughby (1969) suggested methods with weighted first and second derivatives in the form

$$y_{n+1} = y_n + h[(1 - \beta_1)y'_{n+1} + \beta_1 y'_n] \tag{2.27}$$

and

$$y_{n+1} = y_n + \frac{h}{2}[(1 + a)y'_{n+1} + (1 - a)y'_n] - \frac{h^2}{12}[(1 + 3a)y''_{n+1} + (1 - 3a)y''_n] \tag{2.28}$$

[†]J. H. Seinfeld, L. Lapidus, and M. Hwang, *Ind. Eng. Chem. Fund.* 9, 266 (1970).
[‡]H. H. Rosenbrock, *Computer J.* 5, 320 (1963).
[§]D. A. Calahan, *Proc. IEEE (Letters)* 56, 744 (1964).
[‖]J. B. Caillaud and L. Padmanabhan, *Chem. Eng. J.* 2, 227 (1971).

TABLE 2-4

COEFFICIENTS OF A-STABLE SEMI-IMPLICIT RUNGE-KUTTA METHODS

	Order	a_1	a_2	b_1	c_1	w_1	w_2
Rosenbrock	2	$1 - \sqrt{2}/2$	$1 - \sqrt{2}/2$	$(\sqrt{2} - 1)/2$	0	0	1
	3	1.40824829	0.59175171	0.17378667	0.17378667	−0.41315432	1.41315432
Calahan	3	0.788675134	0.788675134	−1.15470054	0	0.75	0.25
Trapezoidal rule	2	0.5	0.5	0	0	1	0

respectively. The methods are of the first and third orders and both are A-stable for $\beta_1 < \frac{1}{2}$ and $a > 0$. We may note that for $\beta_1 = 0$, (2.27) results in an implicit Euler method. Brandon† proposed an algorithm for an adaptive evaluation of the coefficient β_1 as a function of the shape of the solution.

After an appropriate modification of the implicit Adams–Moulton procedures, Gear (1971) developed a multistep A-stable algorithm. This method is provided with an automatic step-size adjustment. The integration package written by Gear represents very powerful and useful software. A detailed description of the algorithm is presented in the literature (see the Bibliography).

A simple L-stable explicit method of second order provided with an automatic step-size adjustment is described in *Chem. Eng. Commun. 1*, 291 (1974). In the last decade a great attention has been devoted to stiff systems, and the reader is referred to the references presented in the Bibliography.

2.2 Dependence of Solution of Initial Value Problems on the Initial Condition and on a Parameter

Often, the dependence of the solution of an initial value problem on the initial conditions or governing parameters is required. On using the shooting method for solution of a nonlinear boundary value problem, the missing initial conditions are guessed. To correct the values of the guessed initial conditions, the derivatives of the residual with respect to guessed initial conditions are necessary (see Section 4.6). The dependence of the solution of an initial value problem on parameters is utilized in the GPM concept for parametric investigation of nonlinear boundary value problems (see Section 5.5) and for evaluation of branching points (see Section 5.6).

We shall study the dependence of the solution of the initial value problem

$$\frac{dy_i}{dx} = f_i(x, y_1, \ldots, y_n) \qquad i = 1, 2, \ldots, n \qquad (2.29)$$

$$y_i(a) = y_{i0}$$

on the initial conditions y_{i0}. Suppose that all first derivatives of the right-hand sides, $\partial f_i/\partial y_j$, are continuous. After differentiation of (2.29) with respect to y_{k0} (k is fixed), we have

$$\frac{\partial}{\partial y_{k0}} \frac{dy_i}{dx} = \frac{\partial}{\partial y_0} f_i(x, y_1, \ldots, y_n) = \sum_{j=1}^{n} \frac{\partial f_i(x, y_1, \ldots, y_n)}{\partial y_j} \frac{\partial y_j}{\partial y_{k0}} \qquad (2.30)$$

†D.M. Brandon, 'IMP—A Software System for the Direct or Iterative Solution of Large Differential and/or Algebraic Systems," *General Manual*, University of Connecticut, Dept. of Chem. Eng., Storrs, CT, 1972.

On denoting $p_{ik}(x) = \partial y_i(x)/\partial y_{k,0}$ and after some rearrangement, we may write

$$\frac{dp_{ik}}{dx} = \sum_{j=1}^{n} \frac{\partial f_i(x, y_1, \ldots, y_n)}{\partial y_j} p_{jk} \qquad i = 1, 2, \ldots, n \qquad (2.31)$$

subject to the initial conditions

$$p_{ik}(a) = \delta_{ik} \qquad (2.32)$$

Here δ_{ik} is the Kronecker delta, $\delta_{ii} = 1$ and $\delta_{ik} = 0$ for $i \neq k$.

Equation (2.31) is referred to as the variational equation for variational variables $p_{ik}(x)$. The solution of (2.31) and (2.32) $p_{ik}(x)$ represents the sensitivity of the solution $y_i(x)$ to changes of the initial condition y_{k0}.

Let $\bar{y}_i(x)$ and $\bar{p}_{ik}(x)$ denote the solution of a set of $2n$ differential equations (2.29) and (2.31) subject to (2.32). After a perturbation of $y_k(a)$ from y_{k0} to $y_{k0} + \epsilon$ the solution of (2.29) with this new initial condition is

$$y_i(x) = \bar{y}_i(x) + \epsilon \bar{p}_{ik}(x) + O(\epsilon^2) \qquad (2.33)$$

where ϵ is a small perturbation.

Supposing that f_i may be differentiated twice and that the derivatives are continuous, (2.31) can be differentiated with respect to y_{m0} (m is fixed). After denoting

$$q_{ikm} = \frac{\partial^2 y_i}{\partial y_{k0} \, \partial y_{m0}} \qquad (2.34)$$

the following variational equations result:

$$\frac{dq_{ikm}}{dx} = \sum_{j=1}^{n} \left[\sum_{s=1}^{n} \frac{\partial^2 f_i(x, y_1, \ldots, y_n)}{\partial y_j \, \partial y_s} p_{sm} p_{jk} + \frac{\partial f_i(x, y_1, \ldots, y_n)}{\partial y_j} q_{jkm} \right]$$
$$i = 1, 2, \ldots, n \qquad (2.35)$$

with the initial conditions

$$q_{ikm}(a) = 0 \qquad i = 1, 2, \ldots, n \qquad (2.36)$$

After perturbation of y_{k0} and y_{m0},

$$y_{k0} + \epsilon \longrightarrow y_{k0}, \qquad y_{m0} + \eta \longrightarrow y_{m0}$$

the solution can be written in the form

$$y_i(x) = \bar{y}_i(x) + \epsilon \bar{p}_{ik}(x) + \eta \bar{p}_{im}(x) + \tfrac{1}{2}(\epsilon^2 \bar{q}_{ikk} + \eta^2 \bar{q}_{imm})$$
$$+ \epsilon \eta \bar{q}_{ikm} + O(\epsilon^3 + \eta^3) \qquad (2.37)$$

Here the bar variables are calculated for the original initial conditions y_{k0} and y_{m0}. In this particular case a set of $6n$ differential equations for y_i, p_{ik}, p_{im}, q_{ikk}, q_{ikm}, q_{imm}, $i = 1, \ldots, n$ must be integrated. The variational variables are used in the Newton–Fox shooting method (see Section 4.6). An application of variational equations is presented next.

Example 2.1

Behavior of a tubular recycle reactor with piston flow under steady-state conditions may be described by the set of ordinary differential equations

$$\frac{dy}{dx} = \text{Da}(1 - y) \exp\left(\frac{\Theta}{1 + \Theta/\gamma}\right)$$

$$\frac{d\Theta}{dx} = B\,\text{Da}(1 - y) \exp\left(\frac{\Theta}{1 + \Theta/\gamma}\right) - \beta(\Theta - \Theta_c)$$

(2.38)

subject to mixed boundary conditions

$$y(0) = (1 - \lambda)y(1), \qquad \Theta(0) = (1 - \lambda)\Theta(1) \qquad (2.39)$$

The problem can be solved by the shooting method (see Section 4.6) on choosing

$$y(0) = y_0, \qquad \Theta(0) = \Theta_0 \qquad (2.40)$$

Let us suppose that the solution is already known; that is, integration of (2.38) and (2.40) yields solutions for (2.38) and (2.39). The values of $y(1)$ and $\Theta(1)$ will be denoted y_1 and Θ_1. Variational equations make it possible to investigate the stability of a particular steady state. Stability may be analyzed by testing the transient behavior of an imposed perturbation. To perform this task an initial value problem (2.38) and (2.40) must be integrated together with the variational equations for the variables $\partial y/\partial y_0$, $\partial y/\partial \Theta_0$, $\partial \Theta/\partial y_0$, and $\partial \Theta/\partial \Theta_0$; that is, six first-order ordinary differential equations have to be solved simultaneously.

The inlet perturbations, Δy_0 and $\Delta \Theta_0$,

$$y(0) = y_0 + \Delta y_0, \qquad \Theta(0) = \Theta_0 + \Delta\Theta_0 \qquad (2.41)$$

give rise to the outlet values $y(1)$ and $\Theta(1)$:

$$y(1) \doteq y_1 + \frac{\partial y}{\partial y_0}(1)\,\Delta y_0 + \frac{\partial y}{\partial \Theta_0}(1)\,\Delta\Theta_0$$

$$\Theta(1) \doteq \Theta_1 + \frac{\partial\Theta}{\partial y_0}(1)\,\Delta y_0 + \frac{\partial\Theta}{\partial\Theta_0}(1)\,\Delta\Theta_0$$

(2.42)

By virtue of the recycle stream, a new inlet perturbation is given by

$$\begin{pmatrix} \Delta y_0^{\text{new}} \\ \Delta\Theta_0^{\text{new}} \end{pmatrix} = (1 - \lambda)A \begin{bmatrix} \Delta y_0 \\ \Delta\Theta_0 \end{bmatrix} \qquad (2.43)$$

where the matrix A is

$$A = \begin{pmatrix} \dfrac{\partial y}{\partial y_0}(1) & \dfrac{\partial y}{\partial\Theta_0}(1) \\[2mm] \dfrac{\partial\Theta}{\partial y_0}(1) & \dfrac{\partial\Theta}{\partial\Theta_0}(1) \end{pmatrix}$$

If the spectral radius of the matrix $(1 - \lambda)A$ is lower than 1, the imposed perturbation is damped and the steady state under question is stable. On the other hand, the inequality

$$(1 - \lambda)\rho(A) > 1 \qquad (2.44)$$

indicates instability.

Now the dependence of the solution of an initial value problem

$$y_i' = f_i(x, y_1, \ldots, y_n, \mu), \qquad y_i(a) = y_{i0} \qquad i = 1, \ldots, n \qquad (2.45)$$

on the parameter μ will be investigated. The functions f_i possess all the necessary continuous derivatives.

After differentiation of (2.45) with respect to μ and on denoting

$$r_i(x) = \frac{\partial y_i(x)}{\partial \mu} \qquad (2.46)$$

a set of differential equations results:

$$\frac{dr_i(x)}{dx} = \sum_{j=1}^{n} \frac{\partial f_i(x, y_1, \ldots, y_n, \mu)}{\partial y_j} r_j(x) + \frac{\partial f_i(x, y_1, \ldots, y_n, \mu)}{\partial \mu}$$

$$i = 1, 2, \ldots, n \qquad (2.47)$$

with the initial conditions

$$r_i(a) = 0 \qquad i = 1, 2, \ldots, n \qquad (2.48)$$

On solving the set of $2n$ differential equations (2.45), (2.47), and (2.48) for $\mu = \bar{\mu}$, we may construct a solution for a perturbed parameter $\mu = \bar{\mu} + \epsilon$:

$$y_i(x) = \bar{y}_i(x) + \epsilon \bar{r}_i(x) + O(\epsilon^2) \qquad (2.49)$$

2.3 Numerical Solution of Linear and Nonlinear Algebraic Equations

Boundary value problems are problems in a certain infinite-dimensional space of functions. To calculate a numerical solution, a problem in this space is approximated by a problem in a finite-dimensional space. This consideration reveals that numerical solution of linear and nonlinear algebraic equations is important for the solution of boundary value problems. Since this part discusses only supplemental material, the text will be very brief and only the most important result will be presented.

2.3.1 Gaussian Elimination and Its Modifications

Often, numerical solution of boundary value problems requires that we solve a set of linear equations

$$Ax = b \qquad (2.50)$$

For example, linear equations must be solved for linear boundary value problems approximated by finite-difference methods (see Section 3.1) or in one iteration step by the Newton method (see below). To solve (2.50) two classes of methods—finite and iterative—may be used. While the finite-difference approximations arising after discretization of linear or linearized partial differential equations of the elliptic type may be handled by iterative methods, the linear equations resulting by finite-difference approximation of boundary value problems for ordinary differential equations are solved via finite methods. Elliptic problems in partial differential equations are not dealt with in this book.

Gaussian elimination is the best known of the class of finite methods. A number of modifications of this method exists. With regard to the possibility of occurrence of ill-conditioned matrices, an algorithm with pivoting is recommended in this text.

2.3.2   Band and Almost-Band Matrices

Very often, the matrix A is tridiagonal; that is, the nonzero elements are only in the main and the two adjacent diagonals. Based on Gaussian elimination, the Thomas algorithm can be developed. The Thomas algorithm can be modified for tridiagonal and multidiagonal (band) matrices with a low number of nonzero off-diagonal elements.†

2.3.3   Solution of a Single Transcendental Equation

Often, solution of a nonlinear boundary value problem requires to solve a single transcendental equation

$$f(x) = 0 \tag{2.51}$$

In a general stationary one-step iteration method a sequence $\{x_k\}$ is constructed which converges to a solution x^* of (2.51), according to

$$x_{k+1} = \varphi(x_k) \tag{2.52}$$

If $|\varphi'(x^*)| < 1$, then $x_k \longrightarrow x^*$, supposing that an initial guess x_0 is close to x^*. Sometimes the function $\varphi(x)$ is chosen in the form $\varphi(x) = x - \psi(x)f(x)$, so that (2.52) may be written

$$x_{k+1} = x_k - \psi(x_k)f(x_k) \tag{2.53}$$

For the regular falsi method, the function $\psi(x)$ is

$$\psi(x) = \frac{x - x_0}{f(x) - f(x_0)} \tag{2.54}$$

where the points x_0 and x_1 are the appropriately selected points in the vicinity of x^*.

One of the most popular methods is the Newton method; here $\psi(x) = 1/f'(x)$; that is,

$$x_{k+1} = x_k - \frac{f(x_k)}{f'(x_k)} \tag{2.55}$$

We may easily find that $\varphi'(x^*) = 0$ and hence that

†M. Kubíček,: *Commun. ACM 16*, 760 (1973).

$$|x_{k+1} - x^*| \leq \alpha |x_k - x^*|^2$$

Here α is a constant depending on both f' and f'' (in the vicinity of x^*). It is said that the Newton method is of the second order of convergence.

For a function $f(x)$ with continuous derivatives of higher orders, the Chebyshev iteration formulas may be developed. The method of second order is identical with the Newton procedure; the method of the third order is

$$x_{k+1} = x_k - \frac{f(x_k)}{f'(x_k)} - \frac{f''(x_k)f^2(x_k)}{2[f'(x_k)]^3} \qquad (2.56)$$

The second group of methods with a high order of convergence is the method of tangent hyperbolas, here represented by the Richmond method of the third order.

$$x_{k+1} = x_k - \frac{2f(x_k)f'(x_k)}{2[f'(x_k)]^2 - f(x_k)f''(x_k)} \qquad (2.57)$$

The method of sequential backward interpolation is a multistep procedure that does not require us to evaluate the derivatives of f. By means of the function f, a map

$$y = f(x) \qquad (2.58)$$

will be studied. This map transforms a certain region M on N. We wish to find such x^* for which $f(x^*) = 0$. Supposing that the inverse function f^{-1} exists near x^*, then

$$x^* = f^{-1}(0) \qquad (2.59)$$

Furthermore, we assume that f^{-1} is continuously differentiable in the vicinity of x^* up to a certain order. The inverse function may be approximated by an interpolation polynomial, assuming that a map of an appropriate discrete set $S_1 \subset N$ to M is known. This map can be calculated in the following way. For a certain discrete mesh $S_2 \subset M$ from the vicinity of x^*, a map f [i.e., $S_1 = f(S_2)$] is performed. The function f^{-1} is then approximated by means of the interpolation polynomial from mesh S_1 with the values S_2. The value of this polynomial for $y = 0$ is supposed to be a new approximation of x^*. For a Lagrangian interpolation polynomial the Aitken schema may be used. For a new point the Aitken schema requires that we calculate only one additional row in the schema. The flowchart of the calculation process is shown in Fig. 2-2.

2.3.4 Solution of Systems of Nonlinear Equations

A number of procedures have been developed to solve a set of nonlinear algebraic equations

$$f_1(x_1, x_2, \ldots, x_n) = 0$$

$$\cdots \cdots$$

$$\qquad (2.60)$$

$$f_n(x_1, x_2, \ldots, x_n) = 0$$

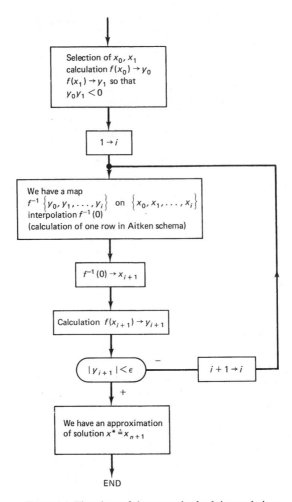

Figure 2-2 Flowchart of the successive back interpolation.

For detailed information the reader is referred to the book by Ortega and Rheinboldt (1970). In this section we restrict ourselves to methods that appear important from the point of view of nonlinear boundary value problems.

In the Newton method a sequence of vectors $\{x^k\}$ is constructed according to the iteration scheme

$$x^{k+1} = x^k + \lambda_k \, \Delta x^k \tag{2.61}$$

$$\Gamma_f(x^k) \, \Delta x^k = -f(x^k) \tag{2.62}$$

Here $\lambda_k \in (0, 1 >$ is an appropriate parameter and $\Gamma_f(x^k)$ is the Jacobian matrix

$$\Gamma_f(x) = \begin{pmatrix} \dfrac{\partial f_1}{\partial x_1} & \dfrac{\partial f_1}{\partial x_2} & \cdots & \dfrac{\partial f_1}{\partial x_n} \\ \cdots\cdots\cdots\cdots\cdots\cdots\cdots \\ \dfrac{\partial f_n}{\partial x_1} & \cdots\cdots\cdots & \dfrac{\partial f_n}{\partial x_n} \end{pmatrix} \tag{2.63}$$

One step in the Newton method consists of determination of the Jacobian matrix and the vector of residuals f for $x = x^k$, solution of a set of linear equations (2.62), and selection of an appropriate value of λ_k. In the vicinity of the solution, $\lambda_k = 1$ is usually adopted, while far from the solution, $\lambda_k < 1$ is used to prevent divergence.

With respect to the fact that the enumeration of the Jacobian matrix for large systems may be time consuming, a modified version of the Newton method may be used:

$$\begin{aligned} x^{k+1} &= x^k - \Gamma_f^{-1}(x^k)f(x^k) & k &= 0, 1, 2, \ldots, s \\ x^{k+1} &= x^k - \Gamma_f^{-1}(x^s)f(x^k) & k &= s+1, s+2, \ldots \end{aligned} \tag{2.64}$$

Here sometimes $s = 0$, assuming that the initial guess x^0 is in the neighborhood of the solution. The rate of convergence of the modified Newton method is comparable with that of the original Newton method. Sometimes Newton-like methods are used:

$$x^{k+1} = x^k - A_k^{-1}f(x^k) \tag{2.65}$$

where the matrix A_k^{-1} is an approximation of $\Gamma_f^{-1}(x^k)$.

The second group of methods is represented by methods of the secant type. These procedures do not require us to evaluate the partial derivatives, and they exhibit good convergence properties. Methods of this type are based on linear interpolation of the functions f_i by planes which are identical with f_i at $(n + 1)$ points.

Warner developed this procedure in the following way. Solution x^* of (2.60) is required. Consider the map $f(R^n \rightarrow R^n)$:

$$y = f(x) \tag{2.66}$$

For an inverse map f^{-1},

$$x = f^{-1}(y) \tag{2.67}$$

or

$$x^* = f^{-1}(0) \tag{2.68}$$

Equation (2.67) rewritten into components yields

$$\begin{aligned} x_1 &= g_1(y_1, y_2, \ldots, y_n) \\ &\;\;\vdots \\ x_n &= g_n(y_1, y_2, \ldots, y_n) \end{aligned} \tag{2.69}$$

Suppose that these functions possess all partial derivatives $\partial g_i / \partial y_j$. On denoting $X_i = g_i(0, 0, \ldots, 0)$, $X'_{ij} = \partial g_i(0, 0, \ldots, 0)/\partial y_j$, the Taylor series yields for (2.69)

$$x_i = X_i + \sum_{j=1}^{n} y_j X'_{ij} \qquad i = 1, 2, \ldots, n \qquad (2.70)$$

Here only the first term in the expansion was considered.

For $(n + 1)$ different vectors $x^1, x^2, \ldots, x^{n+1}$, the vectors $y^1, y^2, \ldots, y^{n+1}$ are calculated according to (2.66). After inserting these vectors into (2.70), n sets of linear algebraic equations of the order $n + 1$ for the unknowns X_i, X'_{ij}, $i = 1, 2, \ldots, n, j = 1, 2, \ldots, n$, result

$$x_i^1 = X_i + \sum_{j=1}^{n} y_j^1 X'_{ij}$$

$$. \qquad (2.71)$$

$$x_i^{n+1} = X_i + \sum_{j=1}^{n} y_j^{n+1} X'_{ij}$$

for $i = 1, 2, \ldots, n$. Assuming that the sets (2.71) possess a solution (which depends on the guesses $x^1, x^2, \ldots, x^{n+1}$), then for each i, (2.71) can be solved independently. The matrices of all sets are identical except that the right-hand sides are different. The unknowns $X_i, i = 1, 2, \ldots, n$, may be calculated simultaneously by the Gaussian elimination for a set $(n + 1) \times (n + 1)$ for n different right-hand sides. Moreover, only the forward elimination may be used if the variables X_i are ordered in the vector of variables as the last, since the unknowns X'_{ij} are not required. The calculated values of $X_i, i = 1, 2, \ldots, n$, can be considered as a new approximation of solution. The "worst" point from the initial guess $x^1, \ldots, x^{n+1}$ is replaced by this new approximation. The flowchart of this algorithm is shown in Fig. 2-3.

For a case of two equations, the Warner interpolation is, in a simple form,

$$X_i = \frac{\begin{vmatrix} x_i^1 & y_1^1 & y_2^1 \\ x_i^2 & y_1^2 & y_2^2 \\ x_i^3 & y_1^3 & y_2^3 \end{vmatrix}}{\begin{vmatrix} 1 & y_1^1 & y_2^1 \\ 1 & y_1^2 & y_2^2 \\ 1 & y_1^3 & y_2^3 \end{vmatrix}} \qquad i = 1, 2$$

A third-order method which is a multidimensional analogy of the Chebyshev method is developed next.[†] Consider the set (2.60) and define the map (2.66). Let x^* be a solution to (2.60). The inverse function f^{-1} will be denoted F and thus in the vicinity of x^*

$$x = F(y) = F[f(x)] \qquad (2.72)$$

†M. Kubíček and V. Hlaváček,: *Chem. Eng. J.* 2, 100 (1971).

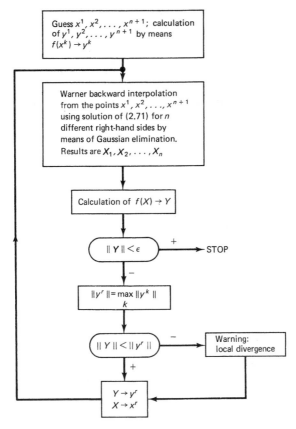

Figure 2-3 Flowchart of a variant of a secant method (Warner's backward interpolation).

Clearly, the following identity is valid:

$$x^* = F[f(x^*)] = F[0] \qquad (2.73)$$

After expansion of (2.73) into a Taylor series, we have for the ith component x_i^* $(i = 1, 2, \ldots, n)$:

$$x_i = F_i(y) - \sum_{j=1}^{n} \frac{\partial F_i}{\partial y_j} y_j + \frac{1}{2} \sum_{j=1}^{n} \sum_{m=1}^{n} \frac{\partial^2 F_i}{\partial y_j \partial y_m} y_j y_m + \cdots \qquad (2.74)$$

All partial derivatives are to be calculated for $y = f(x)$. With regard to (2.66) and (2.72), we may write (2.74) in the form

$$x_i = x_i - \sum_{j=1}^{n} \frac{\partial F_i}{\partial y_j} f_j(x) + \frac{1}{2} \sum_{j=1}^{n} \sum_{m=1}^{n} \frac{\partial^2 F_i}{\partial y_j \partial y_m} f_j(x) f_m(x) \qquad (2.75)$$

31

Let us define, for the sake of brevity,

$$a_{ij} = \frac{\partial F_i}{\partial y_j}, \qquad b_{ijm} = \frac{\partial^2 F_i}{\partial y_j\, \partial y_m}, \qquad i, j, m = 1, 2, \ldots, n$$

Now (2.75) may be rewritten to yield the following iterative procedure:

$$x_i^{k+1} = x_i^k - \sum_j a_{ij} f_j + \frac{1}{2} \sum_j \sum_m b_{ijm} f_j f_m \qquad (2.76)$$

where the coefficients $a_{ij}, b_{ijm}, f_i, f_m$ must be calculated for $x = x^k$. The coefficients a_{ij} and b_{ijm} can be determined from (2.72), which yields for component i:

$$x_i = F_i[f(x)] \qquad i = 1, 2, \ldots, n$$

After differentiation of this equation with respect to x_m and defining c_{jm} by

$$c_{jm} = \frac{\partial f_j}{\partial x_m} \qquad (2.77)$$

one may write

$$\sum_j a_{ij} c_{jm} = \delta_{im} \qquad m, i = 1, 2, \ldots, n \qquad (2.78)$$

where δ_{im} is the Kronecker delta. Since the coefficients c_{jm} are known, (2.78) can be used for calculation of unknown coefficients a_{ij}. The matrix c_{ij} is identical with the Jacobian matrix of (2.60) and, therefore, after omitting the coefficients b_{ijm}, the Newton method results. Furthermore, differentiation of (2.78) with respect to x_s after defining

$$d_{j,m,s} = \frac{\partial^2 f_j}{\partial x_m\, \partial x_s} \qquad (2.79)$$

yields

$$\sum_j \sum_r b_{ijr} c_{rs} c_{jm} = -\sum_j a_{ij} d_{jms} \qquad s, m = 1, 2, \ldots, n; \quad i = 1, 2, \ldots, n \qquad (2.80)$$

The set of equations (2.80) may be utilized for evaluation of the unknown coefficients b_{ijr}. Let us consider the set of equations (2.80) for a given value of i; it is obvious that a matrix of the type $n^2 \times n^2$ is not dependent on i and, moreover, only the right-hand sides are variable. Therefore, instead of solving the system of n^3 equations (2.80) it is possible to solve successively the $n^2 \times n^2$ sets for different i. However, simultaneous solution of the $n^2 \times n^2$ set subject to n right-hand sides by means of Gaussian elimination is done prior to all other methods. One iteration step for given values x^k then requires evaluation of the matrix of first derivatives c_{jm} and a three-dimensional matrix of second derivatives d_{jms} of the function f_j. Therefore, we must invert a matrix ($n \times n$) [see (2.78)] and further, a set of linear equations ($n^2 \times n^2$) with n different right-hand sides has to be solved [see (2.80)]. After inserting the values f_i, a_{ij}, and b_{ijm} in (2.76), a new approximation can be obtained.

2.3.5 Dependence of the Solution of a Set of Nonlinear Equations on a Parameter

Sometimes in a set of nonlinear equations a parameter of physical importance appears. The goal of the computation is to find the dependence of the solution on the value of this parameter. Consider a set of equations

$$f_1(x_1, \ldots, x_n, \alpha) = 0$$
$$\cdot$$
$$\cdot \tag{2.81}$$
$$\cdot$$
$$f_n(x_1, \ldots, x_n, \alpha) = 0$$

where α is the parameter under question. The dependence $x(\alpha)$ is required. This problem was analyzed by Daviděnko (1953) and based on the implicit function theorem, the following differential equations were developed† :

$$\frac{dx}{d\alpha} = -\Gamma^{-1}(x, \alpha) \frac{\partial f}{\partial \alpha} \tag{2.82}$$

subject to an initial condition

$$x(\alpha_0) = x_0 \tag{2.83}$$

Here $\Gamma = \{\partial f_i / \partial x_j\}$ is the Jacobian matrix and

$$\frac{\partial f}{\partial \alpha} = \left(\frac{\partial f_1}{\partial \alpha}, \ldots, \frac{\partial f_n}{\partial \alpha} \right)^T$$

It is supposed that

$$f(x_0, \alpha_0) = 0 \tag{2.84}$$

The differential equation (2.82) possesses a solution $x(\alpha)$ which fulfills (2.81) if $x(\alpha)$ is continuously differentiable and Γ is regular. If these conditions are not satisfied, and if $x(\alpha)$ is a smooth curve in the space (x, α), a modified integration may be performed:

$$\frac{d(x', \alpha)}{dx_k} = -\Gamma_1^{-1}(x, \alpha) \frac{\partial f}{\partial x_k} \tag{2.85}$$

with an initial condition

$$x'(x_{k0}) = x_0', \qquad \alpha(x_{k0}) = \alpha_0 \tag{2.86}$$

where we have denoted $x' = (x_1, \ldots, x_{k-1}, x_{k+1}, \ldots, x_n)^T$ and k is fixed, $1 \leq k \leq n$. The whole continuous dependence $x(\alpha)$ may be calculated by an alternated integration of (2.82) and (2.85), and the transition from (2.82) to (2.85) occurs if Γ becomes almost singular. The subroutine DERPAR‡ is

---

†Equation (2.82) may be developed easily. Differentiation of $f(x, \alpha) = 0$ results in

$$\Gamma(x, \alpha) \, dx + \frac{\partial f}{\partial \alpha} d\alpha = 0$$

For a regular matrix Γ, (2.82) immediately results.

‡M. Kubíček, *ACM Trans. Math. Software* 2, 98 (1976).

33

suitable for determination of parametric dependences. In this subroutine the integration is performed with respect to the arc length of the solution locus.

2.4 Boundary Value Problems

Unlike the initial value problem, where the initial conditions are specified at one point, boundary value problems are subjected to conditions that are given at more points. These conditions are referred to as the boundary conditions. A general m-point boundary value problem may be written as a set of equations

$$\frac{dy_i}{dx} = f_i(x, y_1, \ldots, y_n) \qquad i = 1, 2, \ldots, n \qquad (2.87)$$

and boundary conditions

$$g_j[y_1(x_1), \ldots, y_1(x_m), y_2(x_1), \ldots, y_2(x_m), \ldots, y_n(x_1), \ldots y_n(x_m)] = 0$$
$$j = 1, 2, \ldots, n \qquad (2.88)$$

A set of equations of higher order may always be rewritten as a set of equations of first order. Thus the above-mentioned formulation is quite general.

The most common case is the two-point boundary value problem (where $m = 2$). In physical and engineering applications the boundary conditions are frequently linear. For a discussion of boundary conditions, see Section 2.4.2.

The solution of (2.87) depends generally on n integration constants, which must be determined from given boundary conditions (2.88). As a result, it is necessary to specify n boundary conditions. For this particular case the problem is properly specified. If the number of boundary conditions is lower than the number of equations, the problem is underdetermined. In this situation new boundary conditions must be formulated in order to get a determined problem. In turn, if the number of boundary conditions is higher than the number of equations, the problem is overspecified and we may find a solution only if the overspecification is based on a sound physical model. In this case the extra surplus boundary conditions are results of the remaining conditions and may be omitted.

2.4.1 Existence and Uniqueness of Solution

While the problem of existence and uniqueness of solution for initial value problems has been carefully investigated and a detailed analysis has been published, for boundary value problems this question has been analyzed only for particular situations.

Some authors have investigated the problem of existence and uniqueness of solution for a case of a single second-order differential equation

$$y'' = f(x, y, y') \tag{2.89}$$

with boundary conditions

$$y(0) = A, \qquad y(1) = B \tag{2.90}$$

Sometimes a simpler form of (2.89) is considered:

$$y'' = f(x, y) \tag{2.91}$$

subject to boundary conditions (2.90). Suppose that f possesses continuous second derivatives in the region

$$S = \{(x, y), 0 \leq x \leq 1, -\infty < y < \infty\}$$

and satisfies

$$\frac{\partial f(x, y)}{\partial y} \geq \eta > -\pi^2 \tag{2.92}$$

for each $(x, y) \in S$. Then the problem (2.90)–(2.91) has a unique solution which is two times continuously differentiable. For the sake of illustration, consider a simple linear problem

$$y'' + \alpha y = 0 \tag{2.93a}$$

subject to the boundary conditions

$$y(0) = 0, \qquad y(1) = B \tag{2.93b}$$

A solution to (2.93a) which vanishes at $x = 0$ is

$$y(x) = C \sin (\sqrt{\alpha}\, x)$$

To be able to fulfill the second boundary condition also, we must require that

$$C \sin \sqrt{\alpha} = B \tag{2.94}$$

For $\alpha \neq (k\pi)^2$ the constant C may be evaluated from (2.94) and a unique solution results:

$$y(x) = \frac{B}{\sin \sqrt{\alpha}} \sin (\sqrt{\alpha}\, x) \tag{2.95}$$

For $\alpha = (k\pi)^2$ no solution of the boundary value problem (2.93) exists if $B \neq 0$. On the other hand, for $\alpha = (k\pi)^2$ and $B = 0$, an infinite number of solutions results, which are in the form $y(x) = C \sin (\sqrt{\alpha}\, x)$. Here C is arbitrary.

We shall see below that for a number of physical nonlinear boundary value problems, multiple solutions exist or no solution exists. Here the problem of existence and uniqueness may be analyzed a posteriori (i.e., on the basis of calculated results). Moreover, frequently a set of equations must be analyzed even though the theory is more or less oriented to a single equation.

Example 2.2

The explosion of a solid explosive material may be described as a nonlinear boundary value problem. For a sample in the form of an infinite cylinder, we may write

$$\frac{d^2y}{dx^2} + \frac{1}{x}\frac{dy}{dx} = -\delta e^y \tag{2.96}$$

subject to the boundary conditions

$$x = 0: \quad \frac{dy}{dx} = 0 \tag{2.97}$$

$$x = 1: \quad y = 0 \tag{2.98}$$

The solution of nonlinear equation (2.96) satisfying BC (2.97) can be found in closed analytical form:

$$y = \ln\frac{8B/\delta}{(Bx^2 + 1)^2} \tag{2.99}$$

The value of the integration constant B is given by (2.100):

$$\frac{8B/\delta}{(B + 1)^2} = 1 \tag{2.100}$$

From (2.100) it can readily be shown that for a given δ from the range $0 < \delta < 2$, two distinct real roots of (2.100) exist and thus two solutions of (2.96) with (2.97) and (2.98) occur. For $\delta = 2$, there is only one root of (2.100), $B = 1$, and only one solution of the problem (2.96)–(2.98) exists. Finally, for $\delta > 2$, the problem possesses no solution.

Example 2.3

Heat and mass transfer within a porous catalyst particle is described by

$$\frac{d^2y}{dx^2} + \frac{a}{x}\frac{dy}{dx} = \phi^2 y \exp\left[\frac{\gamma\beta(1 - y)}{1 + \beta(1 - y)}\right] \tag{2.101}$$

$$x = 0: \quad \frac{dy}{dx} = 0 \tag{2.102}$$

$$x = 1: \quad y = 1$$

Here ϕ^2, a, γ, and β are parameters of the problem. This equation cannot be integrated analytically and only the numerical integration is possible.

For a given value of the parameter γ and for a low value of the product $\gamma\beta$, only one solution of (2.101)–(2.102) exists. For higher values of $\gamma\beta$ in the range $\phi \in (\phi_1, \phi_2)$, three solutions occur. All of these observations are summarized in Table 2-5 and Fig. 2-4.

TABLE 2-5
RESULTS FOR EXAMPLE 2.3

Parameters	Number of solutions	Solutions
$a = 0$, $\gamma = 20$, $\gamma\beta = 2$, $\phi = 1$	1	$y(0) = 0.3745$
$a = 2$, $\gamma = 20$, $\gamma\beta = 4$, $\phi = 2$	1	$y(0) = 0.0028$
$a = 0$, $\gamma = 20$, $\gamma\beta = 14$, $\phi = 0.16$	3	See Fig. 2-4

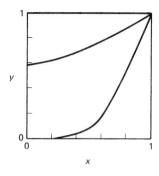

Figure 2-4 Three different solutions of Example 2.3 (see Table 2-5).

2.4.2 Two-Point Boundary Value Problem of the pth-Order

Boundary conditions (2.88) were formulated above quite generally. Define now an occurrence schema of variables in boundary conditions ($x_1 = a$, $x_2 = b$) in a form of a table, for instance:

$$y_1(a), y_2(a), y_3(a), \ldots, y_n(a), y_1(b), y_2(b), y_3(b), \ldots, y_n(b)$$

$$
\begin{array}{l|cccc|cccc}
g_1 & \times & \times & & & & & & \\
g_2 & & & \times & & & & & \\
g_3 & & & & \times & \times & \times & & \\
\cdot & & & & & & & & \\
\cdot & & & & & & & & \\
\cdot & & & & & & & & \\
g_n & & & & & & & \times & \times
\end{array}
\qquad (2.103)
$$

Here a cross indicates that the variable $y_i(a)$ or $y_i(b)$ occurs explicitly in the boundary condition for g_j. For an occurrence schema in the form ($n = 5$)

$$
\begin{array}{l|ccc|cc}
\times & & & & \\
& \times & & & \\
& & \times & & \\
& & & \times & \\
& & & & \times
\end{array}
\qquad (2.104)
$$

37

or

$$
\left[
\begin{array}{ccccc|}
\times & \times & \times & \times & \times \\
\times & \times & \times & \times & \times \\
\times & \times & \times & \times & \times \\
\times & \times & \times & \times & \times \\
\times & \times & \times & \times & \times \\
\end{array}
\right] \qquad (2.105)
$$

an initial value problem results. For (2.105) we must solve before integration a set of five equations $g_i = 0$, $i = 1, \ldots, 5$, for five unknowns, $y_1(a), \ldots, y_5(a)$.

The boundary condition g_3 in (2.103) is of the mixed type since it involves both $y(a)$ and $y(b)$. In practical applications this type of boundary condition is not very common; usually the conditions are separated; that is, in the function g_i either $y(a)$ or $y(b)$ occurs. This means that in the occurrence schema (2.103) in each row the first or the last n elements are blank. For example, for $n = 5$:

$$
\left[
\begin{array}{ccccc|ccccc}
\times & \times & \times & \times & \times & & & & & \\
\times & & & \times & \times & & & & & \\
 & & & & & \times & \times & \times & & \times \\
 & & & & & & & \times & & \\
 & \times & & \times & & & & & & \\
\end{array}
\right] \qquad (2.106)
$$

The boundary value problem with separated boundary conditions is at the point $x = a$ of the pth order if $p = n - r$, where r is the number of functions g_i in (2.88) which depend on $y(a)$. Evidently, if the problem is of the pth order at the point $x = a$, it is of the $n - p$ order at the point $x = b$. For instance, the problem given by Eq. (2.106) is of second order at $x = a$ and of third order at $x = b$. The order of the problem indicates the number of unknown boundary conditions at the given point which it is necessary to specify in order to convert the boundary value problem to an initial value problem. This fact is very important for numerical solution of boundary value problems via shooting procedures. Problems of conversion to an initial value problem are reported in Section 4.6.

BIBLIOGRAPHY

Since this text is supplementary in nature and it is necessary for readers to understand the numerical techniques associated with the solution of nonlinear boundary value problems, the bibliography is supportive in character. As a result, we have restricted the literature to the most important books and papers.

Basic information on numerical solution of initial value problems may be found in books by:

HENRICI, P.: *Discrete Variable Methods in Ordinary Differential Equations*. J. Wiley, New York, 1962.

LAPIDUS, L., AND SEINFELD, J. H.: *Numerical Solution of Ordinary Differential Equations*. Academic Press, New York, 1971.

LAMBERT, J. D.: *Computational Methods in Ordinary Differential Equations*. J. Wiley, London, 1973.

SHAMPINE, L. F., AND GORDON, M. K.: *Computer Solution of Ordinary Differential Equations: The Initial Value Problem*. Freeman, San Francisco, Calif., 1975.

GEAR, C. W.: *Numerical Initial Value Problems in Ordinary Differential Equations*. Prentice-Hall, Englewood Cliffs, N. J., 1971.

Stiff problems are dealt with in:

LAPIDUS, L., AND SEINFELD, J. H.: *Numerical Solution of Ordinary Differential Equations*. Academic Press, New York, 1971.

CAILLAUD, J. B., AND PADMANABHAN, L.: *Chem. Eng. J.* 2, 227 (1971).

LINIGER, W., AND WILLOUGHBY, R. A.: *SIAM J. Numer. Anal.* 6, 47 (1969).

VIŠŇÁK, K., AND KUBÍČEK, M.: *J. Inst. Math. Appl.* 21, 251 (1978).

GEAR, C. W.: *Commun. ACM* 14, 176, 185 (1971).

SHAMPINE, L. F., AND GEAR, C. W.: A User's View of Solving Stiff Ordinary Differential Equations, *SIAM Review* 21, 1 (1979).

GEAR, C. W.: Numerical Solution of Ordinary Differential Equations: Is There Anything Left to Do? *SIAM Review* 23, 10 (1981).

BYRNE, G. D. et al: A Comparison of Two Codes: GEAR and EPISODE, *Comp. Chem. Eng* 1, 133 (1977).

Some problems associated with the theory of dependence of solution of initial value problems on initial conditions and governing parameters may be found in textbooks on differential equations, e.g.:

CODDINGTON, E. A., AND LEVINSON, N.: *Theory of Ordinary Differential Equations*. McGraw-Hill, New York 1955.

For solution of linear algebraic equations, the reader is referred to:

FADDĚJEV, D. K., AND FADDĚJEVA, V. N.: *Numerical Methods of Linear Algebra*. Fizmatgiz, Moscow, 1960 (in Russian).

FORSYTHE, G. E., AND MOLER, C. B.: *Computer Solution of Linear Algebraic Systems*. Prentice-Hall, Englewood Cliffs, N.J., 1967.

ROSE, D. J., AND WILLOUGHBY, R. A., EDS.: *Sparse Matrices and Their Applications*. Plenum Press, New York, 1972. This book is oriented toward sparse systems.

Methods of solution of single transcendental equations are presented in:

TRAUB, J.: *Iterative Methods for the Solution of Equations*. Prentice-Hall, Englewood Cliffs, N.J., 1964.

For solution of systems of nonlinear algebraic equations, see, e.g.:

ORTEGA, J. M., AND RHEINBOLDT, W. C.: *Iterative Solution of Nonlinear Equations in Several Variables*. Academic Press, New York, 1970.

Dependence of solution of a set of algebraic equations on a parameter is discussed by:

DAVIDĚNKO, D. F.: *Dokl. Akad. Nauk SSSR* 88, 601 (1953) (in Russian).

KUBÍČEK, M.: *ACM Trans. Math. Software* 2, 98 (1976).

Certain properties of boundary value problems are discussed in:

BAILEY, P. B., SHAMPINE, L. F., AND WALTMAN, P. E.: *Nonlinear Two Point Boundary Value Problems*. Academic Press, New York, 1968.

A survey of initial value integrators in boundary value problem codes is:

WATTS, H. A.: Initial value integrators in BVP codes, in *Codes for Boundary-Value Problems in Ordinary Differential Equations* (ed. B. Childs et al.), Springer Lecture Notes in Computer Science. Springer-Verlag, Berlin, 1979.

Linear Boundary Value Problems 3

A number of problems in engineering and physics give rise to a linear boundary value problem. Solution of nonlinear boundary value problems is sometimes constructed via solution of linear boundary value problems, an example being the quasi-linearization method. For these reasons our discussion of linear boundary value problems is separated from that dealing with nonlinear problems.

In this chapter two classes of methods are discussed: (1) construction of a finite-difference analogy to the relevant boundary value problem and solution of resulting equations, and (2) methods based on marching procedures for initial value problems. To this class of methods belong the method of superposition of solutions, the method of complementary functions, the method of adjoints, and the factorization and invariant imbedding approach.

3.1 Finite-Difference Methods

To explain the finite-difference methods, a second-order linear two-point boundary value problem

$$p(x)y'' + q(x)y' + r(x)y = s(x) \tag{3.1}$$

subject to the boundary conditions

$$\alpha_0 y(a) + \beta_0 y'(a) = \gamma_0 \tag{3.2a}$$

$$\alpha_1 y(b) + \beta_1 y'(b) = \gamma_1 \tag{3.2b}$$

will be used. The region $\langle a, b \rangle$ will be divided into N equidistant parts ($h = (b - a)/N$) by the mesh points

$$x_0 = a, x_1 = a + h, \ldots, x_k = a + kh, \ldots, x_N = b \qquad (3.3)$$

Consider the unknown function $y(x)$ at these mesh points. After replacing the derivatives in (3.1) by finite-difference approximations, we have

$$y''(x_i) \sim y_i'' = \frac{y_{i-1} - 2y_i + y_{i+1}}{h^2} \qquad (3.4)$$

$$y'(x_i) \sim y_i' = \frac{y_{i+1} - y_{i-1}}{2h} \qquad (3.5)$$

where we have denoted $y_i \sim y(x_i)$. The finite-difference approximations (3.4) and (3.5) are the simplest (three-point) approximations with the error $O(h^2)$. The more accurate approximations are discussed below.

On substitution of these approximations in (3.1) at the points $x_1, \ldots, x_{N-1}$, the following equations result:

$$p_i \frac{y_{i+1} - 2y_i + y_{i-1}}{h^2} + q_i \frac{y_{i+1} - y_{i-1}}{2h} + r_i y_i = s_i \qquad i = 1, 2 \ldots, N - 1 \quad (3.6)$$

Here we have denoted

$$p_i = p(x_i), \qquad q_i = q(x_i), \qquad r_i = r(x_i), \qquad s_i = s(x_i) \qquad (3.7)$$

After algebraic manipulation, (3.6) yields

$$A_i y_{i-1} + B_i y_i + C_i y_{i+1} = D_i \qquad i = 1, 2, \ldots, N - 1 \qquad (3.8)$$

where

$$A_i = \frac{p_i}{h^2} + \frac{q_i}{2h}, \qquad B_i = \frac{-2p_i}{h^2} + r_i, \qquad C_i = \frac{p_i}{h^2} - \frac{q_i}{2h}, \qquad D_i = s_i \quad (3.9)$$

The boundary conditions (3.2a) and (3.2b) may be rewritten

$$\alpha_0 y_0 + \beta_0 \frac{(y_1 - y_0)}{h} = \gamma_0$$
$$\alpha_1 y_N + \beta_1 \frac{(y_N - y_{N-1})}{h} = \gamma_1 \qquad (3.10)$$

The set of equations (3.8), together with (3.10), represents a set of $N + 1$ linear algebraic equations for $N + 1$ variables $y_0, y_1, \ldots, y_N$.

A more accurate approximation of boundary conditions (3.2a) and (3.2b) is

$$\alpha_0 y_0 + \beta_0 \frac{(-3y_0 + 4y_1 - y_2)}{2h} = \gamma_0$$
$$\alpha_1 y_1 + \beta_1 \frac{(y_{N-2} - 4y_{N-1} + 3y_N)}{2h} = \gamma_1 \qquad (3.11)$$

While the approximation (3.10) is of the order $O(h)$, the approximation (3.11) is $O(h^2)$. To explain the method of calculation, let us draw the structure matrix (see Chapter 2) for $N = 5$. Figures 3-1 and 3-2 display the structure matrix for

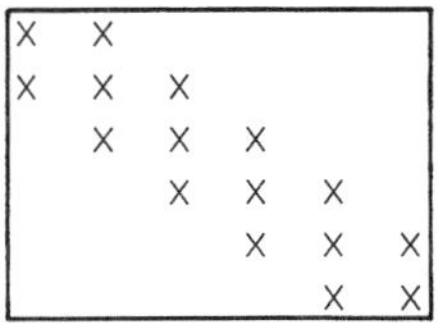

Figure 3-1 Structural matrix for (3.8) and (3.10).

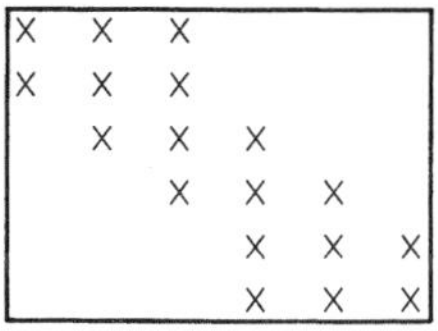

Figure 3-2 Structural matrix for (3.8) and (3.11).

(3.8) and (3.10) and (3.8) and (3.11), respectively. In the former case a set having a tridiagonal matrix results which can be solved easily by the Thomas algorithm (see Chapter 2). The latter case may be rendered tridiagonal after an appropriate linear combination of rows. Sometimes the boundary conditions may be approximated by making use of the Taylor expansion. The method will be shown for the approximation of boundary condition (3.2a). Evidently, for $\beta_0 \neq 0$,

$$y_0' = \frac{\gamma_0 - \alpha_0 y_0}{\beta_0} \tag{3.12}$$

Using the Taylor expansion, we have

$$y_1 - y_0 = h y_0' + \frac{h^2}{2!} y_0'' + \frac{h^3}{3!} y_0''' + \cdots \tag{3.13}$$

Equation (3.1) yields

$$y_0'' = \frac{s_0 - q_0 y_0' - r_0 y_0}{p_0} \tag{3.14}$$

After differentiation of (3.1), we have

$$y_0''' = \frac{s_0' - (p_0' + q_0) y_0'' - (q_0' + r_0) y_0' - r_0' y_0}{p_0} \tag{3.15}$$

On inserting (3.12) into (3.14) and the result and (3.12) into (3.15), explicit relations for y_0'' and y_0''' result which do not contain derivatives of y. After combination of these results with (3.13) a linear relation between y_1 and y_0 results. The order of approximation is higher in comparison with (3.10); however, the structure matrix is the same (see Fig. 3-1). Of course, the functions p, q, r, and s must be analytically differentiated. This type of approximation is very convenient for higher-order finite-difference schemas (see below).

If for approximations of the first and second derivatives, finite-difference formulas are used which make use of more than three mesh points, the resulting structure matrix is multidiagonal. A multidiagonal structure matrix also results if differential equations of higher order are approximated. Any arbitrary finite-difference approximations may be developed by using the method of undetermined coefficients. Assume that the kth derivative at the point $\bar{x}$ should be replaced by a finite-difference formula utilizing the values of the fucntion y at the mesh points $x_0, x_1, \ldots, x_n$. The finite-difference formula is supposed in the form

$$y^{(k)}(\bar{x}) = \sum_{i=0}^{n} c_i y(x_i) = \sum_{i=0}^{n} c_i y_i \qquad (3.16)$$

The unknown coefficients c_i may be evaluated from a condition which requires that (3.16) be satisfied by $y = 1; x; x^2; \ldots; x^n$. This yields a set of linear equations for the unknown coefficients $c_0, c_1, \ldots, c_n$:

$$c_0 + c_1 + \cdots + c_n = 0$$

$$c_0 x_0 + c_1 x_1 + \cdots + c_n x_n = 0 \qquad \left[= \frac{d^k}{dx^k}(x) \Big/ x = \bar{x} \right]$$

$$\cdots \cdots$$

$$c_0 x_0^{k-1} + c_1 x_1^{k-1} + \cdots + c_n x_n^{k-1} = 0$$

$$c_0 x_0^k + c_1 x_1^k + \cdots + c_n x_n^k = k! \qquad \left[= \frac{d^k}{dx^k}(x^k) \Big/ x = \bar{x} \right] \quad (3.17)$$

$$c_0 x_0^{k+1} + c_1 x_1^{k+1} + \cdots + c_n x_n^{k+1} = (k+1)!\bar{x}$$

$$\cdots \cdots$$

$$c_0 x_0^n + c_1 x_1^n + \cdots + c_n x_n^n = n(n-1)\ldots(n-k+1)\bar{x}^{n-k}$$

Equation (3.17) represents a set of $n + 1$ linear equations for unknown coefficients c_i. For approximation of boundary value problems the point $\bar{x}$ is selected in such a way that $\bar{x} = x_j$. The coefficients of such finite-difference formulas for an equidistant mesh ($h = x_{i+1} - x_i$) are presented in Tables 3-1 through 3-3 together with the error of approximation. Here the derivative is approximated at the point where the coefficient is underlined.

For instance, formula 11 in Table 3-1 may be read

$$y_1' = \frac{1}{12h}(-3y_0 - 10y_1 + 18y_2 - 6y_3 + y_4) - \frac{h^4}{20}y^{(V)}(\xi)$$

In consequence of the results presented above, a finite-difference approximation of (3.1) may be constructed by direct use of the method of undetermined coefficients to the original differential equation:

$$[py'' + qy' + ry - s]_{x=x_i} = A_i y_{i-1} + B_i y_i + C_i y_{i+1} - D_i \qquad (3.18)$$

The unknown coefficients A_i, B_i, C_i, and D_i may be calculated from the condition which requires that (3.18) is satisfied exactly by $y = 0; 1; x; x^2$. Of course, the coefficients presented in (3.9) result. Sometimes instead of (3.7) an averaged approximation will be used; for example,

$$p_i = \frac{1}{h} \int_{x_i-h/2}^{x_i+h/2} p(x)\, dx$$

Frequently, this approximation yields more accurate results.

The order of approximation of (3.8) and (3.11) is 2, that is, $O(h^2)$. We are going to present an approximation of higher order for a differential equation of second order which does not contain the first derivative:

$$y'' + r(x)y = s(x) \tag{3.19}$$

subject to boundary conditions of the first kind:

$$y(a) = \gamma_0, \qquad y(b) = \gamma_1 \tag{3.20}$$

TABLE 3-1

FINITE-DIFFERENCE FORMULAS FOR hy' AND EQUIDISTANT GRID

Mult. factor	Coefficients in difference formula							Error		Formula no.
1	-1	1						$-1/2$	$hy''(\xi)$	1
	-1	1						$1/2$		2
1/2	-3	4	-1					$1/3$		3
	-1	0	1					$-1/6$	$h^2 y'''(\xi)$	4
	1	-4	3					$1/3$		5
1/6	-11	18	-9	2				$-1/4$		6
	-2	-3	6	-1				$1/12$	$h^3 y^{(IV)}(\xi)$	7
	1	-6	3	2				$-1/12$		8
	-2	9	-18	11				$1/4$		9
1/12	-25	48	-36	16	-3			$1/5$		10
	-3	-10	18	-6	1			$-1/20$		11
	1	-8	0	8	-1			$1/30$	$h^4 y^{(V)}(\xi)$	12
	-1	6	-18	10	3			$-1/20$		13
	3	-16	36	-48	25			$1/5$		14
1/60	-137	300	-300	200	-75	12		$-1/6$		15
	-12	-65	120	-60	20	-3		$1/30$		16
	3	-30	-20	60	-15	2		$-1/60$	$h^5 y^{(VI)}(\xi)$	17
	-2	15	-60	20	30	-3		$1/60$		18
	3	-20	60	-120	65	12		$-1/30$		19
	-12	75	-200	300	-300	137		$1/6$		20
1/60	-147	360	-450	400	-225	72	-10	$1/7$		21
	-10	-77	150	-100	50	-15	2	$-1/42$		22
	2	-24	-35	80	-30	8	-1	$1/105$		23
	-1	9	-45	0	45	-9	1	$-1/140$	$h^6 y^{(VII)}(\xi)$	24
	1	-8	30	-80	35	24	-2	$1/105$		25
	-2	15	-50	100	-150	77	10	$-1/42$		26
	10	-72	225	-400	450	-360	147	$1/7$		27

TABLE 3-2
FINITE-DIFFERENCE FORMULAS FOR $h^2 y''$ AND EQUIDISTANT GRID

Mult. factor	Coefficients in difference formula					Error			Formula no.
1	1	−2	1			−1		1/6	1
	1	−2	1			0	$hy^{(III)}(\xi_1)+$	−1/12 $\;h^2y^{(IV)}(\xi_2)$	2
	1	−2	1			1		−1/6	3
1/6	12	−30	24	−6		11/12		−1/10	4
	6	−12	6	0		−1/12	$h^2y^{(IV)}(\xi_1)+$	−1/30 $\;h^3y^{(V)}(\xi_2)$	5
	0	6	−12	6		1/12		−1/30	6
	−6	24	−30	12		11/12		−1/10	7
1/24	70	−208	228	−112	22	−5/6		1/15	8
	22	−40	12	8	−2	1/12		−1/60	9
	−2	32	−60	32	−2	0	$h^3y^{(V)}(\xi_1)+$	1/90 $\;h^4y^{(VI)}(\xi_2)$	10
	−2	8	12	−40	22	−1/12		1/60	11
	22	−112	228	−208	70	5/6		1/15	12

TABLE 3-3
FINITE-DIFFERENCE FORMULAS FOR $h^3 y'''$ AND $h^4 y^{(IV)}$ AND EQUIDISTANT GRID

Mult. factor	Coefficients in difference formula					Error		Formula no.
1	−1	3	−3	1		−3/2		1
	−1	3	−3	1		−1/2	$hy^{(IV)}(\xi)$	2
	−1	3	−3	1		1/2		3
	−1	3	−3	1		3/2		4
1/2	−5	18	−24	14	−3	7/4		5
	−3	10	−12	6	−1	1/4		6
	−1	2	0	−2	1	−1/4	$h^2y^{(V)}(\xi)$	7
	1	−6	12	−10	3	1/4		8
	3	−14	24	−18	5	7/4		9
1	1	−4	6	−4	1	−2		10
	1	−4	6	−4	1	−1		11
	1	−4	6	−4	1	1	$hy^{(V)}(\xi)$	12
	1	−4	6	−4	1	2		13
	1	−4	6	−4	1	−1/6	$h^2y^{(VI)}(\xi)$	14

The finite-difference approximation

$$y_{i-1} - 2y_i + y_{i+1} + \frac{h^2}{12}(r_{i-1}y_{i-1} + 10r_iy_i + r_{i+1}y_{i+1}) = \frac{h^2}{12}(s_{i-1} + 10s_i + s_{i+1})$$

$$(3.21)$$

is of the fourth order. In comparison with the finite-difference formula (3.6), the order of approximation is higher; however, from the computational point of view the computer time requirements are almost the same. The structure matrix is again tridiagonal. For another boundary condition it is necessary to use approximations that exhibit the same order of approximation [i.e., $O(h^4)$]. For instance, the procedure described above making use of Taylor expansion may be used. Unfortunately, schema (3.21) may be used only for differential equations of the type (3.19) (i.e., without the first derivative). Equation (3.1) may be transformed to (3.19) by making use of the substitution

$$y(x) = z(x)Y(x) \qquad (3.22)$$

After inserting into (3.1), we have

$$pzY'' + (2pz' + qz)Y' + (pz'' + qz' + rz)Y = s \qquad (3.23)$$

If we put

$$2pz' + qz = 0 \qquad (3.24)$$

that is,

$$z(x) = \exp \int \left(-\frac{q}{2p}\right) dx \qquad (3.25)$$

then (3.23) does not contain the first derivative of the new variable $Y(x)$.

Methods that take advantage of all mesh points to approximate the derivatives at the particular mesh points are discussed in Section 4.1. An application of this procedure to linear boundary value problems does not seem to be as effective as in the nonlinear case.

Example 3.1

For the sake of simplicity, the following boundary value problem will be solved:

$$y'' = y, \qquad y(0) = 1, \qquad y(1) = e = 2.718281829 \qquad (3.26)$$

After using the most simple approximation [see (3.6)], we get

$$y_{i+1} - (2 + h^2)y_i + y_{i-1} = 0, \qquad y_0 = 1, \qquad y_N = e \qquad (3.27)$$

The results for different values of h are presented in Table 3-4. In column "R.E." appear values calculated using the Richardson's extrapolation (deferred approach to the limit) for the order $p = 2$. The asymptotic behavior of the error is in a good agreement with the fact that the approximation is of order $O(h^2)$.

For a more accurate approximation [see (3.21)] the finite-difference approximation is

$$y_{i-1}\left(1 - \frac{h^2}{12}\right) - y_i\left(2 + \frac{5}{6}h^2\right) + y_{i+1}\left(1 - \frac{h^2}{2}\right) = 0 \qquad (3.28)$$

The results of calculation are presented in Table 3-5. From the asymptotic behavior of the error it may be inferred that the approximation is of fourth order.

TABLE 3-4
RESULTING VALUES OF $y_{N/2} \sim y(0.5)$ FOR DIFFERENT VALUES OF h.[a]

h	$y_{N/2}$	Error	R.E.	Error
0.5	1.652569701	0.003848430		
			1.648750938	0.000029667
0.25	1.649705629	0.000984358		
			1.648722502	0.000000231
0.1	1.648879802	0.000158531		

[a]The Exact Solution of Example 3.1 Is $e^{0.5} \doteq 1.648721271$.

TABLE 3-5
RESULTING VALUES OF $y_{N/2} \sim y(0.5)$ FOR THE APPROXIMATION (3.28)

h	$y_{N/2}$	Error	R.E.	Error
0.5	1.648672132	−0.000049139		
			1.648721247	−0.000000024
0.25	1.648718177	−0.000003094		
			1.648721271	0.000000000
0.1	1.648721192	−0.000000079		

Below we are going to discuss the computational aspects of the numerical solution of a set of linear second-order differential equations with boundary conditions. Consider a set

$$p^k(x)y^{k''} + \sum_{j=1}^{M} [q_j^k(x)y^j + r_j^k(x)y^j] = s^k(x) \qquad k = 1, 2, \ldots, M \qquad (3.29)$$

subject to boundary conditions which are analogous to that described by (3.2).

For each $k, k = 1, 2, \ldots, M$, a finite-difference approximation will be constructed which is analogous to (3.6):

$$p_i^k \frac{y_{i-1}^k - 2y_i^k + y_{i+1}^k}{h^2} + \sum_{j=1}^{M} \left(q_{ji}^k \frac{y_{i+1}^j - y_{i-1}^j}{2h} + r_{ji}^k y_i^j \right) = s_i^k$$
$$i = 1, 2, \ldots, N - 1 \qquad (3.30)$$

The following grouping of variables and equations will be used. The variables are grouped in a vector Y,

$$Y = (y_0^1, y_0^2, \ldots, y_0^M, y_1^1, y_1^2, \ldots, y_1^M, \ldots, y_N^1, \ldots, y_N^M)^T \qquad (3.31)$$

The linear algebraic equations are grouped as follows:

(1) Approximation of boundary condition (3.2a) for y^1

(2) Approximation of boundary condition (3.2a) for y^2

.
.
.

(M) Approximation of boundary condition (3.2a) for y^M

48

$(M + 1)$ Approximation (3.30) for $i = 1, k = 1$

$(M + 2)$ Approximation (3.30) for $i = 1, k = 2$

.
.
.

$(2M)$ Approximation (3.30) for $i = 1, k = M$

$(2M + 1)$ Approximation (3.30) for $i = 2, k = 1$

$(2M + 2)$ Approximation (3.30) for $i = 2, k = 2$

.
.
.

$(iM + k)$ Approximation (3.30) for i and k

.
.
.

(NM) Approximation (3.30) for $i = N - 1$ and $k = M$

$(NM + 1)$ Approximation of boundary condition (3.2b) for y^1

.
.
.

$(NM + M)$ Approximation of boundary condition (3.2b) for y^M

For this grouping of variables the structure matrix is of the band type with $4M + 1$ diagonals. The number of diagonals may be reduced if in (3.29) $y^{j'}$ does not appear for $j \neq k$. Recently, a subroutine has been developed that is capable of rapid solving linear equations of this type.†

For mixed and or multipoint boundary conditions, an almost-band matrix results which may be conveniently handled by a subroutine DIAKUB.†

3.2 Superposition of Solution, Method of Complementary Functions

The methods described herein are based on utilization of the idea of replacing the original boundary value problem by an associated initial value problem which can be handled quite easily by making use of efficient integration subroutines. The basic idea of the superposition of solution will be explained on a single second-order differential equation (3.1) subjected to boundary conditions (3.2).

A general analytical solution of (3.1) may be written in the form $[p(x) \neq 0]$

$$y(x) = y_0(x) + A_1 y_1(x) + A_2 y_2(x) \tag{3.32}$$

where y_0 is an arbitrary (particular) solution of (3.1) and $y_1(x)$ and $y_2(x)$ are two

†M. Kubíček: Sci. Pap. Inst. Chem. Technol., Prague, K12, 5 (1977).

linearly independent solutions of a homogeneous equation

$$p(x)y'' + q(x)y' + r(x)y = 0 \tag{3.33}$$

So far the integration constants A_1 and A_2 are not determined and may be evaluated from boundary conditions. For a guess of initial conditions

$$y(a) = \eta_0 \quad \text{and} \quad y'(a) = \eta_0' \tag{3.34}$$

the differential equation (3.1) may be integrated as an initial value problem in the region $\langle a, b \rangle$. Evidently, a particular solution $y_0(x)$ results. Now for

$$y(a) = \eta_1 \quad \text{and} \quad y'(a) = \eta_1' \tag{3.35}$$

and after integration of the homogeneous equation (3.33), the function $y_1(x)$ results. Finally, for an appropriate guess

$$y(a) = \eta_2 \quad \text{and} \quad y'(a) = \eta_2' \tag{3.36}$$

and on numerical integration of (3.33) a new solution $y_2(x)$ results which is linearly independent of $y_1(x)$ if the matrix

$$\begin{pmatrix} \eta_1 & \eta_1' \\ \eta_2 & \eta_2' \end{pmatrix}$$

is regular. For the values $y(a)$ and $y'(a)$, we have according to (3.32) and (3.34)–(3.36),

$$y(a) = \eta_0 + A_1\eta_1 + A_2\eta_2$$
$$y'(a) = \eta_0' + A_1\eta_1' + A_2\eta_2'$$

After inserting these relations into boundary conditions (3.2a), we get

$$(\alpha_0\eta_1 + \beta_0\eta_1')A_1 + (\alpha_0\eta_2 + \beta_0\eta_2')A_2 = \gamma_0 - \alpha_0\eta_0 - \beta_0\eta_0' \tag{3.37a}$$

In an analogous way for $x = b$, (3.2b) yields

$$(\alpha_1 y_1(b) + \beta_1 y_1'(b))A_1 + (\alpha_1 y_2(b) + \beta_1 y_2'(b))A_2 = \gamma_1 - \alpha_1 y_0(b) - \beta_1 y_0'(b) \tag{3.37b}$$

Since the values of $y_i(b)$ and $y_i'(b)$, $i = 0, 1, 2$, have been calculated via numerical integration of relevant initial value problems, (3.37a) and (3.37b) represent a set of two linear equations for two unknowns A_1 and A_2. The procedure described above will be illustrated by a simple example.

Example 3.2

Solve a linear boundary value problem

$$xy'' + 2y' - xy = e^x \tag{3.38a}$$

$$2y(1) + y'(1) = e, \qquad y(2) = 2 \tag{3.38b}$$

Assume that $y(1) = \eta_0 = y'(1) = \eta_0' = 0$. Equation (3.38a) may be integrated by the Runge–Kutta method, and a particular solution $y_0(x)$ results. The homogeneous equation

$$xy'' + 2y' - xy = 0$$

will be integrated for $y_1(1) = \eta_1 = 0$, $y_1'(1) = \eta_1' = 1$ and $y_2(1) = \eta_2 = 1$, $y_2'(1) = \eta_2' = 0$. The results of these three integrations are presented in Table 3-6. After inserting into (3.37a) and (3.37b), two linear algebraic equations result:

$$A_1 + 2A_2 = e = 2.71828$$

$$0.58760A_1 + 1.35914A_2 = 0.95137$$

TABLE 3-6
METHOD OF SUPERPOSITION OF SOLUTIONS FOR THE PROBLEM
(3.38)

x	y_0	y_1	y_2	y
1.0	0.00000	0.00000	1.00000	−3.51143
1.2	0.04864	0.16778	1.01784	−1.89107
1.4	0.18055	0.29339	1.06559	−0.70324
1.6	0.38788	0.39791	1.13882	0.26509
1.8	0.67378	0.49339	1.23641	1.13839
2.0	1.04863	0.58760	1.35914	2.00000

The solution is $A_1 = 9.74114$ and $A_2 = -3.51143$. The final solution, $y = y_0 + A_1 y_1 + A_2 y_2$, is presented in Table 3-6. This solution agrees with the exact analytical solution

$$y^* = \frac{e^x}{2} + \frac{1}{x}\left[-\frac{e^x}{4} + \left(4e^2 - \frac{3}{4}e^4\right)e^{-x}\right]$$

within five significant digits.

Obviously, it has been necessary to solve for the particular problem three initial value problems and a set of two linear equations for two unknowns A_1 and A_2. However, the amount of computational work can be reduced. For instance, the solution to (3.1) and (3.2) may be found in the form

$$y(x) = y_0(x) + A_1 y_1(x) \tag{3.39}$$

where $y_0(x)$ is a solution of (3.1) for the conditions

$$y_0(a) = \eta_0, \qquad y_0'(a) = \eta_0'$$

so that

$$\alpha_0 \eta_0 + \beta_0 \eta_0' = \gamma_0 \tag{3.40a}$$

and $y(x)$ is a solution of the homogeneous equation (3.33) for the conditions

$$y_1(a) = \eta_1, \qquad y_1'(a) = \eta_1'$$

which satisfy

$$\alpha_0 \eta_1 + \beta_0 \eta_1' = 0 \tag{3.40b}$$

It is obvious that for any arbitrary constant A_1, the solution (3.39) satisfies the boundary condition (3.2a) at the point $x = a$. The unknown constant A_1 may be evaluated from the condition at the point $x = b$:

$$\alpha_1(y_0(b) + A_1 y_1(b)) + \beta_1(y_0'(b) + A_1 y_1'(b)) = \gamma_1 \tag{3.41}$$

Evidently, this procedure saves one integeration of the homogeneous equation.

The procedure presented above may also be adopted for an nth-order differential equation

$$\sum_{i=0}^{n} p_i(t)y^{(i)} = r(t) \tag{3.42}$$

For this particular case the solution may be found in the form

$$y(t) = y_0(t) + \sum_{i=1}^{n} A_i y_i(t) \tag{3.43}$$

where $y_0(t)$ is an arbitrary solution of the inhomogeneous problem (3.42) and $y_i(t)$ are linearly independent solutions of the pertinent homogeneous equation. The coefficients A_i can be evaluated from the boundary conditions.

Now the algorithm discussed above will be generalized for a general set of n first-order linear differential equations

$$y' = A(t)y + f(t) \tag{3.44}$$

subject to general two-point boundary conditions

$$B_1 y(a) + B_2 y(b) = c \tag{3.45}$$

where B_1 and B_2 are $(n \times n)$ matrices and $y = (y_1, y_2, \ldots, y_n)^T$. A general solution to (3.44) may be written in a form

$$y(t) = y^0(t) + \beta_1 y^1(t) + \ldots + \beta_n y^n(t) \tag{3.46}$$

where $y^0(t)$ is a solution of (3.44) and $y^1(t), \ldots, y^n(t)$ are linearly independent solutions of a homogeneous system

$$y' = A(t)y \tag{3.47}$$

A fundamental matrix $Y(t)$ of (3.47) is composed of columns $y^1(t), \ldots, y^n(t)$ and also satisfies the matrix differential equation

$$\frac{dY}{dt} = A(t)Y \tag{3.48}$$

subject to the initial conditions

$$Y(a) = D \tag{3.49}$$

If the matrix D is regular, the matrix solution $Y(t)$ is fundamental. Equation (3.46) may be rewritten in the form

$$y(t) = y^0(t) + Y(t)\beta \tag{3.50}$$

where $\beta = (\beta_1, \ldots, \beta_n)^T$ is a vector that can be determined from the boundary conditions (3.45):

$$(B_1 D + B_2 Y(b))\beta = c - B_1 y^0(a) - B_2 y^0(b) \tag{3.51}$$

Equation (3.51) represents a set of linear algebraic equations for the unknowns $\beta_i, i = 1, \ldots, n$. To evaluate the coefficients in this set it is necessary to integrate one times the inhomogeneous set (3.44) from a to b and n times the homogeneous set (3.47) from a to b with initial conditions given by the matrix D [see (3.49)]. These profiles may be either stored and the solution found by making use of

(3.46) or we can store only the initial conditions used [i.e., $y^0(a)$ and D]; the initial conditions for the solution $y(t)$ [i.e., $y(a)$] may be calculated from (3.46). Of course, to find this solution we must integrate (3.44) once more with these initial conditions. The results of this integration may be used to test the accuracy of satisfying boundary conditions.

This formulation is rather general and, as a result, a further generalization to multipoint boundary value problems, that is, for the boundary conditions

$$\sum_{i=1}^{s} B_i y(t_i) = c \qquad (3.52)$$

may be performed easily. The set (3.51) may be then written

$$[B_1 D + \sum_{i=2}^{s} B_i Y(t_i)]\beta = c - \sum_{i=1}^{s} B_i y^0(t_i) \qquad (3.53)$$

where toward solution of a homogeneous set (3.48) an initial condition

$$Y(t_1) = D \qquad (3.54)$$

has been chosen. The integration is performed within the region where the points t_i, $i = 1, \ldots, s$, are specified. For a majority of boundary value problems this formulation is too general and a great deal of integrations result.

Now assume separated boundary conditions

$$B_1 y(a) = c \qquad (3.55)$$

$$B_2 y(b) = d \qquad (3.56)$$

where B_1 and B_2 are the $r \times n$ and $(n - r) \times n$ full-rank matrices, respectively. The initial condition for solution of the inhomogeneous equation (3.44), $y^0(a)$, has been chosen to satisfy (3.55), that is,

$$B_1 y^0(a) = c \qquad (3.57)$$

The initial conditions for integration of homogeneous equations will be chosen in such a way as to satisfy the homogeneous boundary conditions (3.55), that is,

$$B_1 y^i(a) = 0 \qquad i = 1, 2, \ldots, n - r \qquad (3.58)$$

To meet (3.58) $n - r$ components of the vectors $y^i(a)$ will be chosen (with respect to the required linear independence) and the resulting r components will be calculated from (3.58). The homogeneous differential set (3.47) must be integrated only $n - r$ times for these initial conditions, and the solution of the original problem (3.44), (3.55), and (3.56) will be found in a form

$$y(t) = y^0(t) + \sum_{i=1}^{n-r} \beta_i y^i(t) \qquad (3.59)$$

The boundary condition (3.55) is satisfied for an arbitrary β_i as a result of (3.57) and (3.58); (3.56) yields

$$\sum_{i=1}^{n-r} \beta_i B_2 y^i(b) = d - B_2 y^0(b) \qquad (3.60)$$

Equation (3.60) represents a set of $n - r$ linear algebraic equations for the unknowns $\beta_1, \ldots, \beta_{n-r}$.

Below we describe a method where only the inhomogeneous set (3.44) is to be integrated. The solution is supposed as a linear combination of solutions of an inhomogeneous problem, $y^i(t)$:

$$y(t) = \sum_{i=1}^{n+1} \beta_i y^i(t) \tag{3.61}$$

Evidently to fulfill (3.45), we obtain

$$\sum_{i=1}^{n+1} \beta_i [B_1 y^i(a) + B_2 y^i(b)] = c \tag{3.62}$$

Equation (3.62) represents a set of n linear algebraic equations. In addition, if we require that the linear combination given by (3.61) satisfy (3.44), the following relation must be valid:

$$\sum_{i=1}^{n+1} \beta_i [(y^i(t))' - A(t) y^i(t)] = f(t) \tag{3.63}$$

Since $y^i(t)$ satisfies (3.44), we have

$$\sum_{i=1}^{n+1} \beta_i f(t) = f(t)$$

which gives rise to a missing condition for the evaluation of β_i:

$$\sum_{i=1}^{n+1} \beta_i = 1 \tag{3.64}$$

The set of equations (3.62) and (3.64) yields the unknown constants β_i which after insertion into (3.61) result in $y(t)$. The algorithm described above is sometimes referred to as the method of complementary functions.

For solution of any arbitrary set of linear differential equations, a linear dependence is valid:

$$y(b) = By(a) + d \tag{3.65}$$

Here, of course, the matrix B and the vector d are dependent on a and b as well as on $A(t)$ and $f(t)$. Equation (3.65) may be easily verified if a general solution of (3.50) is adopted for $t = a$:

$$y(a) = y^0(a) + Y(a)\beta$$

that is,

$$\beta = Y(a)^{-1}[y(a) - y^0(a)] \tag{3.66}$$

On the other hand, for $t = b$

$$y(b) = y^0(b) + Y(b)\beta = Y(b)Y(a)^{-1}y(a) + y^0(b) - Y(b)Y(a)^{-1}y^0(a) \tag{3.67}$$

Relation (3.67) is identical to (3.65).

The coefficients in (3.65) (i.e., the matrix B and the vector d) may be calculated after $n + 1$ integrations of (3.44) for different initial conditions $y^i(a)$. However, it is obvious that we need not calculate these coefficients, and the fact that this dependence is linear yields sufficient information. For instance, for the boundary conditions (3.55) and (3.56), $n - r + 1$ initial conditions $y^i(a)$, $i = 1, 2, \ldots, n - r + 1$, must be guessed so that (3.55) is fulfilled; that is,

$$B_1 y^i(a) = c \tag{3.68}$$

These initial conditions are used toward integration of the inhomogeneous set (3.44). The results calculated at $t = b$ are inserted into (3.56), and for each initial condition a residuum vector R^i results:

$$R^i = B_2 y^i(b) - d \qquad i = 1, 2, \ldots, n - r + 1 \tag{3.69}$$

Making use of a generalized secant method (or the Warner's scheme; see Chapter 2) an approximation of $y(a)$ results which is an exact solution since there is a linear dependence between R and $y(a)$.

3.3   Method of Adjoints

Let us introduce an adjoint system to (3.44):

$$x' = -A^T(x)x \tag{3.70}$$

and investigate the relation

$$\frac{d}{dt} \sum_{i=1}^{n} x_i y_i = \frac{d}{dt} x^T y = x^T y' + (x^T)' y$$

After combination with (3.44) and (3.70), we get

$$\frac{d}{dt} \sum_{i=1}^{n} x_i y_i = x^T(Ay + f) - (A^T x)^T y = x^T A y + x^T f - x^T A y = x^T f$$

or in a scalar form

$$\frac{d}{dt} \sum_{i=1}^{n} x_i y_i = \sum_{i=1}^{n} x_i f_i \tag{3.71}$$

On integration of (3.71) from a to b, one may write

$$\sum_{i=1}^{n} x_i(b) y_i(b) - \sum_{i=1}^{n} x_i(a) y_i(a) = \sum_{i=1}^{n} \int_a^b x_i(t) f_i(t) \, dt \tag{3.72}$$

The relation (3.72) is often referred to as Green's identity and represents a fundamental relation for the method of adjoints. It is not restricted by the values of initial conditions $x(a)$ or $x(b)$. Evidently, these relations represent the missing conditions for $y(a)$ or $y(b)$.

Let us choose n different "initial" vectors $x^1(b), \ldots, x^n(b)$ and form a $n \times n$ matrix:

$$X(b) = B^T \tag{3.73}$$

After insertion into (3.72), we have

$$By(b) - X^T(a)y(a) = \int_a^b X^T(t) f(t) \, dt = r \tag{3.74}$$

where $X(a)$ has been calculated by integration of (3.70) from b to a with the initial condition (3.73). The integrals on the right-hand side of (3.74) may be evaluated simultaneously with this integration; for example, together with the integration of (3.70) we can integrate

$$r' = -\sum x_i f_i, \qquad r(b) = 0$$

In addition to (3.74), the boundary conditions (3.45) must be also satisfied, that is,

$$B_2 y(b) + B_1 y(a) = c \tag{3.45}$$

Equations (3.74) and (3.45) represent a set of $2n$ linear algebraic equations for $2n$ unknown variables $y(a)$ and $y(b)$.

The initial conditions for the adjoint variables can also be chosen at $x = a$. The selection of the matrix B is quite general; however, some special formulation may lower the dimension of the set of linear algebraic equations or decrease the number of integrations, supposing that the boundary conditions are not of a completely general form given by (3.45).

For a special form of chosen terminal conditions

$$X(b) = B_2^T \tag{3.75}$$

(3.74) yields

$$B_2 y(b) - X^T(a) y(a) = r$$

After substitution for $B_2 y(b)$ from (3.45), we get

$$[B_1 + X^T(a)] y(a) = c - r \tag{3.76}$$

Equation (3.76) is a set of n linear algebraic equations. In an analogous way we may choose

$$X(a) = B_1^T \tag{3.77}$$

Here again a set of n equations for the unknowns $y(b)$ results:

$$[B_2 + X^T(b)] y(b) = r + c \tag{3.78}$$

If a certain row of the matrix B_2 is zero, the pertinent column $X(b)$ [and also $X(t)$] is also zero and the integration of the adjoint set need not be performed for these initial conditions. As a result, the number of necessary integrations decreases. Assuming that the matrix B_1 contains n_1 zero rows and the matrix B_2 contains n_2 zero rows (evidently $n_1 + n_2 \leq n$), it is necessary to integrate the adjoint equations (3.70) either

$$(n - n_1) \text{ times from } a \text{ to } b$$

or

$$(n - n_2) \text{ times from } b \text{ to } a$$

For both alternatives we determine the values of initial or terminal conditions, $y(a)$ or $y(b)$, respectively. Of course, the direction of the integration process will be chosen in such a way that a lower number of integrations of adjoint equations results. It is obvious that we must also take into consideration the possible inherent instability of these systems in a given direction. The profiles $y(t)$ can be calculated by integration of the original system (3.44) for already known initial conditions.

For the multipoint boundary conditions the Green's identity (3.72) can be considered for all regions $\langle t_i, t_{i+1}\rangle$; here the points t_i are such points where the boundary conditions are specified.

3.4 Method of Splitting the Differential Operator: The Factorization Method

The factorization method takes advantage of the solution of initial value problems associated with the original boundary value problem. The term "factorization" has been developed from the decomposition of the differential operator

$$L^{(n)} = \frac{d^n}{dt^n} - q$$

to differential operators of the first order, so called "factors," that is,

$$L^{(n)}y = L_1(L_2(\ldots L_n y)\ldots)$$

The basic idea of this method will be explained for a second-order differential equation in the form ($' = d/dt$)

$$(p(t)y')' - q(t)y(t) = f(t) \tag{3.79}$$

subject to the boundary conditions (3.2)

$$\alpha_0 y(a) + \beta_0 y'(a) = \gamma_0 \tag{3.2a}$$

$$\alpha_1 y(b) + \beta_1 y'(b) = \gamma_1 \tag{3.2b}$$

In operator form (3.79) may be rewritten

$$(DpD - q)y = f \tag{3.80}$$

Here D denotes the operator of differentiating. Our goal is to decompose the operator $DpD - q$ in the form

$$DpD - q = \frac{1}{\varphi}(D + \varphi q)(\varphi pD - 1) \tag{3.81}$$

Here φ is an as yet unknown function of the variable t. After some rearrangement we get

$$DpD - q = \frac{1}{\varphi}D\varphi pD + q\varphi pD - \frac{1}{\varphi}D - q$$

$$p'D + pDD = \frac{1}{\varphi}[\varphi'p + \varphi p']D + \frac{1}{\varphi}\varphi pDD + q\varphi pD - \frac{1}{\varphi}D$$

Finally, we may write

$$\left(\frac{\varphi'}{\varphi}p + q\varphi p - \frac{1}{\varphi}\right)D = 0$$

To guarantee the validity of the last equation, we must require that

$$\varphi' + q\varphi^2 = \frac{1}{p} \tag{3.82}$$

The decomposition (3.81) is valid if the function $\varphi(t)$ obeys (3.82). Now (3.80) may be rewritten

$$\frac{1}{\varphi}(D + \varphi q)(\varphi p D - 1)y = f \tag{3.83}$$

The required function y can be found by a successive solution of equations

$$\frac{1}{\varphi}(D + \varphi q)\xi = f \tag{3.84}$$

$$(\varphi p D - 1)y = \xi \tag{3.85}$$

which can be rewritten

$$\xi' + \varphi q\xi = \varphi f \tag{3.86}$$

$$y' - \frac{1}{\varphi p}y = \frac{\xi}{\varphi p} \tag{3.87}$$

Let us find for three first-order differential equations (3.82), (3.86), and (3.87) such conditions which assure that the resulting $y(t)$ would meet (3.79) subject to (3.2a) and (3.2b). At $t = a$ the differential equation (3.87) yields

$$\varphi(a)p(a)y'(a) - y(a) = \xi(a) \tag{3.88}$$

Assuming that $\alpha_0 \neq 0$, we can fulfill (3.2a) in the following way:

$$\varphi(a) = -\frac{1}{p(a)}\frac{\beta_0}{\alpha_0}, \qquad \xi(a) = -\frac{\gamma_0}{\alpha_0} \tag{3.89}$$

Integration of (3.82) and (3.86) from a to b yields $\varphi(t)$ and $\xi(t)$. At $t = b$, (3.87) yields

$$\varphi(b)p(b)y'(b) - y(b) = \xi(b) \tag{3.90}$$

Since we have calculated $\varphi(b)$ and $\xi(b)$ by integration, (3.90) and (3.2b) give rise to a set of two linear algebraic equations for the unknowns $y(b)$ and $y'(b)$. On solving, we have

$$y(b) = \frac{\beta_1\xi(b) + \gamma_1\varphi(b)p(b)}{\alpha_1\varphi(b)p(b) + \beta_1} \tag{3.91}$$

After integration of differential equations (3.87) from b to a, the solution of the original problem (3.79) with boundary conditions (3.2a) and (3.2b) may be obtained.

To develop (3.89) we have supposed that $\alpha_0 \neq 0$. Now the case $\alpha_0 = 0$ will be dealt with:

$$p(DpD - q) = (pD + \varphi)(pD - \varphi) \tag{3.92}$$

On using $\varphi D - D\varphi = -(D\varphi)$ and after rearrangement, we may develop for φ:

$$\varphi' + \frac{\varphi^2}{p} = q \tag{3.93}$$

The equations for ξ and y result from (3.92):

$$p\xi' + \varphi\xi = pf \tag{3.94}$$

$$py' - \varphi y = \xi \tag{3.95}$$

After inserting (3.95) into (3.2a) at $t = a$ and assuming that $\beta_0 \neq 0$, we get for φ and ξ:

$$\varphi(a) = -\frac{\alpha_0}{\beta_0}p(a), \qquad \xi(a) = \frac{\gamma_0}{\beta_0}p(a) \tag{3.96}$$

The profiles $\xi(t)$ and $\varphi(t)$ may be calculated by integration of (3.93) and (3.94) subject to (3.96) from a to b. At $t = b$, (3.95) is valid:

$$p(b)y'(b) - \varphi(b)y(b) = \xi(b)$$

In an analogous way to (3.91) we may develop

$$y(b) = \frac{\gamma_1 p(b) - \beta_1 \xi(b)}{\beta_1 \varphi(b) + \alpha_1 p(b)} \tag{3.97}$$

The solution of the original problem can be calculated by integration of (3.95) from $t = b$ to $t = a$ by using the initial condition (3.97).

Evidently, we have replaced the integration of the original boundary value problem for a second-order differential equation (3.79) by an integration of an initial value problem for two first-order differential equations from a to b and one first-order differential equation from b to a. To perform the latter integration it is necessary to store the profiles $\xi(t)$ and $\varphi(t)$. This shortcoming may be overcome in two different ways. The first takes advantage of "composed factorization." However, this procedure is not economical if the solution $y(t)$ is to be known at more than one point t, $t \in (a, b)$. The second possibility makes use of the integration of the original second-order differential equation from $t = b$ since the terminal values of $y(b)$ and $y'(b)$ are known. In this particular case the fulfillment of boundary conditions at $t = a$ may be checked. Unfortunately, this procedure may be spoiled by the instability of the original equations.

Without any development we will show the factorization method for a set of $2n$ first-order differential equations:

$$\begin{pmatrix} x_1(t) \\ x_2(t) \end{pmatrix}' + \begin{pmatrix} A_1(t), A_2(t) \\ A_3(t), A_4(t) \end{pmatrix}\begin{pmatrix} x_1(t) \\ x_2(t) \end{pmatrix} = \begin{pmatrix} f_1(t) \\ f_2(t) \end{pmatrix} \tag{3.98}$$

subject to the boundary conditions

$$x_1(a) + U_1 x_2(a) = u_1 \tag{3.99}$$

$$x_1(b) + U_2 x_2(b) = u_2 \tag{3.100}$$

Here x_1, x_2, f_1, f_2, u_1, and u_2 are vectors of n components and A_1, A_2, A_3, A_4, U_1, and U_2 are $n \times n$ matrices. The algorithm is as follows:

1. Integrate the differential equations

$$G' = GA_4 - A_1 G - GA_3 G + A_2$$
$$g' = -(A_1 + GA_3)g + f_1 + Gf_2 \tag{3.101}$$

with the initial conditions

$$G(a) = U_1, \qquad g(a) = u_1 \tag{3.102}$$

from $t = a$ to $t = b$. Here $G(t)$ is an $n \times n$ matrix and $g(t)$ is a vector of n components. For each t we may write

$$x_1(t) + G(t)x_2(t) = g(t) \tag{3.103}$$

2. From a set

$$\begin{aligned} x_1(b) + G(b)x_2(b) &= g(b) \\ x_1(b) + U_2 x_2(b) &= u_2 \end{aligned} \tag{3.104}$$

the values of $x_2(b)$ and $x_1(b)$ are calculated.
3. Integrate the differential equation

$$x_2' = -(A_4 - A_3 G)x_2 + f_2 - A_3 g \tag{3.105}$$

from $t = b$ to $t = a$. The components of $x_1(t)$ are calculated in the course of integration from the identity (3.103). Instead of this integration we can integrate (3.98) from $t = b$ to $t = a$; this alternative does not require us to store the profiles $G(t)$ and $g(t)$.

For the reverse direction of integration, the following algorithm may be developed:

1. Integrate (3.101), where $G = \bar{G}$ and $g = \bar{g}$, with the initial conditions

$$\bar{G}(b) = U_2, \qquad \bar{g}(b) = u_2 \tag{3.106}$$

For each t, (3.103) is valid:

$$x_1(t) + \bar{G}(t)x_2(t) = \bar{g}(t) \tag{3.107}$$

2. From the set

$$\begin{aligned} x_1(a) + \bar{G}(a)x_2(a) &= \bar{g}(a) \\ x_1(a) + U_1 x_2(a) &= u_1 \end{aligned} \tag{3.108}$$

the values $x_2(a)$ and $x_1(a)$ can be calculated.
3. Integrate differential equation (3.105) for $G = \bar{G}$ and $g = \bar{g}$ from $t = a$ to $t = b$. The variable $x_1(t)$ is calculated from (3.107).

If we do not intend to integrate the differential equation (3.105) or (3.98) we may take advantage of (3.103) and (3.107) to determine $x_1(t)$ and $x_2(t)$. For this particular case we must integrate (3.101) and (3.102) from a to b and (3.101) and (3.106) from b to a. However, the overall number of first-order differential equations to be integrated is $2(n^2 + 1)$, in comparison with the former case, which requires integration only of $n^2 + n + 1$ differential equations.

The procedure presented is formulated below for more general boundary conditions. Consider a set of n differential equations

$$x'(t) + A(t)x(t) = f(t) \tag{3.109}$$

subject to the boundary conditions

$$U_1 x(a) = u_1$$
$$U_2 x(b) = u_2 \tag{3.110}$$

To solve (3.109) an identity like (3.103) or (3.107) will be developed. For an $m \times n$ matrix $R(t)$ and an m-component vector $r(t)$, the following differential equations will be defined:

$$R'(t) = R(t)A(t) + Z(t)R(t) \tag{3.111}$$

and

$$r'(t) = R(t)f(t) + Z(t)r(t) \tag{3.112}$$

Here $Z(t)$ is an arbitrary $m \times m$ matrix. If

$$R(t_0)x(t_0) = r(t_0) \tag{3.113}$$

is valid, then we may write for all t from an appropriate region

$$R(t)x(t) = r(t) \tag{3.114}$$

The identity (3.114) may be developed easily. On left multiplication of (3.109) by a matrix $R(t)$ and after inserting for $R(t)A(t)$ and $R(t)f(t)$ from (3.111) and (3.112), we have

$$[R(t)x(t) - r(t)]' = Z(t)[R(t)x(t) - r(t)]$$

Letting

$$y(t) = R(t)x(t) - r(t)$$

it is obviously

$$y'(t) = Z(t)y(t)$$

and as a result of (3.113) also,

$$y(t_0) = 0$$

The last two relations yield $y(t) \equiv 0$ and prove the validity of (3.114). For different selection of the matrix $Z(t)$, different functions $R(t)$ and $r(t)$ result which are interconnected by a relation

$$R_2(t) = K(t)R_1(t)$$
$$r_2(t) = K(t)r_1(t) \tag{3.115}$$

if for a certain $t = t_0$, (3.115) is valid. Here we have denoted R_1, r_1 and R_2, r_2 the functions resulting after integration of (3.111)–(3.113) for two different matrices $Z(t)$.

The algorithm toward solution of the boundary value problem given by (3.109) and (3.110) may be performed as follows. The solutions $R(t)$ and $r(t)$ of the initial value problem (3.111)–(3.112) for the initial conditions

$$R(a) = K_1 U_1, \qquad r(a) = K_1 u_1 \tag{3.116}$$

must be found by integration from a to b for an arbitrary regular matrix K_1. In an analogous way the solutions $\bar{R}(t)$ and $\bar{r}(t)$ of the initial value problem

$$\bar{R}'(t) = \bar{R}(t)A(t) + \bar{Z}(t)\bar{R}(t) \qquad (3.117)$$

$$\bar{r}'(t) = \bar{R}(t)f(t) + \bar{Z}(t)\bar{r}(t) \qquad (3.118)$$

with the initial conditions

$$\bar{R}(b) = K_2 U_2, \qquad \bar{r}(b) = K_2 u_2 \qquad (3.119)$$

must be determined by integration of (3.117) and (3.118) from b to a for an arbitrary regular matrix K_2. A solution of (3.109) $x(t)$ which fulfils

$$R(t)x(t) = r(t) \qquad (3.120)$$

satisfies at $t = a$ the boundary condition (3.110), while a solution $x(t)$ which meets

$$\bar{R}(t)x(t) = \bar{r}(t) \qquad (3.121)$$

obeys at $t = b$ the second boundary condition (3.110). As a result, a solution of (3.109), $x(t)$, which satisfies both (3.120) and (3.121), that is,

$$\begin{bmatrix} R(t) \\ \bar{R}(t) \end{bmatrix} x(t) = \begin{bmatrix} r(t) \\ \bar{r}(t) \end{bmatrix} \qquad (3.122)$$

matches both boundary conditions (3.110) and, evidently, is a solution of the original boundary value problem (3.109)–(3.110). Let us note that the number of rows of $R(t)$ or $\bar{R}(t)$ is the same as the number of rows U_1 or U_2, respectively; that is, the matrix of the set (3.122) is $n \times n$.

The function $Z(t)$ may be dependent on R, so we may choose the function $Z(t, R(t), r(t))$ in such a way that integration of (3.111) and (3.112) does not result in an overflow. The same conclusions are valid for $\bar{Z}(t)$. The method that we have described above is referred to as composed factorization. For a two-point boundary value problem it can be shown that the integrations used for evaluation of $\bar{R}$ and $\bar{r}$ can be performed rather economically. Suppose that we have already calculated $R(t)$ and $r(t)$ by integration of (3.111) and (3.112), which satisfy the initial conditions (3.116). Then, for an arbitrary solution of (3.109) which fulfills the first boundary condition (3.110) at $t = a$, one can show that, at $t = b$,

$$R(b)x(b) = r(b) \qquad (3.123)$$

This condition, together with the resulting boundary conditions at $t = b$, that is,

$$U_2 x(b) = u_2 \qquad (3.124)$$

yields a sufficient number of equations for determination of the unknown "terminal" conditions at $t = b$. The original equation (3.109) can be then integrated from $t = b$ to $t = a$. We may proceed in an analogous way for $\bar{R}(t)$, $\bar{r}(t)$ at $t = a$. It can be easily shown that this procedure is, for $Z(t) \equiv 0$, identical with the method of adjoints for a special selection of initial conditions for adjoint variables. The reader who is interested in details of the factorization method is referred to the Bibliography at the end of the chapter. There, for instance, can be found procedures that lower the number of necessary integrations for some types of boundary conditions. Below only the case of composed factorization with multipoint boundary conditions will be dealt with. Suppose that the boundary conditions are in the form

$$U_i x(t_i) = u_i \qquad i = 1, 2, \ldots, k \tag{3.125}$$

The algorithm is as follows:

The initial value problems

$$\begin{aligned} R_i'(t) &= R_i(t)A(t) + Z_i(t)R_i(t) \\ r_i'(t) &= R_i(t)f(t) + Z_i(t)r_i(t) \end{aligned} \tag{3.126}$$

with the initial conditions

$$R_i(t_i) = K_i U_i, \qquad r_i(t_i) = K_i u_i \qquad i = 1, \ldots, k \tag{3.127}$$

are to be solved. The solution $x(t)$ can be then obtained from the equation

$$R(t)x(t) = r(t) \tag{3.128}$$

where $R = (R_1^T, R_2^T, \ldots, R_k^T)^T$, $r = (r_1^T, r_2^T, \ldots, r_k^T)^T$, and K_i are regular matrices.

3.5 Invariant Imbedding Approach

The invariant imbedding approach frequently incorporates such methods as the field method, the factorization method, the method of sweeps, and the Riccati transformation method. A detailed discussion of this approach for the linear BVP may be found in the literature, for instance in Scott's book (1973). In this section only an algorithm developed by Scott will be presented.

Consider a set of differential equations

$$u'(t) = A(t)u(t) + B(t)v(t) + e(t) \tag{3.129}$$

$$-v'(t) = C(t)u(t) + D(t)v(t) + f(t) \tag{3.130}$$

in the interval $t \in \langle 0, x \rangle$ subject to mixed boundary conditions

$$\begin{aligned} \alpha_1 u(0) + \beta_1 v(0) + \gamma_1 u(x) + \delta_1 v(x) &= \eta_1 \\ \alpha_2 u(0) + \beta_2 v(0) + \gamma_2 u(x) + \delta_2 v(x) &= \eta_2 \end{aligned} \tag{3.131}$$

Here $u(t)$ and $v(t)$ are m- and n-dimensional vectors, respectively; A, B, C, D, α_i, β_i, γ_i, and δ_i are matrices; and e, f, and η_i are vectors. The invariant imbedding approach operates with the interval length x and the solution obtained depends on this variable. Evidently, a whole class of solutions is constructed. Consider two transformations: the Riccati transformation,

$$u(t) = R_1(t)v(t) + R_2(t)u(0) + R_3(t) \tag{3.132}$$

and the recovery transformation,

$$v(0) = S_1(t)v(t) + S_2(t)u(0) + S_3(t) \tag{3.133}$$

The first term on the right-hand side of (3.132) represents a contribution of a homogeneous system, the second term includes a contribution arising from boundary conditions at $t = 0$, and the last term is a contribution because of the source term in (3.129) and (3.130). After differentiation of (3.132) and (3.133) with respect to t and on using (3.129) and (3.130), the following equations may be constructed:

$$R_1'(t) = B(t) + A(t)R_1(t) + R_1(t)D(t) + R_1(t)C(t)R_1(t) \qquad (3.134)$$

$$R_2'(t) = [A(t) + R_1(t)C(t)]R_2(t) \qquad (3.135)$$

$$R_3' = [A + R_1C]R_3 + R_1f + e \qquad (3.136)$$

$$S_1' = S_1[D + CR_1] \qquad (3.137)$$

$$S_2' = S_1CR_2 \qquad (3.138)$$

$$S_3' = S_1[CR_3 + f] \qquad (3.139)$$

with the initial conditions

$$R_1(0) = 0, \qquad R_2(0) = I, \qquad R_3(0) = 0$$
$$S_1(0) = I, \qquad S_2(0) = 0, \qquad S_3(0) = 0 \qquad (3.140)$$

The calculation procedure is then as follows:

1. Integration of (3.134)–(3.139) with initial conditions (3.140).
2. Evaluation of (3.132) and (3.133) for $t = x$. This operation gives rise to

$$[I - R_2(x)]u(0) - R_1(x)v(x) = R_3(x)$$
$$-S_2(x)u(0) - S_1(x)v(x) + v(0) = S_3(x) \qquad (3.141)$$

These relations, together with the boundary conditions (3.131), result in a set of linear algebraic equations for determination of $u(0)$, $v(0)$, $u(x)$, and $v(x)$.

3. The values are inserted into (3.132) and (3.133). We have for an arbitrary t, $t \in \langle 0, x \rangle$,

$$v(t) = [S_1(t)]^{-1}[v(0) - S_2(t)u(0) - S_3(t)] \qquad (3.142)$$

After inserting $v(t)$ into (3.132), we calculate $u(t)$.

This procedure possesses some interesting features:

1. The results of integration of a set (3.134)–(3.139) must be stored only at the point t where the solution is required.
2. Supposing that the dependence of the solution on the length is required, then for a new value of x only steps 2 and 3 must be repeated. A new integration of the initial value problem is not necessary. The same conclusion is valid if the boundary conditions (3.131) are changed.
3. The algorithm proposed may result in diminishing the number of significant digits. Modifications of this algorithm must be adopted to prevent this effect.

4. For the mixed boundary conditions (3.131) the number of integrations of the particular initial value problems is in an agreement with the method of superposition of solution, which is also quite flexible with respect to the different length of the interval x and boundary conditions as well.

5. For a different source term [$e(t)$, $f(t)$ in (3.129) and (3.130)] it is necessary to integrate again only (3.136) and (3.139).

3.6 Discussion

Next, the methods presented above are compared with respect to the following criteria:

1. The extent of the computer time expenditure (i.e., the number of integrations for methods described in Sections 3.2, 3.3, and 3.4) and the necessary computer memory to store the results.

2. The possibility of improving the effectiveness of the algorithm for repeated integration for a number of different right-hand terms $f(t)$ in (3.109).

3. The possibility of improving the effectiveness for repeated calculations for different boundary conditions (3.110) and interval length.

4. Applicability for nonlinear boundary conditions and mixed boundary conditions.

We will now elaborate on these criteria.

1. A general set of n differential equations of first order, (3.109), subject to boundary conditions (3.110) will be discussed. Here the number of rows of the matrices U_1 and U_2 is r and $n - r$, respectively. A comparison of different approaches is presented in Table 3-7. Based on this table the reader can easily compare the advantages and shortcomings of methods discussed with respect to the number of necessary integrations as well as computer memory.

2. For repeated calculations with different right-hand terms $f(t)$, Table 3-7 reveals that a reasonable method for presolving such problems is the first row (computer time is important) and the second row (the storage capacity criterion is important) in the modified superposition of solution method. A number of other procedures can compete with these methods (see Table 3-7).

3. Table 3-7 also contains information on how to presolve the problem if the matrix U_1 and the vector u_1 or the matrix U_2 and the vector u_2 change but the number of rows is constant. The advantages of particular procedures may be inferred from Table 3-7. The case with different boundary conditions in both points is also included. Here r also changes.

TABLE 3-7
COMPARISON OF METHODS

Method	Number of integrated first-order differential equations in direction: $a \to b$	$b \to a$	Number of profiles to be stored	Necessity to solve the linear algebraic equations: number × order	Additional work for different $f(t)$ — Integrations $a \to b$	$b \to a$	Solution of lin. alg. eqs.: number × order
Superposition of solutions (general)	$(n+1)n$ $(n+2)n$	0 0	$(n+1)n$ 0 [a]	$1 \times n$ $1 \times n$	n $2n$	0 0	$1 \times n$ $1 \times n$ [a]
Modified superposition of solutions [see (3.59)]	$(n-r+1)n$ $(n-r+2)n$	0 0	$(n-r+1)n$ 0 [a]	$1 \times (n-r)$ and $(n-r+1) \times r$ —"—	n $2n$	0 0	$1 \times (n-r)$ and $1 \times r$ $1 \times (n-r)$ and $1 \times r$ [a]
Method of complementary functions [see (3.61)]	$(n+1)n$ $(n+2)n$	0 0	$(n+1)n$ 0 [a]	$1 \times (n+1)$ $1 \times (n+1)$	repeat repeat		
Method of adjoints	$(n+1)n$ n	n $(n-r+1)n$	0 0 0 0	$1 \times 2n$ $1 \times 2n$ $1 \times n$ $1 \times n$	$n+1$ $n+2$ $r+1$ n	n 0 n $n-r+1$	$1 \times 2n$ [e] $1 \times 2n$ [e] $1 \times n$ [e] $1 \times n$ [e]
Combined factorization [see (3.122)]	$(n+1)r$ $(n+1)r$ n	$(n-r)(n+1)$ n $(n-r)(n+1)$	$(n+1)r$ or $(n-r+1)n$ 0 0	$p \times n$ [e] $1 \times n$ $1 \times n$	r r n	$n-r$ n $n-r$	$p \times n$ [e,f] $1 \times n$ [g] $1 \times n$ [h]
Factorization for (3.98)–(3.100)	$(r+1)r$ $(r+1)r$	r n	$(r+1)r$ 0	$1 \times n$ $1 \times n$	r r	$n-r$ n	$1 \times n$ $1 \times n$ [i]

Here $r = \dfrac{n}{2}$

TABLE 3-7
(Continued)

Method	Additional work for different BC at $t = a$ — Integrations $a \rightarrow b$	$b \rightarrow a$	Solution of lin. alg. eqs.: number × order	Additional work for different BC at $t = b$ — Integrations $a \rightarrow b$	$b \rightarrow a$	Solution of lin. alg. eqs.: number × order	Additional work for different structure of BC — Integrations $a \rightarrow b$	$b \rightarrow a$	Solution of lin. alg. eqs.: number × order
Superposition of solutions (general)	0	0	$1 \times n$	0	0	$1 \times n$	0	0	$1 \times n$
	n	0	$1 \times n$[a]	n	0	$1 \times n$[a]	n	0	$1 \times n$[a]
Modified superposition of solutions [see (3.59)]	repeat		[b]	repeat		$1 \times (n-r)$ and $1 \times r$	repeat		[b]
	repeat		[b]	repeat		—"—	repeat		[b]
Method of complementary functions [see (3.61)]	0	0	$1 \times (n+1)$	0	0	$1 \times (n+1)$	0	0	$1 \times n$
	n	0	$1 \times (n+1)$[a]	n	0	$1 \times (n+1)$[a]	n	0	$1 \times n$[a]
	0	n	$1 \times 2n$[d]	0	n	$1 \times 2n$[d]	0	n	$1 \times 2n$[d]
Method of adjoints	n	0	[b]	n	0	$1 \times 2n$[d]	n	0	$1 \times 2n$[d]
	0	n	$1 \times n$[d]	0	n	$1 \times n$[d]	0	n	[b]
Combined factorization [see (3.122)]	$(n+1)r$	0	$p \times n$[h]	0	$(n-r)(n+1)$	$p \times n$[g]	repeat	0	[b]
	repeat	0	$1 \times n$[d]	0	n	$1 \times n$[d]	repeat	0	[b]
Factorization for (3.98)–(3.100)	n	repeat		repeat	r	$1 \times n$	repeat	0	
	n	repeat		repeat	n	$1 \times n$[d]	repeat	0	

Here $r = \dfrac{n}{2}$

[a] Initial and terminal conditions for computed profiles are stored.
[b] More effective strategy exists which utilizes the profiles already obtained.
[c] The profiles of adjoint variables are stored.
[d] The coefficients of linear systems are stored.
[e] p, number of points where solution $y(t)$ is requested.
[f] R and $\bar{R}$ are stored.
[g] R is stored.
[h] $\bar{R}$ is stored.
[i] G is stored.

4. Generally speaking, the methods presented, with the exception of factorization, can also be readily adopted for mixed boundary conditions. For nonlinear boundary conditions the method of superposition of solution or the method of adjoints may be used. We are going to show the method of superposition of solution for (3.44) and the boundary conditions

$$F_i(y(a), y(b)) = 0 \qquad i = 1, 2, \ldots, n \qquad (3.143)$$

The solution is supposed in the form (3.46) and $y^0(t), y^1(t), \ldots, y^n(t)$ can be calculated by making use of (3.47)–(3.49). On inserting from (3.50) into (3.129), we have

$$F_i(y^0(a) + Y(a)\beta, y^0(b) + Y(b)\beta) = 0 \qquad i = 1, \ldots, n \qquad (3.144)$$

Evidently, (3.144) represents a set of nonlinear algebraic equations for unknowns $\beta_1, \ldots, \beta_n$. After solving these equations by the Newton–Raphson method, $y(t)$ can be calculated from (3.50).

There is a number of other special procedures which can be used to solve linear boundary value problems. However, these procedures cannot be easily adopted to solve a general nonlinear boundary value problem but are tailored to specific linear problems. We do not discuss such procedures in this book.

PROBLEMS

1. Solve the linear boundary value problem

$$y'' - 400y = 400 \cos^2 \pi z + 2\pi^2 \cos 2\pi z$$

subject to the boundary conditions

$$y(0) = y(1) = 0$$

Show that the solution is given by

$$y(z) = \frac{e^{-20}}{1 + e^{-20}} e^{20z} + \frac{1}{1 + e^{-20}} e^{-20z} - \cos^2 \pi z$$

Try to solve the problem by making use of superposition and show that this problem is difficult for superposition. Compare superposition with the imbedding algorithm.

Reference: Scott, M. R.: Numerical solution of boundary value problems for O.D.E., in *Numerical Solutions of Boundary Value Problems for Ordinary Differential Equations* (ed. A. K. Aziz). Academic Press, New York, 1975.

2. Consider

$$y'' - (z^2 - 1)y = 0$$

subject to

$$y(0) = 1, \qquad y(10) = e^{-50}$$

The general solution of the differential equation is

$$y(z) = Ae^{-z^2/2} + Be^{-z^2/2} \int_0^z e^{t^2} dt$$

With the prescribed initial conditions, we have

$$y(z) = e^{-z^2/2}$$

Show that (a) forward integration with exact initial values is successful only over an interval of about $z = 6$; (b) backward integration is successful. Test also the finite-difference approach and the imbedding algorithm.

Reference: SCOTT, M. R.: ibid.

3. Consider the linear boundary value problem

$$y'' - 11y' - 12y + 22e^z = 0$$

subject to the boundary conditions

$$y(0) = 1, \quad y(15) = e^{15}$$

The general solution of the differential equation is

$$y(z) = Ae^{-z} + Be^{12z} + e^z$$

With the given boundary conditions, the solution is

$$y(z) = e^z$$

Show that neither forward nor backward integration of the original is appropriate. Prove by a numerical experiment that even if the exact initial values are prescribed, the forward integration can be successfully performed only for about $z = 3.0$. Show that the imbedding algorithm is successful over the interval $0 \le z \le 15.0$. Solve this equation by the finite-difference approach.

Reference: SCOTT, M. R.: ibid.

4. Stress distribution in a spherical membrane with normal and tangential loads is described by the linear boundary value problem

$$y'' + (3 \cot z + 2 \tan z)y' + 0.7y = 0$$

subject to the boundary conditions

$$y\left(\frac{\pi}{6}\right) = 0, \quad y\left(\frac{\pi}{3}\right) = 5$$

The solution curve has a sharp spike approximately at 0.535, with the magnitude of the solution at this point approximately 283. Test the procedures described for linear boundary value problems on this example.

Reference: LENTINI, M., AND PEREYRA, V.: Boundary problem solvers for first order systems based on deferred correction, in A. K. Aziz, ed., *Numerical Solutions of Boundary Value Problems for Ordinary Differential Equations*. Academic Press, New York, 1975.

5. Mass transfer in tubular reactors with significant axial mixing effects for a consecutive first-order reaction is described by a set of differential equations:

$$-\frac{1}{Pe}\frac{d^2 f_1}{dz^2} + \frac{df_1}{dz} + Da_1 f_1 = 0$$

$$-\frac{1}{Pe}\frac{d^2 f_2}{dz^2} + \frac{df_2}{dz} + Da_2 f_2 - Da_1 f_1 = 0$$

$$z = 0: \qquad f_1 = 1 + \frac{1}{\text{Pe}}\frac{df_1}{dz}$$

$$f_2 = 1 + \frac{1}{\text{Pe}}\frac{df_2}{dz}$$

$$z = 1: \qquad \frac{df_1}{dz} = \frac{df_2}{dz} = 0$$

Solve these differential equations for $\text{Pe} = 100$, $\text{Da}_1 = \text{Da}_2 = 0.2$.

Show that the forward integration of these equations is difficult unless an orthogonalization procedure is used. Prove by a numerical experiment that the backward integration can be successfully performed.

Try the finite-difference approach and suggest an appropriate ordering of variables in order to get a band matrix.

BIBLIOGRAPHY

Finite-difference methods are discussed in:

KELLER, H. B.: Numerical solution of boundary value problems for ordinary differential equations: survey and some recent results on difference methods, in *Numerical Solution of Boundary Value Problems for Ordinary Differential Equations* (ed. A. K. Aziz). Academic Press, New York, 1975, pp. 27–88.

FOX, L.: *Numerical Solution of Two-Point Boundary Problems in Ordinary Differential Equations.* Oxford University Press, London, 1957.

HENRICI, P.: *Discrete Variable Methods in Ordinary Differential Equations.* J. Wiley, New York, 1962.

ISAACSON, E., AND KELLER, H. B.: *Analysis of Numerical Methods.* J. Wiley, New York, 1966.

KELLER, H. B.: *Numerical Methods for Two-Point Boundary Value Problems.* Blaisdell, Waltham, Mass., 1968.

COLLATZ, L.: *The Numerical Treatment of Differential Equations.* Springer-Verlag, Berlin, 1960.

The methods of superposition of solution, complementary functions, and adjoints are discussed in:

ROBERTS, S. M., AND SHIPMAN, J. S.: *Two Point Boundary Value Problems: Shooting Methods.* Elsevier, New York, 1971.

OSBORNE, M. R.: On shooting methods for boundary value problems. *J. Math. Anal. Appl.* 27, 417–433 (1969).

GOODMAN, T. R., AND LANCE, G. N.: The numerical integration of two-point boundary value problems. *Math. Tables Other Aids Comput.* 10, 54, 82–86 (1956).

BELLMAN, R. E., AND KALABA, R. E.: *Quasilinearization and Nonlinear Boundary Value Problems.* American Elsevier, New York, 1965.

Orthonormalization technique combined with integration of initial value problems which eliminates the case of "almost linearly dependent" solutions is discussed in:

SCOTT, M. R., AND WATTS, H. A.: Superposition, orthonormalization, quasilinearization and two-point boundary-value problems, in *Codes for Boundary-Value Problems in Ordinary Differential Equations* (ed. B. Childs et al.), Lecture Notes in Computer Science. Springer-Verlag, Berlin, 1979.

BELLMAN AND KALABA: see above.

Details on factorization and invariant imbedding methods may be found in:

BABUŠKA, I., PRÁGER, M., AND VITÁSEK, E.: *Numerical Processes in Differential Equations.* Wiley-Interscience, London/SNTL, Prague, 1966.

TAUFER, J.: *Apl. Mat.* 11, 427–451 (1966); 13, 199–200, 201–202, 191–198 (1968); 17, 209–224 (1972).

SCOTT, M. R.: *Invariant Imbedding and Its Applications to Ordinary Differential Equations. An Introduction.* Addison-Wesley, Reading, Mass., 1973.

LEE, E. S.: *Quasilinearization and Invariant Imbedding.* Academic Press, New York, 1968.

SCOTT, M. R.: On the conversion of BVPs into stable initial-value problems via several invariant imbedding algorithms, in *Numerical Solution of Boundary Value Problems for Ordinary Differential Equations* (ed. A. K. Aziz). Academic Press, New York, 1975, pp. 89–196.

Numerical Methods for Nonlinear Boundary Value Problems **4**

This and subsequent chapters represent the heart of the book. In this part, methods for solution of nonlinear boundary value problems are discussed. Section 4.1 is devoted to finite-difference approximations of the problem and to solution of the resulting sparse set of nonlinear algebraical equations. In Section 4.2 a representation of nonlinear boundary value problems in terms of integral equations is described. The quasi-linearization which takes advantage of the linearization of original nonlinear equations is discussed in Section 4.3. Here a sequence of linear boundary value problems must be solved. Section 4.4 is devoted to an analysis of the false transient method. This approach takes advantage of the idea that the solution of a nonlinear boundary value problem may be found by solving the pertinent transient equation. Integration of this equation proceeds through time until the solution ceases to change significantly. In Section 4.5 a new class of methods, one-parameter imbedding techniques, is discussed. The methods described in Sections 4.1 through 4.5 are compared. In Section 4.6 the shooting methods, when the boundary value problem is replaced by an initial value problem, are dealt with. Methods such as invariant imbedding and variational techniques are omitted because they are not universal and their applicability depends strongly on the problem in question.

4.1 Finite-Difference Methods

Solution of nonlinear two-point boundary value problems can often be found by the finite-difference approach. For problems where shooting methods fail (see Section 4.6) the finite-difference method can be adopted successfully. In contradistinction to shooting methods, where the right-hand sides of differential equations are written as a subroutine which, connected with an appropriate correction algorithm, may easily yield solutions for the majority of "stable" problems, the finite-difference methods sometimes require several weeks of engineering and programming effort. For many physical and engineering problems, integration of the relevant initial value problem gives rise to a "blow-up" effect, that is, to profiles that diverge beyond a prescribed limit (overflow in the digital computer). For a great number of practical problems, however, even a very accurate initial guess does not allow us to integrate across and the blow-up effect is unavoided. On the other hand, the finite-difference technique may be successful for such types of problems, and usually no substantial difficulties are encountered. Of course, for large and strongly nonlinear systems, the development of the appropriate finite-difference equations and their solution is a tedious task even for an experienced programmer.

The finite-difference technique will be outlined largely for a simple second-order differential equation

$$F[x, y(x), y'(x), y''(x)] = 0 \qquad (4.1)$$

subject to the boundary conditions

$$y(a) = A \qquad (4.2a)$$

$$y(b) = B \qquad (4.2b)$$

The goal of this section is to present some methods to replace nonlinear differential equations together with the boundary conditions by the finite-difference analogy. Finite-difference methods using approximations of higher order based on Hermitian approximations, as well as the compact difference scheme, are discussed. A connection between a compact high-order difference scheme and orthogonal collocation is shown. For strongly sensitive problems, the advantages of the finite-difference approach over shooting methods are discussed. Furthermore, problems of grouping of finite-difference equations for a set of second-order equations will be dealt with. Finally, numerical methods for solution of finite-difference equations will be suggested.

4.1.1 Construction of the Finite-Difference Analogy

The differential equation is replaced in the selected meshpoints $x_0 = a$, $x_1, \ldots, x_{n-1}, x_n = b$ by the finite-difference analogy. Usually, the equidistant mesh is used (i.e., $x_{i+1} - x_i = h$). After replacing the first and second derivatives in (4.1) by the finite-difference formulas

$$y'(x_i) = \frac{y(x_{i+1}) - y(x_{i-1})}{2h} + O(h^2) \qquad (4.3)$$

and

$$y''(x_i) = \frac{y(x_{i+1}) - 2y(x_i) + y(x_{i-1})}{h^2} + O(h^2) \qquad (4.4)$$

the finite-difference analogy to (4.1) is

$$F\left[x_i, y_i, \frac{y_{i+1} - y_{i-1}}{2h}, \frac{y_{i+1} - 2y_i + y_{i-1}}{h^2}\right] = 0 \qquad (4.5)$$

for $i = 1, 2, \ldots, n - 1$. The boundary conditions are

$$y_o = A, \qquad y_n = B \qquad (4.6)$$

Equations (4.5) and (4.6) form a set of $n + 1$ nonlinear algebraic equations for $n + 1$ unknowns $y_o, y_1, \ldots, y_n$. In fact, for the simple form of boundary conditions (4.6), only $n - 1$ equations result. This set of equations is solved by methods that were discussed in Chapter 2 (e.g., by the Newton–Raphson method). Note that the structural matrix is tridiagonal (Fig. 4-1). For the general boundary conditions

$$g_o(y(a), y'(a)) = 0$$
$$g_1(y(b), y'(b)) = 0 \qquad (4.7)$$

the tridiagonal nature of the structural matrix can be retained for the approximation

$$g_o\left(y_o, \frac{y_1 - y_o}{h}\right) = 0, \qquad g_1\left(y_n, \frac{y_n - y_{n-1}}{h}\right) = 0 \qquad (4.8)$$

Unfortunately, the first derivative approximation is of the first order, so that the approximation to (4.5) and (4.7) is of first order only. To overcome this drawback, two different procedures can be used:

1. The finite-difference analogy is considered for $i = 0, 1, \ldots, n$ and the approximation to (4.7) is

$$g_o\left(y_o, \frac{y_1 - y_{-1}}{2h}\right) = 0$$
$$g_1\left(y_n, \frac{y_{n+1} - y_{n-1}}{2h}\right) = 0 \qquad (4.9)$$

We have now a set of $n + 3$ equations for $n + 3$ unknowns.

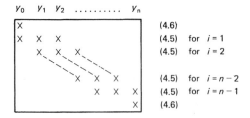

Figure 4-1 Structural matrix for (4.5) and (4.6).

2. The first derivatives $y'(a)$ and $y'(b)$ in (4.7) are replaced by asymmetrical formulas:

$$g_0\left(y_o, \frac{-3y_o + 4y_1 - y_2}{2h}\right) = 0$$

$$g_1\left(y_n, \frac{3y_n - 4y_{n-1} + y_{n-2}}{2h}\right) = 0 \tag{4.10}$$

Now a set of $n + 1$ equations for $n + 1$ unknowns results. In both cases, however, the structure of the original structural matrix is destroyed; for the second case the structural matrix is as shown in Fig. 4-2.

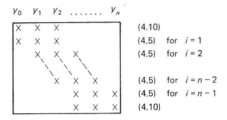

Figure 4-2 Structural matrix for (4.5) and (4.10).

A generalization to a set of differential equations is obvious; however, only for an appropriate sequence of variables does a matrix of the band type result.

4.1.2 Finite-Difference Methods
Using Approximations of Higher Order

So far we have considered only the simple three-points approximation to differential equations of second order. In this section we address ourselves to approximations of higher order.

Collatz (1960) presented the Hermitian method, which exhibits a higher order of approximation than that of the common finite-difference formulas. This approximation makes use of the fact that the differential equation must be satisfied at each grid point. This approximation can be readily developed for differential equations of simple form. For a special case of differential equations of second order without a first derivative

$$y'' = f(x, y)$$

the approximation of fourth order is

$$y_{i+1} - 2y_i + y_{i-1} = \frac{h^2}{12}[f(x_{i-1}, y_{i-1}) + 10f(x_i, y_i) + f(x_{i+1}, y_{i+1})]$$

Fox (1957) has recommended taking advantage of approximations using more than three grid points. This type of approximation yields a multidiagonal structural matrix.

For a number of engineering problems where the profile is expected to be sufficiently smooth, a low number of mesh points may be used, the resulting finite-difference approximation makes use of all mesh points. Such type of finite-difference schema will be referred to as a compact high-order difference scheme. The development for a set of first-order differential equations is presented next.

Let us consider a set of first-order differential equations

$$(y^i)' = f^i(x, y^1, y^2, \dots, y^N) \qquad i = 1, 2, \dots, N \tag{4.11}$$

subject to two-point boundary conditions,

$$\sum_{i=1}^{N} a_i^j y^i(a) + \sum_{i=1}^{N} b_i^j y^i(b) = c^j \qquad j = 1, 2, \dots, N \tag{4.12}$$

For the sake of simplicity, linear two-point boundary conditions have been chosen. Now the points $x_0 = a, x_1, \dots, x_{n-1}, x_n = b$ will be selected on $\langle a, b \rangle$. Since N boundary conditions (4.12) are given, one has to construct nN relations in order to establish $N(n + 1)$ values of $y^i(x_j), i = 1, 2, \dots, N; j = 0, 1, \dots, n$. Therefore, the difference approximation will be "centered" at the points $\tilde{x}_j$ defined by the relations

$$\tilde{x}_j = \alpha x_j + (1 - \alpha) x_{j+1} \qquad j = 0, 1, \dots, n - 1 \tag{4.13}$$

where $\alpha \in \langle 0, 1 \rangle$. Usually, $\alpha = 0.5$ will be used. Taking advantage of the method of undetermined coefficients, the difference formulas for the first derivative of y can be written:

$$\frac{dy(\tilde{x}_j)}{dx} \sim \tilde{y}_j' = \sum_{k=0}^{n} D_{jk} y_k \qquad j = 0, 1, \dots, n - 1 \tag{4.14}$$

where y_k denotes $y(x_k)$. The coefficients D_{jk} can be calculated from the systems of linear algebraic equations

$$XD^T = P_1 \tag{4.15}$$

where

$$X = \begin{pmatrix} 1 & 1 & \cdots & 1 \\ x_0 & x_1 & & x_n \\ x_0^2 & x_1^2 & & x_n^2 \\ \cdot & & & \\ \cdot & & & \\ \cdot & & & \\ x_0^n & x_1^n & & x_n^n \end{pmatrix} \tag{4.16}$$

$$D = \begin{pmatrix} D_{00} & D_{01} & \cdots & D_{0n} \\ D_{10} & D_{11} & \cdots & D_{1n} \\ \cdot & & & \\ \cdot & & & \\ \cdot & & & \\ D_{n-1,0} & D_{n-1,1} & \cdots & D_{n-1,n} \end{pmatrix}$$

$$P_1 = \begin{pmatrix} 0 & 0 & \cdots & 0 \\ 1 & 1 & \cdots & 1 \\ 2\tilde{x}_0 & 2\tilde{x}_1 & \cdots & 2\tilde{x}_{n-1} \\ \cdot & & & \\ \cdot & & & \\ \cdot & & & \\ n\tilde{x}_0^{n-1} & n\tilde{x}_1^{n-1} & \cdots & n\tilde{x}_{n-1}^{n-1} \end{pmatrix}$$

The Lagrange interpolation formula for any arbitrary y takes the form

$$y(\tilde{x}_j) \sim \tilde{y}_j = \sum_{k=0}^{n} C_{jk} y_k \qquad j = 0, 1, \ldots, n-1 \tag{4.17}$$

where the coefficients C_{jk} obey the system of equations

$$XC^T = P_2 \tag{4.18}$$

where $C = \{C_{jk}\}$ and

$$P_2 = \begin{pmatrix} 1 & 1 & \cdots & 1 \\ \tilde{x}_0 & \tilde{x}_1 & \cdots & \tilde{x}_{n-1} \\ \tilde{x}_0^n & \tilde{x}_1^n & \cdots & \tilde{x}_{n-1}^n \end{pmatrix} \tag{4.19}$$

For the application of method III (see below), a quadrature formula will be constructed:

$$\int_{x_j}^{x_{j+1}} y(x)\, dx \sim \sum_{k=0}^{n} A_{jk} y_k \qquad j = 0, 1, \ldots, n-1 \tag{4.20}$$

The coefficients A_{jk} satisfy

$$XA^T = P_3 \tag{4.21}$$

where $A = \{A_{jk}\}$ and

$$P_3 = \begin{pmatrix} \mu_{00} & \mu_{01} & \cdots & \mu_{0,n-1} \\ \mu_{10} & \mu_{11} & \cdots & \mu_{1,n-1} \\ \cdot & & & \\ \cdot & & & \\ \cdot & & & \\ \mu_{n0} & \mu_{n1} & \cdots & \mu_{n,n-1} \end{pmatrix} \tag{4.22}$$

Here

$$\mu_{ij} = \int_{x_j}^{x_{j+1}} x^i\, dx$$

Using the approximations, three distinct difference procedures result. It is obvious that the following relations are valid ($i = 1, 2, \ldots, N, j = 0, 1, \ldots, n-1$):

Method I

$$\sum_{k=0}^{n} D_{jk} y_k^i = f^i\left[\tilde{x}_j, \sum_{k=0}^{n} C_{jk} y_k^1, \sum_{k=0}^{n} C_{jk} y_k^2, \ldots, \sum_{k=0}^{n} C_{jk} y_k^N\right] \tag{4.23}$$

Method II

$$\sum_{k=0}^{n} D_{jk} y_k^i = \sum_{k=0}^{n} C_{jk} f^i[x_k, y_k^1, \ldots, y_k^N] \tag{4.24}$$

Method III

$$y_{j+1}^i - y_j^i = \sum_{k=0}^{n} A_{jk} f^i[x_k, y_k^1, \ldots, y_k^N] \tag{4.25}$$

In the first method the difference approximation is centered at the point $\tilde{x}_j$; that is, the nonlinearities f^i are evaluated at these points from the values $\tilde{y}^i$ calculated from interpolation. When using the second method, the nonlinear functions f^i are established at $\tilde{x}_j$ by interpolating from the values f^i enumerated in the mesh points x_j. The last procedure differs substantially from the first two. Instead of an approximation of the derivative, a substitution for an integral is made:

$$\int_{x_j}^{x_{j+1}} [y^i(x)]' \, dx = \int_{x_j}^{x_{j+1}} f^i(x, y^1, y^2, \ldots, y^N) \, dx$$

In order to enable a facile application of the methods presented, the coefficients D, C, and A for $b - a = 1$ and $n = 2 - 5$ are given in Tables 4-1 through 4-4.

TABLE 4-1
COEFFICIENTS D, C, AND A: $n = 2$, $\alpha = 0.5$, $x_1 = 0.5$

		k			
		0	1	2	
j	$\tilde{x}_j$	0.0	0.5	1.0	x_k
0	0.25	−2.00000	2.00000	0.00000	D_{jk}
1	0.75	0.00000	−2.00000	2.00000	
0	0.25	0.37500	0.75000	−0.12500	C_{jk}
1	0.75	−0.12500	0.75000	0.37500	
0		0.20833	0.33333	−0.04167	A_{jk}
1		−0.04167	0.33333	0.20833	

TABLE 4-2
COEFFICIENTS D, C, AND A: $n = 3$, $\alpha = 0.5$, $x_1 = 0.333$, $x_2 = 0.666$

	k				
j	0	1	2	3	
0	−2.87800	2.62819	0.37425	−0.12444	D_{jk}
1	0.12500	−3.37782	3.37725	−0.12444	
2	0.12575	−0.37707	−2.61751	2.86882	
0	0.31256	0.93722	−0.31194	0.06216	C_{jk}
1	−0.06256	0.56278	0.56194	−0.06216	
2	0.06288	−0.31410	0.93900	0.31222	
0	0.12489	0.26356	−0.06925	0.01380	A_{jk}
1	−0.01389	0.18004	0.18025	−0.01380	
2	0.01400	−0.06994	0.26475	0.12519	

TABLE 4-3

COEFFICIENTS D, C, AND A: $n = 4$, $\alpha = 0.5$, $x_1 = 0.25$, $x_2 = 0.50$, $x_3 = 0.75$

j	k 0	1	2	3	4	
0	−3.66667	2.83333	1.50000	−0.83333	0.16667	
1	0.16667	−4.50000	4.50000	−0.16667	0.00000	
2	0.00000	0.16667	−4.50000	4.50000	−0.16667	D_{jk}
3	−0.16667	0.83333	−1.50000	−2.83333	3.66667	
0	0.27344	1.09375	−0.54688	0.21875	−0.03906	
1	−0.03906	0.46875	0.70313	−0.15625	0.02344	
2	0.02344	−0.15625	0.70312	0.46875	−0.03906	C_{jk}
3	−0.03906	0.21875	−0.54688	1.09375	0.27344	
0	0.08715	0.22431	−0.09167	0.03681	−0.00660	
1	−0.00660	0.12014	0.15833	−0.02569	0.00382	
2	0.00382	−0.02569	0.15833	0.12014	−0.00660	A_{jk}
3	−0.00660	0.03681	−0.09167	0.22431	0.08715	

TABLE 4-4

COEFFICIENTS D, C, AND A: $n = 5$, $\alpha = 0.5$

j	k 0	1	2	3	4	5	
0	−4.39844	2.61719	3.72396	−2.89062	1.13281	−0.18490	
1	0.18490	−5.50781	5.39062	0.02604	−0.11719	0.02344	
2	−0.02344	0.32552	−5.85938	5.85938	−0.32552	0.02344	D_{jk}
3	−0.02344	0.11719	−0.02604	−5.39062	5.50781	−0.18490	
0	0.24609	1.23047	−0.82031	0.49219	−0.17578	0.02734	
1	−0.02734	0.41016	0.82031	−0.27344	0.08203	−0.01172	
2	0.01172	−0.09766	0.58594	0.58594	−0.09766	0.01172	C_{jk}
3	−0.01172	0.08203	−0.27344	0.82031	0.41016	−0.02734	
4	0.02734	−0.17578	0.49219	−0.82031	1.23047	0.24609	
0	0.06597	0.19819	−0.11083	0.06694	−0.02403	0.00375	
1	−0.00375	0.08847	0.14194	−0.03583	0.01069	−0.00153	
2	0.00153	−0.01292	0.11139	0.11139	−0.01292	0.00153	A_{jk}
3	−0.00153	0.01069	−0.03583	0.14194	0.08847	−0.00375	
4	3.00375	−0.02403	0.06694	−0.11083	0.19819	0.06597	

To match the boundary conditions (4.12),

$$\sum_{i=1}^{n} a_i^j y_0^i + \sum_{i=1}^{n} b_i^j y_n^i = c^j \qquad j = 1, 2, \ldots, n \qquad (4.26)$$

must hold. Equations (4.26), together with, for instance, (4.23), represent a set of $N(n + 1)$ nonlinear algebraic equations for the unknown values y_0^i $[= y^i(a)]$, $y_1^i, \ldots, y_n^i$ $[= y^i(b)]$, $i = 1, 2, \ldots, N$. It can be solved by the Newton–Raphson method when the f^i are continuously differentiable. The Newton–Raphson method converges fast to the solution if the initial guess is not too far from it. The choice of the mesh points is generally a difficult task. It appears to be convenient to place the mesh points uniformly for solving problems without any preliminary information regarding the nature of the solution. In attempting to obtain a solution one makes use, as a rule, of four to nine mesh points, as is discussed below. The most appropriate choice of α proved to be $\alpha = 0.5$.

In the following example a comparison of all three methods mentioned will be made from the point of view of their applicability, speed of convergence, and accuracy attained.

Example 4.1

A nonadiabatic tubular piston-flow recycle reactor in which a single first-order reaction takes place is described by dimensionless enthalpy and mass balances

$$\frac{d\theta}{dx} = B\mathrm{Da}(1 - y) \exp\left(\frac{\theta}{1 + \theta/\gamma}\right) - \beta(\theta - \theta_c) = R_1(\theta, y) \qquad (4.27)$$

$$\frac{dy}{dx} = \mathrm{Da}(1 - y) \exp\left(\frac{\theta}{1 + \theta/\gamma}\right) = R_2(\theta, y) \qquad (4.28)$$

subject to the boundary conditions

$$\begin{aligned} \theta(0) &= (1 - \lambda)\theta(1) \\ y(0) &= (1 - \lambda)y(1) \end{aligned} \qquad (4.29)$$

where the following parameters and variables are used: θ, dimensionless temperature; y, conversion; x, dimensionless space coordinate, $x \in \langle 0, 1 \rangle$; B, dimensionless adiabatic temperature rise; Da, Damköhler number; γ, dimensionless activation energy; β, dimensionless heat-transfer coefficient; θ_c, dimensionless temperature of cooling medium; λ, recycle ratio. Construct a uniform mesh of grid points $x_i = i/n$ and $\alpha = 0.5$. One can find for method I ($n = 2$) the values of coefficients in Table 4-1; a simple approximation results:

$$\begin{aligned} -2\theta_0 + 2\theta_1 &= R_1(\tilde{\theta}_0, \bar{y}_0) \\ -2y_0 + 2y_1 &= R_2(\tilde{\theta}_0, \bar{y}_0) \\ -2\theta_1 + 2\theta_2 &= R_1(\tilde{\theta}_1, \bar{y}_1) \\ -2y_1 + 2y_2 &= R_2(\tilde{\theta}_1, \bar{y}_1) \end{aligned} \qquad (4.30)$$

where according to Table 4-1, for example,

$$\tilde{\theta}_0 = 0.375\theta_0 + 0.750\theta_1 - 0.125\theta_2$$

The boundary conditions may be written

$$-\theta_0 + (1 - \lambda)\theta_2 = 0 \qquad (4.31)$$
$$-y_0 + (1 - \lambda)y_2 = 0$$

At each step, the values of θ_2 and y_2 can be expressed through (4.31) as a function of θ_0 and y_0, so that four unknowns θ_0, y_0, θ_1, and y_1 are involved in (4.30).

Table 4-5 shows the course of iteration procedure when the Newton–Raphson method for the solution of algebraic equations is used. For most problems of that type, three to six iterations are sufficient in order to obtain the tolerance 10^{-6}; all three methods require approximately the same number of iterations. In Table 4-6 the deviation of the outlet temperature θ_n from its correct value

TABLE 4-5
COURSE OF ITERATIONS FOR METHOD I:
$B = 4$, $Da = 0.1$, $\gamma = 20$, $\lambda = 0.5$, $\beta = 0$, $n = 2$, $\alpha = 0.5$

Iteration	θ_0	y_0	θ_1	y_1	$\theta_2{}^a$	$y_2{}^a$
0[b]	0.7259	0.1815	1.0520	0.2630	1.4518	0.3631
1	0.9555	0.2389	1.3767	0.3442	1.9111	0.4778
2	0.9640	0.2410	1.3898	0.3474	1.9281	0.4820
3	0.9640	0.2410	1.3898	0.3474	1.9281	0.4820
Accurate value[c]	0.9705	0.2426	1.3996	0.3499	1.9409	0.4852

[a]Calculated from the boundary conditions.
[b]The initial values are the solution for $Da = 0.09$.
[c]Results obtained by the shooting method.

TABLE 4-6
DEPENDENCE OF DEVIATIONS $\theta_n - \theta^a(1)$ ON THE NUMBER OF MESH POINTS:
$B = 4$, $Da = 0.12$, $\gamma = 20$, $\beta = 0$, $\lambda = 0.5$, $\theta^a(1) = 3.05728$

$n + 1$	Method I	Method II	Method III
2	$25{,}952 \times 10^{-5}$	$-11{,}790 \times 10^{-5}$	$-11{,}790 \times 10^{-5}$
3	5,193	4,804	51
4	159	−689	−453
5	233	292	55
6	−39	43	37
9	2	2	2
10	2	2	2

[a]Exact solution.

$\theta(1)$, obtained by using the shooting method, is presented as a dependence of the number of the mesh points involved. It is obvious that the results for $n = 9$ are practically identical for all three methods and the difference is of the order 10^{-5}. The same dependence is plotted in Figs. 4-3 and 4-4 for two different sets of parameters B and Da. In the latter case, where conversion approaches to 1.0 and an inflection point occurs on the profile, the error is evidently larger. The results of particular methods differ from each other and method II diverges.

When the dependence of the error of methods I and III on the number of mesh points is compared, it can be concluded that method III is the most convenient from the point of view of both absolute value of its error and its almost monotone convergence. For $N = 9$ the error never exceeded 0.3% when using method III.

The high-order difference method can also be used for differential equations of higher order. For the mth derivative, the following approximation is used:

$$\frac{d^m y(x_j)}{dx^m} \sim y_j^{(m)} = \sum_{k=0}^{n} D_{jk} y_k \qquad j = 1, 2, \ldots, n-1 \qquad (4.32)$$

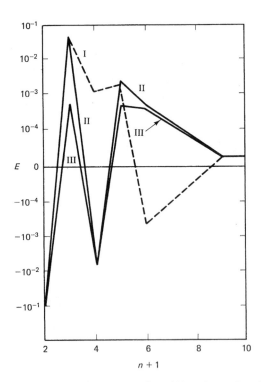

Figure 4-3 Dependence of the error $E = \theta_n - \theta(1)$ on the number of mesh points ($B = 4$, $Da = 0.12$, $\gamma = 20$, $\beta = 0$, $\lambda = 0.5$). Example 4.1.

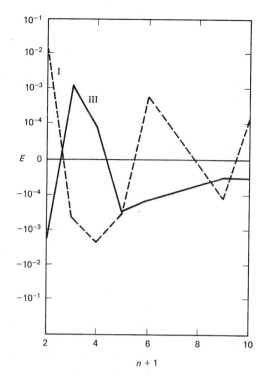

Figure 4-4 Dependence of the error $E = \theta_n - \theta(1)$ on the number of mesh points. Method II has diverged ($B = 6$, Da $= 0.08$, $\gamma = 20$, $\beta = 0$, $\lambda = 0.5$). Example 4.1.

where the coefficients D_{jk} obey (4.33)

$$XD^T = P_4 \qquad (4.33)$$

The matrix P_4 is

$$P_4 = \begin{pmatrix} 0 & 0 & 0 \\ \cdot & \cdot & \cdot \\ \cdot & \cdot & \cdot \\ \cdot & \cdot & \cdot \\ 0 & 0 & 0 \\ m! & m! & m! \\ m!\,x_1 & m!\,x_2 & m!\,x_{n-1} \\ \cdot & \cdot & \cdot \\ \cdot & \cdot & \cdot \\ \cdot & \cdot & \cdot \\ \dfrac{n!}{(n-m)!}\,x_1^{n-m} & \dfrac{n!}{(n-m)!}\,x_2^{n-m} & \dfrac{n!}{(n-m)!}\,x_{n-1}^{n-m} \end{pmatrix} \qquad (4.34)$$

Equation 4.33 can be developed after inserting $y = x^i$, $i = 0, 1, \ldots, n$, into (4.32).

Usually, this procedure is used for differential equations of first and second order. The second-order equations are frequently "diffusion-like" differential equations, for example of the type

$$\frac{d^2y}{dx^2} + \frac{a}{x}\frac{dy}{dx} = R(y) \tag{4.35}$$

subject to the boundary conditions

$$x = 0: \qquad \frac{dy(0)}{dx} = 0 \quad \text{(symmetry)} \tag{4.36}$$

$$x = 1: \qquad \alpha y(1) + \beta \frac{dy(1)}{dx} = \gamma \tag{4.37}$$

Two ways to approximate (4.35)–(4.37) are possible:

1. Equation (4.35) will be approximated by virtue of (4.32) at the mesh points $x_1, x_2, \ldots, x_{n-1}$. The boundary condition (4.36) will be approximated by (4.32) for $m = 1$ and $j = 0$. The first derivative $dy(1)/dx$ will be approximated by means of (4.32) for $m = 1$ and $j = n$. After these approximations a set of $n + 1$ nonlinear algebraic equations for $n + 1$ unknowns $y_0, y_1, \ldots, y_n$ results.

2. A set of $2n$ mesh points: $-1 = -x_n, -x_{n-1}, \ldots, -x_1, x_1, \ldots, x_{n-1}$, $x_n = 1$ will be selected. Equation (4.35) will be approximated by formula (4.32) taking into consideration $2n$ mesh points located at $x_1, x_2, \ldots, x_{n-1}$. It is obvious, due to symmetry [see (4.36)], that $y(-x_k) = y(x_k)$. Hence the symmetry enables us to eliminate one-half of the unknowns. The boundary condition (4.37) will be replaced analogously, as mentioned above. This second approach is for the boundary condition $y(1) = \gamma$ identical with the orthogonal collocation method, proposed by Villadsen and Stewart.[†] The orthogonal collocation method takes advantage of the approximation in the form

$$y(x) = y(1) + (1 - x^2) \sum_{i=0}^{n-2} a_i P_i(x^2) \tag{4.38}$$

where a_i are constant coefficients and $P_i(x^2)$ are polynomials of the order i having the argument x^2. At the chosen mesh points $x_1, x_2, \ldots, x_{n-1}$, $x_n = 1$ the function $y(x)$ has to obey the differential equation in question. It appears to be convenient to select these points as the roots of some orthogonal polynomials. Villadsen and Stewart recommended the Jacobi polynomials and reported the values of the roots $x_1, x_2, \ldots, x_n$ as well as the coefficients A_{ij} and B_{ij} for the approximations

†J. V. Villadsen and W. E. Stewart, *Chem. Eng. Sci.* 22, 1483 (1967).

$$\frac{dy(x_i)}{dx} = \sum_{j=1}^{n} A_{ij} y(x_j)$$

$$\frac{d^2 y(x_i)}{dx^2} + \frac{a}{x} \frac{dy(x_i)}{dx} = \sum_{j=1}^{n} B_{ij} y(x_j)$$

(4.39)

These values for a slab ($a = 0$), a cylinder ($a = 1$), and a sphere ($a = 2$), are reported in Tables 4-7 through 4-9 for $n = 2$ and $n = 3$.

TABLE 4-7
COLLOCATION CONSTANTS FOR SLAB GEOMETRY ($a = 0$)

n	i	x_i	A_{i1}	A_{i2}	A_{i3}	B_{i1}	B_{i2}	B_{i3}
2	1	0.44721	−1.1180	1.1180		−2.5	2.5	
	2	1	−2.5000	2.5000		−2.5	2.5	
3	1	0.285235	−1.7530	2.5076	−0.7547	−4.7399	5.6771	−0.9373
	2	0.765055	−1.3706	−0.6535	2.0241	8.3229	−23.2601	14.9373
	3	1	1.7915	−8.7915	7.0000	19.0719	−47.0789	28.0000

TABLE 4-8
COLLOCATION CONSTANTS FOR CYLINDRICAL GEOMETRY ($a = 1$)

n	i	x_i	A_{i1}	A_{i2}	A_{i3}	B_{i1}	B_{i2}	B_{i3}
2	1	0.57735	−1.7320	1.7320		−6.0000	6.0000	
	2	1	−3.0000	3.0000		−6.0000	6.0000	
3	1	0.393765	−2.5396	3.8256	−1.2860	−9.9024	12.2997	−2.3973
	2	0.803087	−1.3777	−1.2452	2.6229	9.0337	−32.7649	23.7306
	3	1	1.7155	−9.5155	8.0000	22.7575	−65.4241	42.6667

TABLE 4-9
COLLOCATION CONSTANTS FOR SPHERICAL GEOMETRY ($a = 2$)

n	i	x_i	A_{i1}	A_{i2}	A_{i3}	B_{i1}	B_{i2}	B_{i3}
2	1	0.654654	−2.2913	2.2913		−10.5000	10.5000	
	2	1	−3.5000	3.5000		−10.5000	10.5000	
3	1	0.46889	−3.1993	5.0152	−1.8158	−15.6700	20.0349	−4.3649
	2	0.830224	−1.4087	−1.8067	3.2154	9.9651	−44.3300	34.3649
	3	1	1.6968	−10.6968	9.0000	26.9329	−86.9329	60.0000

Example 4.2

The collocation method has been applied to an eigenvalue problem of the form

$$\frac{d^2 y}{dx^2} + \frac{2}{x} \frac{dy}{dx} + 2\lambda(1 - x^2)y = 0$$

(4.40)

$$y(1) = 0, \qquad \frac{dy(0)}{dx} = 0$$

(4.41)

The nontrivial solutions occur only for certain values of the parameter λ; an approximation of the first (lowest value of λ) is desired. This case arises in the Graetz–Nusselt problem of heat transfer to a Newtonian fluid in laminar flow through a tube. The smallest eigenvalue, λ_1, is the asymptotic value of the local Nusselt number for large distances into the heat-transfer region.

Approximation of (4.40) gives the collocation equations

$$\sum_{j=1}^{n} B_{ij}y(x_j) + 2\lambda(1 - x_i^2)y(x_i) = 0 \qquad i = 1, 2, \ldots, n \qquad (4.42)$$

or in a matrix notation

$$MY = \lambda Y \qquad (4.43)$$

where M is an $n \times n$ matrix with the elements given by

$$m_{ij} = \frac{-B_{ij}}{2(1 - x_i^2)} \qquad i, j = 1, 2, \ldots, n \qquad (4.44)$$

and Y is a column vector with the elements $y(x_1), \ldots, y(x_n)$. Equation (4.43) yields nontrivial solutions Y only when λ is an eigenvalue of the matrix M. The eigenvalues of M can be easily calculated by standard methods. The results of calculation are reported in Table 4-10. The orthogonal collocation approximations of λ_1 converge rapidly with increasing n.

TABLE 4-10
FIRST EIGENVALUE λ_1 OF (4.40)
SUBJECT TO (4.41)

Collocation	λ_1
$n = 2$	4.5
$n = 3$	3.679
$n = 4$	3.65714
Exact	3.656793

Source: Result taken from the paper by J. V. Villadsen and N. E. Stewart, *Chem. Eng. Sci.*, 22, 1483 (1967) Pergamon Press, Inc.

4.1.3 Aspects of the Effective Use of Finite-Difference Methods

This section has as a principal goal the discussion of problems connected with the type of finite-difference approximations, structural matrix representation of the finite-difference approximations, and strategy of evaluation of an appropriate initial guess. Finally, the approximation of boundary conditions is dealt with. The emphasis is on procedures for effective handling of different types of nonlinear boundary value problems by finite-difference methods.

1. Seeking a guide to the evaluation of an appropriate finite-difference approximation to the original nonlinear boundary value problem, we are going to discuss various approximations in the context of engineering applications. Frequently, for a second-order differential equation, three-point approximations

to first- and second-order derivatives are employed. The number of mesh points ranges from some tens to some hundreds, depending largely on the shape of the profile examined. The set of first-order differential equations $y = f(x, y)$ is usually replaced by the simplest approximation

$$\frac{y_{i+1} - y_i}{h} = \frac{f(x_{i+1}, y_{i+1}) + f(x_i, y_i)}{2} \tag{4.45}$$

which is "centered" at the point $(x_{i+1} + x_i)/2$. For "smooth" profiles the high-order difference approximation or the orthogonal collocation procedure should be applied because of low computer storage requirements and high speed of calculation. For profiles that exhibit a sudden change in the first derivative and are "flat", these methods may fail and the three-point approximations should be preferred. The coefficients in the finite-difference approximations making use of more than three points to replace the particular derivatives can be easily calculated.

With typical engineering problems, however, one will often have some idea of the shape of the solution and this will facilitate the introduction of a non-equidistant mesh. Such a procedure is virtually a necessity if the profiles exhibit a sudden change in the first derivative. This refinement of finite-difference technique may entail for difficult problems a considerable decrease in computational time.

2. Based on the results of finite-difference approximations (4.5) and (4.6) we have noticed that a simple tridiagonal matrix results. For a set of differential equations we attempt always to achieve a reduction of the sparse structural matrix to the band type. This depends largely on the ordering of particular variables. If the variables are grouped according to mesh points, then for separated boundary conditions a band matrix results. This idea can be demonstrated on a set of two second-order differential equations:

$$y'' = f_1(x, y, z, y', z') \tag{4.46}$$

$$z'' = f_2(x, y, z, y', z') \tag{4.47}$$

A three-point approximation will be used. Now we have two possibilities for grouping the finite-difference equations:

(a) Equation (4.46) will be approximated at $x_0, x_1, \ldots, x_n$ and afterwards the same is done with (4.47) at $x_0, x_1, \ldots, x_n$.
(b) Equation (4.46) will be approximated at x_0, (4.47) at x_0, (4.46) at x_1, and so on.

If we choose the latter possibility, we can group the variables y_i and z_i essentially into two different sequences: $y_0, y_1, \ldots, y_n$, $z_0, z_1, \ldots, z_n$ and $y_0, z_0, y_1, z_1, \ldots, y_n, z_n$. The structural matrix of both groupings is drawn in Fig. 4-5. The former grouping of the variables gives rise to a sparse matrix, whereas the latter ordering results in a band seven-diagonal matrix. If, for

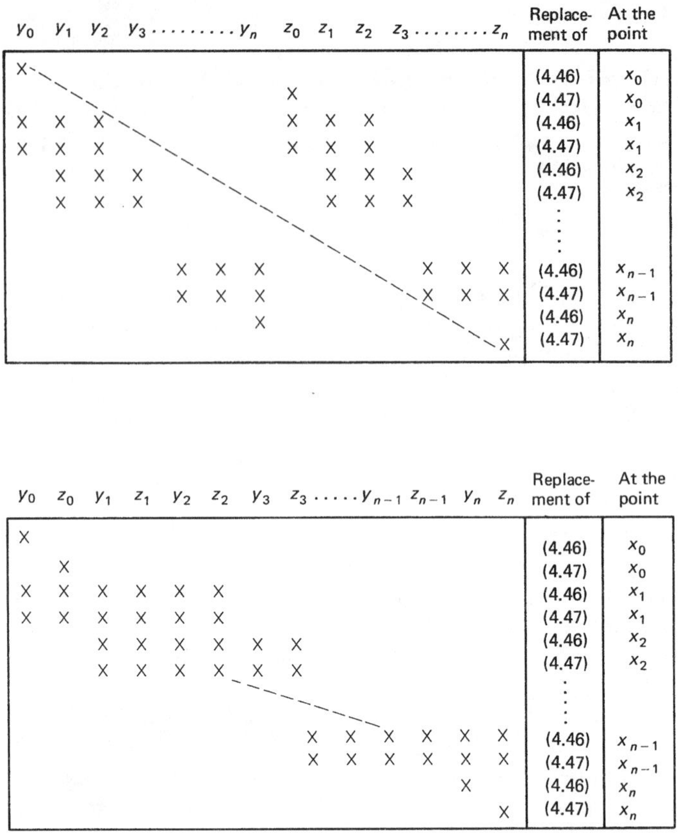

y_0	y_1	y_2	$y_3 \cdots\cdots y_n$	z_0	z_1	z_2	$z_3 \cdots\cdots z_n$	Replacement of	At the point
X								(4.46)	x_0
				X				(4.47)	x_0
X	X	X		X	X	X		(4.46)	x_1
X	X	X		X	X	X		(4.47)	x_1
	X	X	X		X	X	X	(4.46)	x_2
	X	X	X		X	X	X	(4.47)	x_2
			$\vdots$						$\vdots$
			X X X				X X X	(4.46)	x_{n-1}
			X X X				X X X	(4.47)	x_{n-1}
			X					(4.46)	x_n
							X	(4.47)	x_n

y_0	z_0	y_1	z_1	y_2	z_2	y_3	$z_3 \cdots y_{n-1}$	z_{n-1}	y_n	z_n	Replacement of	At the point
X											(4.46)	x_0
	X										(4.47)	x_0
X	X	X	X	X	X						(4.46)	x_1
X	X	X	X	X	X						(4.47)	x_1
		X	X	X	X	X	X				(4.46)	x_2
		X	X	X	X	X	X				(4.47)	x_2
							$\vdots$					$\vdots$
						X	X X X X	X	X		(4.46)	x_{n-1}
						X	X X X X	X	X		(4.47)	x_{n-1}
									X		(4.46)	x_n
										X	(4.47)	x_n

Figure 4-5 Structural matrix of finite-difference approximations for two second-order equations.

instance, the Newton–Raphson procedure is used to solve the nonlinear finite-difference equations, the resulting band matrix can be economically handled from the viewpoint of simplicity of algorithm, computer storage requirements and computer time expenditure. The number of diagonals in the band matrix depends on the size of the system and on the type of finite-difference approximations to be used. For mixed and integral boundary conditions, grouping according to mesh points yields the band matrix with some off-diagonal elements. To handle this type of structural matrix effectively, a special algorithm can be used: for example, the Crout procedure or modified Gaussian elimination.

3. In employing the finite-difference technique, a considerable complication may arise for large systems of differential equation because of the dimensions of the set of nonlinear equations. Of course, the size of the set of nonlinear equations is also large for a very short step size h. To avoid the difficulty of

estimating a "reasonable" first guess for the Newton–Raphson method, the following procedure can be used:

(a) A rather sparse grid of mesh points is selected. The problem of low dimensionality can be solved easily. The first rough guess is obtained.

(b) The number of grid points is doubled and the unknown values of dependent variables at the new grid points are evaluated by interpolation. This is the first approximation for the Newton–Raphson technique. The halving of step h is repeated as long as the preassigned tolerance is met.

(c) This scheme possesses an attractive simplicity and when employed in conjunction with the Richardson extrapolation procedure (deferred approach to the limit), the accuracy can be substantially improved without additional iterations.

For some very difficult problems with extreme gradients, the number of grid points may increase to such an extent that a numerical attack on the nonlinear finite-difference equations, with regard to round-off errors, computer time expenditure, and storage requirements, becomes a limiting factor. To handle these equations by finite-difference techniques, a nonequidistant mesh must be employed.

4. The technique for handling boundary conditions is as important as the approximation of the original differential equation. The boundary conditions must be replaced by the finite-difference formulas of the same (or higher) order of approximation as those used for approximation of differential equations. This fact is often overlooked by inexperienced programmers. As we have already shown, the mixed boundary conditions give rise to the off-diagonal elements in the structural band matrix. The integral boundary conditions must be replaced by appropriate quadrature formulas such as Newton–Cotes. There are physical problems where the differential equation is singular on the boundary. For instance, let us mention the diffusion and heat conduction problems in spherical and cylindrical geometry, where the term $(a/x)(dy/dx)$ occurs. Such problems require special care; usually, a modified equation is solved in the vicinity of the singular point.

For further work on a number of the cases discussed in this section, the reader is referred to the Problems.

4.1.4 Numerical Solution of Finite-Difference Equations

The problem we face is that of selecting an appropriate procedure toward solution of a large set of nonlinear algebraic equations. To establish the solution, the following methods can be used: the Newton–Raphson technique, the modified Newton method (the Jacobi matrix is evaluated only in the first iteration), the method of successive approximations or the modified linearization according to Holt (1964) (if convergence is guaranteed), and finally, for second-order differential equations, the relaxation technique (Pope 1960).

To demonstrate the procedures presented above, a simple second-order differential equation

$$p(x)y'' + q(x)y' + r(x)y = R(x, y) \tag{4.48}$$

subject to linear boundary conditions for $x = a$ and $x = b$ will be considered. After replacing (4.48) by the finite-difference analogy, we can write

$$p_i \frac{y_{i-1} - 2y_i + y_{i+1}}{h^2} + q_i \frac{y_{i+1} - y_{i-1}}{2h} + r_i y_i = R(x_i, y_i) \tag{4.49}$$

On denoting $Y = (y_0, y_1, \ldots, y_n)^T$ and after including the approximation of boundary conditions in the set of finite-difference equations, a matrix equation results:

$$AY = f(Y) \tag{4.50}$$

where $f_i(Y) = R(x_i, y_i)$, $i = 1, 2, \ldots, n - 1$. The structural matrix A is for (4.49) tridiagonal. The simplest method of solution of (4.50) is the simple iteration (successive approximation)

$$AY^{k+1} = f(Y^k) \tag{4.51}$$

or

$$Y^{k+1} = A^{-1}f(Y^k) \tag{4.52}$$

Here the Thomas algorithm can be used for solution of the tridiagonal system (4.51). The conditions that assure convergence of the iteration process (4.51) can usually be developed only with serious difficulties; moreover, for practical engineering problems, they are too pessimistic. The iteration process may be modified to speed up convergence; for example,

$$(A + B)Y^{k+1} = f(Y^k) + BY^k \tag{4.53}$$

where B is an appropriately chosen matrix. Equation (4.51) can be also linearized in some fashion; for example, $[y_i^k]^\alpha$ on the right-hand side is replaced by $(y_i^k)^{\alpha-1}y_i^{k+1}$. This product can now be incorporated in the left-hand side (i.e., in A); the matrix A contains in coefficients the last iteration, y_i^k.

The nonlinear successive relaxation method (NSOR) can be used to establish the solution of (4.50). We may notice that a change in the variable y_i gives rise to a change in three residua for equations $i - 1$, i, and $i + 1$ only. The "alternating corrections" method, proposed by Pope (1960), takes advantage of this fact. Let us denote by D the diagonal matrix defined $D = \text{diag}(A)$. Equation (4.50) can be written in a modified form,

$$Y = (E - D^{-1}A)Y + D^{-1}f(Y) \tag{4.54}$$

or for components,

$$y_i = a_i y_{i-1} + b_i y_{i+1} + c_i R(x_i, y_i) \qquad i = 1, 2, \ldots, n-1 \qquad (4.55)$$

For n grid points (n even) the procedure is as follows:

1. A first guess of y_i (i.e., y_i^0; $i = 0, 1, \ldots, n$) is chosen.
2. New values y_i^{k+1} are calculated for $i = 1, 3, 5, \ldots, n-1$ according to (4.55):

$$y_i^{k+1} = a_i y_{i-1}^k + b_i y_{i+1}^k + c_i R(x_i, y_i^k) \qquad (4.56)$$

3. New values y_i^{k+1} are calculated for $i = 2, 4, 6, \ldots, n-2$ by means of

$$y_i^{k+1} = a_i y_{i-1}^{k+1} + b_i y_{i+1}^{k+1} + c_i R(x_i, y_i^k) \qquad (4.57)$$

4. The values of y_0^{k+1} and y_n^{k+1} are calculated from the boundary conditions.
5. Check whether $\| Y^{k+1} - Y^k \| < \epsilon$. If this condition is violated, the calculation procedure returns back to step 2. If this condition is satisfied, the computation is either finished, if the step size $h = (b-a)/n$ is short enough owing to the accuracy required, or the iteration process is repeated when h is too large. Usually under such circumstances, the number of mesh points is doubled, the unknown values of y at the new mesh points are interpolated, and the calculation procedure returns to step 2. Unfortunately, this procedure sometimes does not converge, or if multiple solutions occur, for example, only one solution can be obtained. Despite these facts, the technique may be used for simple problems because it is not necessary to evaluate derivatives and to solve linear equations as well.

The most conventional method for the solution of nonlinear finite-difference equations is the Newton–Raphson procedure. To solve the set of nonlinear equations

$$F(Y) = AY - f(Y) = 0 \qquad (4.58)$$

the following sequence is constructed:

$$[A - \Gamma_f(Y^k)] \Delta Y^k = f(Y^k) - AY^k \qquad (4.59)$$

where we have denoted $\Delta Y^k = Y^{k+1} - Y^k$. The Jacobian matrix of first derivatives Γ_f is for (4.48) diagonal. Hence, for this simple case, the matrix $(A - \Gamma_f)$ is tridiagonal and may be solved easily. The structure of the Jacobian matrix for numerous types of nonlinear boundary value problems was discussed in the preceding section.

For a successful application of the method outlined in this section, especially for the Newton–Raphson technique, a "reasonable" initial profile is required. The procedure leading to a satisfactory guess was described in the preceding section. Sometimes we can obtain a sufficiently good initial profile by the shooting method. If the differential equations are strongly sensitive to initial conditions and the "blow-up" effect occurs, we can select as a first guess the solution for parameters across which it is still possible to integrate.

Example 4.3

The nonisothermal diffusion and first-order chemical reaction in a porous catalyst are described by a second-order differential equation:

$$\frac{d^2y}{dx^2} + \frac{a}{x}\frac{dy}{dx} = \phi^2 y \exp\left[\frac{\gamma\beta(1-y)}{1+\beta(1-y)}\right] \tag{4.60}$$

subject to the boundary conditions

$$x = 0: \quad \frac{dy}{dx} = 0$$
$$x = 1: \quad y = 1 \tag{4.61}$$

where y is concentration of the reacting component and a, β, γ, and ϕ are parameters of this problem. After replacing (4.60) according to (4.48)–(4.55) a finite-difference approximation results:

$$y_i = \frac{1}{2}\left[\left(1 + \frac{a}{2i}\right)y_{i+1} + \left(1 - \frac{a}{2i}\right)y_{i-1} - h^2 R(y_i)\right] \quad i = 1, 2, \ldots, n-1 \tag{4.62}$$

Here, for the sake of simplicity, an abridged notation $R(y_i)$ for the right-hand side of (4.60) was used. The boundary conditions are approximated by

$$\frac{-3y_0 + 4y_1 - y_2}{2h} = 0, \qquad y_n = 1 \tag{4.63}$$

The results obtained by the "alternating corrections" technique are reported in Table 4-11.

TABLE 4-11
RESULTS OBTAINED BY THE "ALTERNATING CORRECTION" TECHNIQUE:
$a = 2, \gamma = 20, \beta = 0.2, \phi = 1$

							x					
n	Itera-tion	0.0	0.1	0.2	0.3	0.4	0.5	0.6	0.7	0.8	0.9	1.0
10	0	0	1	1	1	1	1	1	1	1	1	1
	1	0.997	0.995	0.990	0.995	0.990	0.995	0.990	0.995	0.990	0.995	1
	5	0.955	0.953	0.947	0.953	0.949	0.956	0.954	0.962	0.968	0.983	1
	10	0.900	0.899	0.896	0.903	0.904	0.915	0.922	0.937	0.953	0.975	1
	20	0.821	0.821	0.824	0.836	0.846	0.864	0.883	0.908	0.935	0.966	1
	44	0.744	0.747	0.755	0.771	0.791	0.817	0.846	0.881	0.918	0.958	1
20	0	1	1	1	1	1	1	1	1	1	1	1
	1	0.999	0.998	0.998	0.998	0.998	0.998	0.998	0.998	0.998	0.998	1
	5	0.989	0.987	0.987	0.987	0.987	0.987	0.987	0.998	0.998	0.992	1
	10	0.976	0.974	0.974	0.974	0.974	0.974	0.975	0.977	0.980	0.987	1

It is obvious that for more grid points the number of iterations increases. At the early stage of calculation a sparse grid is used. After reaching the vicinity of the solution the grid is refined, and the iteration process is continued. The course of iteration is reported in Table 4-12. The lowest row of this table contains the exact solution obtained by more accurate methods.

TABLE 4-12
COURSE OF ITERATION: $a = 2$, $\gamma = 20$, $\beta = 0.2$, $\phi = 2$, $\epsilon = 0.0001$[a]

		x					
n	Itera-tion	0.0	0.2	0.4	0.6	0.8	1.0
10	0	1	1	1	1	1	1
	10	0.3469	0.2907	0.3563	0.4915	0.7114	1
	20	0.0038	0.0068	0.0255	0.1160	0.4426	1
	35[b]	0.0031	0.0060	0.0228	0.1061	0.4249	1
20	50	0.0028	0.0054	0.0218	0.1071	0.4299	1
	75[b]	0.0029	0.0055	0.0221	0.1078	0.4311	1
40	100	0.0028	0.0054	0.0218	0.1079	0.4319	1
	146[b]	0.0028	0.0053	0.0219	0.1082	0.4324	1
80	162[b]	0.0028	0.0053	0.0219	0.1082	0.4325	1
∞	∞	0.0028	0.0053	0.0219	0.1085	0.4332	1

[a] n is doubled for $(k + 1)$st iteration if $\| Y^k - Y^{k-1} \| < \epsilon$.
[b] n was doubled after this iteration.

Example 4.4

The boundary value problem from Example 4.3 will be considered for $a = 0$. The finite-difference approximation (4.60) is ($h = 1/n$):

$$F_i = y_{i-1} - 2y_i + y_{i+1} - h^2\phi^2 y_i \exp\left[\frac{\gamma\beta(1 - y_i)}{1 + \beta(1 - y_i)}\right] = 0$$
$$i = 0, 1, 2, \ldots, n - 1 \tag{4.64}$$

The approximation of boundary conditions (4.61) yields

$$y_{-1} = y_1, \qquad y_n = 1 \tag{4.65}$$

On inserting (4.65) in (4.64), we have a set of n nonlinear algebraic equations for n unknowns $y_0, y_1, \ldots, y_{n-1}$. These equations can be solved by the Newton–Raphson procedure. The elements of the Jacobian matrix $\Gamma = \{g_{i,j}\}$ are

$$g_{i,i-1} = \frac{\partial F_i}{\partial y_{i-1}} = 1 \qquad i = 1, 2, \ldots, n - 1$$

$$g_{i,i+1} = \frac{\partial F_i}{\partial y_{i+1}} = 1 \qquad i = 0, 1, \ldots, n - 2$$

$$g_{i,i} = \frac{\partial F_i}{\partial y_i} = -2 - h^2\phi^2 \exp\left[\frac{\gamma\beta(1 - y_i)}{1 + \beta(1 - y_i)}\right]\left[1 + \frac{\gamma\beta y_i}{(1 + \beta(1 - y_i))^2}\right]$$
$$i = 0, 1, \ldots, n - 1$$

The other elements of the Jacobian matrix are zero. The Newton–Raphson iteration algorithm can be written

$$\Gamma(Y^k)(Y^{k+1} - Y^k) = -F(Y^k)$$

where $Y = (y_0, y_1, \ldots, y_{n-1})^T$, $F = (F_0, F_1, \ldots, F_{n-1})^T$. The set of linear algebraic equations with tridiagonal matrix has been solved by the Thomas algorithm. The course of iterations is reported in Table 4-13. It is obvious that the rate of convergence is high.

TABLE 4-13
APPLICATION OF THE NEWTON–RAPHSON METHOD:
$a = 0, \gamma = 20, \beta = 0.05, \phi = 1, h = 0.1$[a]

Itera-tion	x										
	0.0	0.1	0.2	0.3	0.4	0.5	0.6	0.7	0.8	0.9	1.0
0	1	1	1	1	1	1	1	1	1	1	1
1	0.5000	0.5050	0.5200	0.5450	0.5800	0.6250	0.6800	0.7450	0.8200	0.9050	1
2	0.5514	0.5557	0.5686	0.5902	0.6206	0.6601	0.7088	0.7669	0.8347	0.9124	1
3	0.5522	0.5565	0.5693	0.5909	0.6212	0.6606	0.7092	0.7673	0.8350	0.9125	1
4	0.5522	0.5565	0.5693	0.5909	0.6212	0.6606	0.7092	0.7673	0.8350	0.9125	1

[a]Exact value of $y(0) = 0.5521$.

4.1.5   Remarks on the Relative Merits of the Finite-Difference Technique

Computational experience with the finite-difference technique has shown that for a practical engineering problem, this method is more difficult to apply than the shooting method. As far as comparison between these techniques is concerned, the competition between the methods appears to be for some mildly sensitive problems. Such problems can be solved with some difficulty by shooting methods. For this technique, however, preliminary information about the behavior of the solution should be at hand, the algorithm of modifying the initial guess of missing conditions must be supported by some heuristic arguments based on physical description, and finally, double-precision arithmetics must be used. The programming effort can be compared with that for finite-difference methods. Treatment of extremely sensitive problems reported in the literature (e.g., some problems in boundary layer theory) will center on utilization of the finite-difference approach. On the other hand, however, problems with extreme gradients (e.g., combustion and explosion problems) may be handled better with procedures based on the shooting method strategy. The reader can infer that for such difficult problems the state of the art in the theory of numerical solution of nonlinear differential equations does not allow us to draw general conclusions and experimentation with the problem is sometimes necessary.

1. Solve the nonlinear differential equation

$$y'' = -1 - \sin(y')$$

subject to the linear boundary conditions

$$y(0) = 0, \qquad y(1) = 1$$

After replacing this equation by its finite-difference analogy and using a simple iteration formula, we have

$$\frac{y_{i+1}^{k+1} - 2y_i^{k+1} + y_{i-1}^{k+1}}{h^2} = -1 - \sin\frac{y_{i+1}^k - y_{i-1}^k}{2h}$$

$$y_0^{k+1} = 0, \qquad y_n^{k+1} = 1$$

The initial profile, y_i^0, can be chosen constant (e.g., $y_i^0 \equiv 1$). For $h = 0.2$, calculate the profile $y(x)$. The first three iterations are reported in the following table:

x	y^0	y^1	y^2	y^3
0.0	1	0	0	0
0.2	1	0.280	0.350	0.349
0.4	1	0.520	0.621	0.618
0.6	1	0.720	0.816	0.810
0.8	1	0.880	0.941	0.934
1.0	1	1	1	1

Solve the finite-difference analogy for $h = 0.05$.

Reference: Fox, L.: *The Numerical Solution of Two-Point Boundary Problems in Ordinary Differential Equations.* Oxford University Press, New York, 1957.

2. Solve the problem given by (4.60) and (4.61) in Example 4.3. The following iteration scheme can be constructed:

$$y_i^{k+1} = \frac{1}{2}\left[\left(1 + \frac{a}{2i}\right)y_{i+1}^k + \left(1 - \frac{a}{2i}\right)y_{i-1}^k - h^2 R(y_i^k)\right]$$

for $i = 1, 2, \ldots, n - 1$. $y_n = 1$, and y_0 can be evaluated from (4.63): $y_0 = (4y_1^{k+1} - y_2^{k+1})/3$. Compare the rate of convergence of this iteration process with the "alternating correction" presented in Table 4-11.

Modify the iteration process by constructing a Gauss–Seidel-like schema:

$$y_i^{k+1} = \frac{1}{2}\left[\left(1 + \frac{a}{2i}\right)y_{i+1}^{k+1} + \left(1 - \frac{a}{2i}\right)y_{i-1}^k - h^2 R(y_i^k)\right]$$

Again compare the rate of convergence of this modification with the "alternating correction" method.

Furthermore, use the simple iteration method given by (4.51) and check the convergence for different values of the parameter ϕ. It can be shown that for higher values of ϕ, the simple iteration method can fail (see Example 4.5 in Section 4.2). To improve the convergence, use the Holt linearization technique and replace

$$\phi^2 y_i \exp\left[\frac{\gamma\beta(1-y_i)}{1+\beta(1-y_i)}\right] \sim \phi^2 y_i^{k+1} \exp\left[\frac{\gamma\beta(1-y_i^k)}{1+\beta(1-y_i^k)}\right]$$

Compare the domain of convergence with the simple iteration method.

3. Velocity and temperature distribution in a viscous magnetohydrodynamic flow past a flat plate is described by virtue of two dimensionless differential equations:

$$(hw')' - M^2 sw + 1 = 0$$
$$(k\theta')' + M^2 sw^2 + h(w')^2 = 0$$

subject to the linear boundary conditions

$$w'(0) = 0, \qquad w(1) = -K$$
$$\theta'(0) = 0, \qquad \theta(1) = 0$$

where the coefficients s, K, and h are functions of the dimensionless temperature θ:

$$s = (1 + N\theta)^\alpha, \qquad k = (1 + N\theta)^\beta, \qquad h = (1 + N\theta)^\gamma$$

Here α, β, γ, N, and M are constant coefficients. Solve these equations by the finite-difference method and the Newton–Raphson technique for the following values of parameters: $M = 1$, $N = 1$, $\alpha = 3/2$, $\beta = \gamma = 5/2$, $K = -0.6$. Discuss for this example the advantages and shortcomings of the finite-difference approach.

4. Mass diffusion and first-order chemical reaction in an isothermal porous catalyst particle, taking into consideration the concentration dependence of the diffusion coefficient, can be described by a simple second-order differential equation:

$$\frac{d^2\theta}{dx^2} + \frac{1}{1-\theta}\left(\frac{d\theta}{dx}\right)^2 = \phi^2\theta(1-\theta)$$

$$x = 0: \qquad \frac{d\theta}{dx} = 0$$

$$x = 1: \qquad \theta = 1$$

Solve this problem by the high-order difference method with an equidistant mesh for $h = 0.25$ and $\phi = 1$.

5. Axial mixing in an isothermal tubular reactor where a second-order reaction occurs is described by

$$\frac{1}{Pe}\frac{d^2y}{dx^2} - \frac{dy}{dx} - Da\, y^2 = 0$$

$$y(0) = 1 + \frac{1}{Pe}y'(0), \qquad y'(1) = 0$$

Solve this equation by the finite-difference method. Use the Holt approximation $(y^{k+1})^2 = y^k y^{k+1}$. To approximate the first and second derivative, use the standard three-point formulas. Compare the accuracy of the solution obtained for $h = 0.02$ with the results obtained by the high-order difference schema using $h = 0.2$. (The parameters have the values Pe $= 5$, Da $= 1$.)

6. The equilibrium of neighboring drops at different potentials is described by a nonlinear differential equation

$$\frac{d^2y}{dx^2} + \frac{1}{x}\frac{dy}{dx} = \alpha + \frac{\beta}{y^2}$$

with the boundary conditions

$$x = 0: \qquad \frac{dy}{dx} = 0$$

$$x = 1: \qquad y = 1$$

This equation contains a singularity for $x = 0$. To handle this problem numerically, it is necessary to develop an equation that does not contain a singularity. On using l'Hospital rule, we have a new equation

$$2\frac{d^2y}{dx^2} = \alpha + \frac{\beta}{y^2}$$

which is valid near $x = 0$. Show that for $\alpha = 0$ and $\beta = 0.77295$, two solutions exist. [$y(0) = 0.62733$ and 0.47935].

7. Combined heat and mass transfer within and outside a porous catalyst particle and a first-order chemical reaction can be described by the nonlinear differential equation

$$\frac{d^2Y}{dx^2} = \phi^2 Y \exp\frac{\gamma\beta(\chi - Y)}{1 + \beta(\chi - Y)}$$

where

$$\chi = \left(1 - \frac{Sh}{Nu}\right)Y(1) + \frac{Sh}{Nu}$$

The boundary conditions are

$$x = 0: \qquad \frac{dY}{dx} = 0$$

$$x = 1: \qquad -\frac{dY}{dx} = Sh(Y - 1)$$

We note that the right-hand side of differential equations contains boundary condition. For the parameter values $\gamma = 20$, $\beta = 0.1$, $Sh = 30$, $Nu = 10$, and $\phi = 0.2$, solve this problem by the finite-difference method and the Newton–Raphson technique.

BIBLIOGRAPHY

Various aspects of finite-difference approach to solving boundary value problems are discussed in:

COLLATZ, L.: *The Numerical Treatment of Differential Equations.* Springer-Verlag, Berlin, 1960.

Methods of solution of a second-order nonlinear boundary value problem are described in:

Fox, L.: *The Numerical Solution of Two-Point Boundary Problems in Ordinary Differential Equations.* Oxford University Press, New York, 1957.

HENRICI, P.: *Discrete Variable Methods in Ordinary Differential Equations.* J. Wiley, New York, 1962.

For the use of orthogonal collocation methods for solution of nonlinear BVPs, see:

VILLADSEN, J., AND MICHELSEN, M. L.: *Solution of Differential Equation Models by Polynomial Approximation.* Prentice-Hall, Englewood Cliffs, N. J., 1978.

For solution of nonlinear BVPs by compact high-order difference schemes, see:

KUBÍČEK, M., HLAVÁČEK, V., AND CAHA, J.: *Chem. Eng. Sci.* 27, 1829.

For the discussion of finite-difference methods and nonuniform grid:

PEARSON, C. E.: *J. Math. Phys.* 47, 123 (1968); 47, 351 (1968).
BROWN, R. R.: *J. Soc. Ind. Appl. Math.* 10, 475 (1962).
DENNY, V. E., AND LANDIS, R. B.: *J. Comput. Phys.* 9, 120 (1972).
LENTINI, M., AND PEREYRA, V.: *SIAM J. Num. Anal.* 15, 59 (1978).
RUSSELL, R. D., AND CHRISTIANSEN, J.: *SIAM J. Num. Anal.* 15, 59 (1978).

For a discussion of the Newton–Raphson method and the linearization technique used toward solution of nonlinear finite-difference equations, see:

HOLT, J.: *Commun. ACM 7*, 366 (1964).

HOLT, J.: *Numerical Solution of Nonlinear Two-Point Boundary Value Problems by Finite Differences Using the Newton Method.* SAMSO-TR-68-385, Aerospace Corp., San Bernardino, Calif., May 1968.

For a description of procedures solving finite-difference approximations with "probability one":

WANG, C. Y., AND WATSON, L. T.: *Appl. Sci. Res.* 35, 195 (1979); *J. Appl. Math. Phys.* 30, 773 (1979).

For the relaxation method for nonlinear finite-difference equations, see:

POPE, D.: *Math. Comput.* 14, 354 (1960).

For the sucessive accelerated replacement method used to solve nonlinear finite-difference equations, see:

LEW, H.: *AIAA J.* 6, 929 (1968).
DELLINGER, T. C.: *AIAA J.* 9, 262 (1971).

For quasi-Newton methods for discretized nonlinear BVP, see:

HART, W. E., AND SOUL, S. O. V.: *J. Inst. Math. Appl.* 11, 351 (1973).
DEY, S. K.: *Numer. Heat Transfer 3*, 505 (1980).

For descriptions of adaptive and general-purpose codes for nonlinear boundary value problems, see:

PEREYRA, V.: PASVA 3—An adaptive finite difference FORTRAN program for first order nonlinear, ordinary boundary problems; and ASCHER, U., CHRISTIANSEN, J., RUSSELL, R. D.: COLSYS—A collocation code for boundary value problems; both in *Codes for Boundary Value Problems in Ordinary Differential Equations* (ed. B. Childs et al.), Lecture Notes in Computer Science. Springer-Verlag, Berlin, 1978.
DIEKOFF, H. J. et al.: *Numer. Math.* 27, 449 (1977).

For the solution of nonlinear equations arising after discretization of differential equations, see:

DEUFLHARD, P.: Nonlinear equation solvers in boundary value problem codes, in *Codes for Boundary Value Problems in Ordinary Differential Equations* (ed. B. Childs et al.), Lecture Notes in Computer Science. Springer-Verlag, Berlin, 1978.

4.2 Method of Green's Functions

We have discussed in preceding sections the situation where the differential operator was directly approximated. Now we are going to construct a Green's function, which converts the original differential equation into its inverse form (i.e., into some integral equation). Since the integral operators are usually bounded, the convergence properties can be easily tested, unlike the differential operators.

4.2.1 Construction of a Green's Function

We shall address ourselves to the linear differential operator

$$\mathcal{L}(y) = \sum_{i=0}^{n} p_i(x) y^{(i)}(x) \qquad n \ge 2 \tag{4.66}$$

where $p_i(x)$, $i = 0, \ldots, n$, as well as $p_n^{-1}(x)$ are continuous functions in the region $\langle a, b \rangle$. The functions $y(x)$ having continuous derivatives up to the nth order which obey $R_i(y) = 0$, $i = 1, 2, \ldots, n$, will be the subject of our study. The boundary conditions R_i are in the form

$$R_i(y) = \sum_{k=0}^{n-1} \alpha_{ik} y^{(k)}(a) + \sum_{k=0}^{n-1} \beta_{ik} y^{(k)}(b) \qquad i = 1, 2, \ldots, n \tag{4.67}$$

The vectors $(\alpha_{io}, \ldots, \alpha_{i,n-1}; \beta_{io}, \ldots, \beta_{i,n-1})$ are linearly independent. For the sake of simplicity we shall write for a homogeneous problem $\mathcal{L}(y) = 0$ and $R_i(y) = 0$ an abridged equation

$$Ly = 0 \tag{4.68}$$

The detailed theory shows that if only the trivial solution $y \equiv 0$ exists, then only one inverse operator given by the Green's function $G(x, t)$ can be constructed. The theory of Green's functions is covered in textbooks on differential equations (see the Bibliography at the end of this section); here we shall summarize only the basic properties of a Green's function $G(x, t)$:

1. $G(x, t)$, $G'(x, t)$, $\ldots$, $G^{(n)}(x, t)$ are continuous functions of x and t in the region

$$a \le x < t \le b \quad \text{and} \quad a \le t < x \le b$$

Here

$$G^{(k)} = \frac{\partial^k G(x, t)}{\partial x^k}$$

2. G and the derivatives up to the $(n - 2)$nd order are continuous for $x = t$ and $a < t < b$; for the derivative $G^{(n-1)}$ a discontinuity occurs:

$$G^{(n-1)}(t_+, t) - G^{(n-1)}(t_-, t) = \frac{1}{p_n(t)} \qquad (4.69)$$

3. In the region $\langle a, t) \cup (t, b \rangle$, $G(x, t)$ obeys the homogeneous differential equation (boundary value problem) given by the relation (4.68).

If L has an inverse operator, the inhomogeneous equation

$$Ly = z \qquad (4.70)$$

where $z(x)$ is continuous on $\langle a, b \rangle$, exhibits only one solution $y(x)$, and it becomes

$$y(x) = \int_a^b G(x, t)z(t)\, dt \qquad (4.71)$$

To evaluate the Green's function we may take advantage of the fundamental system $y_1(x), \ldots, y_n(x)$ calculated for $\mathcal{L}(y) = 0$:

$$G(x, t) = \begin{cases} \displaystyle\sum_{i=1}^n (a_i + b_i)y_i(x) & x \leq t \\ \displaystyle\sum_{i=1}^n (a_i - b_i)y_i(x) & x \geq t \end{cases} \qquad (4.72)$$

where the functions $a_i(t)$ and $b_i(t)$ are not yet determined. The property of continuity of $G, G', \ldots, G^{(n-2)}$ yields for $x = t$:

$$\sum_{i=1}^n b_i y_i^{(k)}(t) = 0 \qquad k = 0, 1, \ldots, n - 2 \qquad (4.73a)$$

On using (4.69) we obtain

$$\sum_{i=1}^n b_i y_i^{(n-1)}(t) = -\frac{1}{2p_n(t)} \qquad (4.73b)$$

Because of the linear independence of functions in the fundamental system, the determinant of (4.73a) and (4.73b) has a nonzero value; the set of linear equations can be solved and the unknowns $b_1(t), \ldots, b_n(t)$ result. The Green's function (4.72) also satisfies the boundary conditions (4.67) and thus

$$R_i(G) = \sum_{k=0}^{n-1} \alpha_{ik} \sum_{j=1}^n (a_j + b_j)y_j^{(k)}(a) + \sum_{k=0}^{n-1} \beta_{ik} \sum_{j=1}^n (a_j - b_j)y_j^{(k)}(b) = 0$$
$$i = 1, 2, \ldots, n \qquad (4.74)$$

Using the known functions $b_j(t)$ we can establish the unknowns $a_1(t), \ldots, a_n(t)$ because the determinant of (4.74) is nonzero with regard to the fact that (4.68) exhibits only a trivial solution.

We present here two simple examples in order to elucidate the construction of Green's functions. For

$$\mathcal{L}(y) = y'' \qquad (4.75)$$

and the linear separable boundary conditions

$$\alpha y(a) + \beta y'(a) = 0$$
$$\gamma y(b) + \delta y'(b) = 0 \tag{4.76}$$

the Green's function is

$$G(x, t) = \begin{cases} \dfrac{(\alpha x - a\alpha - \beta)(\gamma t - b\gamma - \delta)}{\Delta} & x \le t \\[3mm] \dfrac{(\alpha t - a\alpha - \beta)(\gamma x - b\gamma - \delta)}{\Delta} & x \ge t \end{cases} \tag{4.77}$$

Here

$$\Delta = (b - a)\alpha\gamma + \alpha\delta - \beta\gamma$$

For

$$\mathcal{L}(y) = y'' - \lambda y' \tag{4.78}$$

and for boundary conditions (4.76) the Green's function is

$$G(x, t) = \begin{cases} (a_1 + b_1)e^{\lambda x} + (a_2 + b_2) & x \le t \\ (a_1 - b_1)e^{\lambda x} + (a_2 - b_2) & x \ge t \end{cases} \tag{4.79}$$

where

$$b_1 = -\frac{1}{2\lambda}e^{-\lambda t} \qquad b_2 = \frac{1}{2\lambda}$$

$$a_1 = \left\{ -\frac{\alpha\gamma}{a} + \left[\gamma\left(\frac{\alpha}{2\lambda} + \frac{\beta}{2}\right)e^{\lambda a} + \alpha\left(\frac{\gamma}{2\lambda} + \frac{\delta}{2}\right)e^{\lambda b} \right]e^{-\lambda t} \right\} / \Delta_1$$

$$a_2 = \left\{ \frac{(\alpha + \beta\lambda)\gamma}{2\lambda}e^{\lambda a} + \frac{(\gamma + \delta\lambda)}{2}e^{\lambda b} - \left[(\alpha + \beta\lambda)e^{\lambda a}\left(\frac{\gamma}{2\lambda} + \frac{\delta}{2}\right)e^{\lambda b} \right. \right.$$
$$\left. \left. + (\gamma + \delta\lambda)e^{\lambda b}\left(\frac{\alpha}{2\lambda} + \frac{\beta}{2}\right)e^{\lambda a} \right]e^{-t} \right\} / \Delta_1$$

$$\Delta_1 = \gamma(\alpha + \beta\lambda)e^{\lambda a} - \alpha(\gamma + \delta\lambda)e^{\lambda b}$$

For a number of problems, inhomogeneous boundary conditions are encountered [i.e., $R_i(y) = C_i \ne 0$]. To solve this problem the transformation $w = y - v$ is used, which gives rise to homogeneous boundary conditions $R_i(w) = 0$. Hence for v, $R_i(v) = C_i$. Now the Green's function can be determined for $\mathcal{L}(w) = 0$ subject to $R_i(w) = 0$. It is obvious that $\mathcal{L}(w) = \mathcal{L}(y) - \mathcal{L}(v)$, and for an inhomogeneous problem $\mathcal{L}(y) = f(x)$, we have $\mathcal{L}(w) = f(x) - \mathcal{L}(v)$ and hence

$$w(x) = \int_a^b G(x, t)[f(t) - \mathcal{L}(v(t))]\, dt$$

The solution $y(x)$ is

$$y(x) = v(x) + \int_a^b G(x, t)[f(t) - \mathcal{L}(v(t))]\, dt$$

4.2.2 Method of Successive Approximations

The procedure of transforming the differential equation to an integral relation (4.71) can also be generalized for nonlinear problems in the form

$$\mathcal{L}(y) = f(y(x)), \qquad R_i(y) = 0 \qquad i = 1, 2, \ldots, n \qquad (4.80a)$$

or

$$\mathcal{L}(y) = f(x, y(x), \ldots, y^{n-1}(x)), \qquad R_i(y) = 0 \qquad i = 1, 2, \ldots, n \qquad (4.80b)$$

The problem (4.80) can be rewritten in operator form:

$$Ly - F(y) = 0 \qquad (4.81)$$

Assuming the existence of an inverse operator (i.e., a Green's function) $G = L^{-1}$, we may rewrite (4.81):

$$y - GF(y) = 0 \qquad (4.82)$$

The method of successive approximations yields

$$y_{k+1} = GF(y_k) \qquad (4.83)$$

or

$$y_{k+1}(x) = \int_a^b G(x, t) f(y_k(t)) \, dt \qquad (4.84)$$

The problem of solving a nonlinear boundary value problem has been converted to an evaluation of a sequence of integrals. For a contraction mapping operator GF the sequence $\{y_k(x)\}$ converges to the solution of (4.80) if $GF: \Omega \rightarrow \Omega$; here Ω is a region where GF is defined (see Chapter 2). A sufficient condition to guarantee contraction mapping can be written for the case (4.80a), for example, in the form

$$\max_{t, x \in \langle a, b \rangle} |G(x, t)| \max_{y \in (-\infty, \infty)} |f'(y)| < 1 \qquad (4.85)$$

If the requirements of the contraction mapping theorem are fulfilled [see (4.85)], the unicity of the solution is assured. This condition can be used to prove the existence and uniqueness of the solution.

Relation (4.83) can also be rewritten in an equivalent form,

$$Ly_{k+1} = F(y_k) \qquad (4.86)$$

Indeed, the numerical solution of a nonlinear boundary value problem demands to solve a sequence of linear boundary value problems. Hence the numerical realization of two analytically equivalent iteration procedures (4.83) [or (4.84)] and (4.86) is different, the convergence properties can be established by making use of (4.83).

In a completely analogous way the method of successive approximations may be also used for problems given by (4.80b); for example, for $n = 2$:

$$y_{k+1}(x) = \int_a^b G(x, t) f(t, y_k(t), y_k'(t)) \, dt \qquad (4.87)$$

To prove the assumption of the contraction mapping theorem it is reasonable to rewrite (4.87):

$$y_{k+1}(x) = \int_a^b G(x, t)f(t, y_k(t), v_k(t))\, dt$$

$$v_{k+1}(x) = \int_a^b G'_x(x, t)f(t, y_k(t), v_k(t))\, dt \tag{4.88}$$

To enhance the rate of convergence the generalized Aitken–Steffensen method can be used.†

4.2.3 Newton–Kantorovich Linearization

The operator equation (4.81)–(4.82) can be solved by, for example, the Newton–Kantorovich method. The increment $\Delta y = y_{k+1} - y_k$ may be calculated from (see Section 4.3).

$$L\,\Delta y - F'(y_k)\,\Delta y = F(y_k) - Ly_k \tag{4.89}$$

or analogously to (4.82) from

$$\Delta y - GF'(y_k)\,\Delta y = GF(y_k) - y_k \tag{4.90}$$

Using the operator G on (4.89), equation (4.90) results; that is, the Newton–Kantorovich linearization of (4.81) is identical with that applied to (4.82). However, the numerical realization of both processes is different. The Newton–Kantorovich method in the form of (4.89) is the subject of Section 4.3; we will show that a sequence of linear boundary value problems results. Equation (4.90) in turn, is the Fredholm integral equation of the second kind:

$$\Delta y - \int_a^b G(x, t)f'(y_k(t))\Delta y(t)\, dt = \int_a^b G(x, t)f(y_k(t))\, dt - y_k(x) \tag{4.91}$$

This linear integral equation can be solved by making use of the quadrature formulas, for example,

$$\int_a^b \varphi(x)\, dx \doteq \sum_{i=0}^m C_i\varphi(x_i) \tag{4.92}$$

where $x_0 = a$, $x_m = b$, and $x_{i+1} - x_i = (b - a)/m$.

For the corrections $\Delta y(x_i)$ a set of linear algebraic equations results:

$$\sum_{j=0}^m K_{ij}\Delta y(x_j) = s_i \qquad i = 0, 1, \ldots, m \tag{4.93}$$

where

$$K_{ij} = (\delta_{ij} - C_j G(x_i, x_j))f'(y_k(x_j)) \tag{4.94}$$

$$s_i = \sum_{j=0}^m [C_j G(x_i, x_j)f(y_k(x_j))] - y_k(x_i) \tag{4.95}$$

†See examples in J. M. Ortega and W. C. Rheinboldt, *Iterative Solution of Nonlinear Equations in Several Variables* (New York: Academic Press), 1970; see also C. Brezinski, *Algorithms d'Accélération de la Convergence: Étude Numérique* (Paris: Editions Technique, 1978).

Example 4.5

A simultaneous diffusion, heat conduction, and exothermic chemical reaction in a porous catalyst is described by a nonlinear boundary value problem

$$\frac{d^2y}{dx^2} = \phi^2 y \exp\left[\frac{\gamma\beta(1-y)}{1+\beta(1-y)}\right] \tag{4.96}$$

subject to

$$x = 0: \quad \frac{dy}{dx} = 0$$
$$x = 1: \quad y = 1 \tag{4.97}$$

To have homogeneous boundary conditions, we put $z = 1 - y$. For

$$\frac{d^2z}{dx^2} = R(z(x)) = -\phi^2(1-z)\exp\left(\frac{\gamma\beta z}{1+\beta z}\right) \tag{4.98}$$

and homogeneous boundary conditions $x = 0$: $dz/dx = 0$, $x = 1$: $z = 0$, we can write

$$z(x) = \int_0^1 G(x,t)R(z(t))\, dt = T(z(x)) \tag{4.99}$$

obviously with respect to (4.77) the Green's function is

$$G(x,t) = \begin{cases} t-1 & x \le t \\ x-1 & x \ge t \end{cases} \tag{4.100}$$

The method of successive approximations

$$z_{k+1}(x) = T(z_k(x)) \tag{4.101}$$

converges if (4.85) is fulfilled. Clearly, for the Green's function, we have

$$\max_{t, x \in \langle 0, 1\rangle} |G(x,t)| = 1 \tag{4.102}$$

Hence relation (4.85) yields

$$K = \phi^2 \max_{(z)} \left| \left[-1 + \frac{\gamma\beta(1-z)}{(1+\beta z)^2}\right] \exp\left(\frac{\gamma\beta z}{1+\beta z}\right)\right| < 1 \tag{4.103}$$

Of course, this inequality can be easily tested a posteriori (i.e., when the profiles of the governing equation have been evaluated). For instance, for $\phi = 0.16$, $\beta = 0.7$, and $\gamma = 20$, three profiles calculated by the shooting technique have been determined (see Fig. 4-6). For profile I, $K \doteq 0.2$; for II and III we have the values 3.1 and 35.5, respectively. The method of successive approximations converges only to profile I, where the profile is flat and the nonlinear term low while the solutions II and III cannot be calculated by the successive approximations method. To perform the numerical calculation we rewrite (4.99)–(4.101) in the form

$$z_{k+1}(x) = (x-1)\int_0^x R(z_k(t))\, dt + \int_x^1 (t-1)R(z_k(t))\, dt \tag{4.104}$$

The necessary quadratures were performed by making use of the generalized

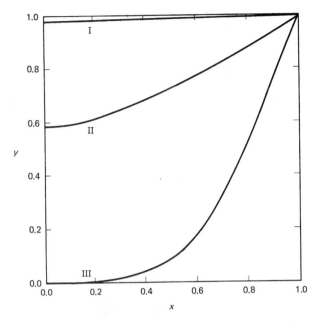

Figure 4-6 Three solutions of the NBVP given by (4.96) and (4.97); $\gamma = 20$, $\beta = 0.7$, $\phi = 0.16$.

trapezoidal rule. An equidistant mesh was used, and the interval $\langle 0, 1 \rangle$ was divided into n parts. The results of the calculation are reported in Tables 4-14 to 4-16.

TABLE 4-14
COURSE OF SUCCESSIVE APPROXIMATIONS:
$\gamma = 20$, $\beta = 0.05$, $n = 20$, $z_0(x) = 0^a$

Iteration	$\phi = 0.1$	$\phi = 1$	$\phi = 2$	$\phi = 4$
0	0.0000	0.0000	0.0000	0.0000
1	0.0050	0.5000	2.0000	8.0000
2	0.0050	0.4337	−7.5137	9,687.1
3		0.4514	0.0845	3.1E13
4		0.4470	1.9920	Divergence
5		0.4481	−7.3979	
6		0.4478	0.0874	
7		0.4479	1.9914	
8			−7.3900	
9			0.0876	
10			1.9914	
11			−7.3895	

^{a}The value of $z(0)$ is presented. Constant initial profiles.

Table 4-14 reveals convergence properties of the method depending on the parameter ϕ. For low values of this parameter satisfactory convergence properties were observed; on the other hand, for higher values of ϕ, divergence occurred. A very interesting phenomenon has been observed for $\phi = 2$, where stationary "cyclic iterations" have been determined. We also note this behavior for a case where as the zeroth iteration is guessed, the solution is accurate to three significant figures. Table 4-15 presents the case where multiple solutions occur. The method of successive approximations enables one to calculate only the profile having low values of z.

TABLE 4-15
METHOD OF SUCCESSIVE APPROXIMATIONS—
A CASE WITH MULTIPLE SOLUTIONS:
$\gamma = 20$, $\beta = 0.7$, $\phi = 0.16$, $n = 20$

Iteration	$z(0)$	$z(0)$	$z(0)$
0	0.0000[a]	0.9999[a]	0.9990[b]
1	0.0128	0.0048	1.047
2	0.0147	0.0135	−1.146
3	0.0150	0.0148	0.0003
4	0.0150	0.0150	0.0128
5		0.0150	0.0147
6			0.0150
7			0.0150

[a]Constant initial profile.
[b]Initial profile was in the vicinity of the profile III (see Fig. 4-6).

The other two profiles cannot be evaluated by this method, even if the exact values are used as the zeroth iteration, the solution with low values of z always results. From Table 4-16 can be inferred the effect of forcing procedures, as for instance Krasnoselskij's method[†] on the rate of convergence. The solution II (see Fig. 4-6) can be calculated using this forcing procedure, unlike the results obtained by the method of successive approximations. The scalar products necessary for evaluation have been calculated according to the formula

$$(u, v) = \int_0^1 u(x)v(x)\, dx$$

[†]M. A. Krasnoselskij et al, *Approximate Solution of Operator Equations* (Moscow: Nauka, 1969).

TABLE 4-16
EFFECT OF KRASNOSELSKIJ'S (AITKEN–STEFFENSEN) METHOD[a]

	k	z_k	$T(z_k)$	$T(T(z_k))$	β_k
$\gamma = 20$	0	0.0000[b]	0.5000	0.4337	0.8883
$\beta = 0.05$	1	0.4411	0.4496	0.4475	0.8007
$\phi = 1$	2	0.4479			
$\gamma = 20$	0	0.0000[b]	2.0000	−7.5137	0.1829
$\beta = 0.05$	1	0.2597	1.8859	−5.9454	0.1713
$\phi = 2$	2	0.5444	1.5963	−2.9320	0.1874
	3	0.7477	1.2318	−0.4801	0.2190
	4	0.8569	0.9660	0.6343	0.2463
	5	0.8843	0.8898	0.8733	0.2551
	6	0.8856			
$\gamma = 20$	0	0.4100[c]	0.4211	0.4380	−1.4993
$\beta = 0.7$	1	0.3958	0.3740	0.3214	−0.7933
$\phi = 0.16$	2	0.4157	0.4228	0.4413	−0.5664
	3	0.4123	0.4131	0.4150	−0.5864
	4	0.4120	0.4121	0.4123	−0.6532
	5	0.4120			

[a]The value of $z(0)$ is presented.
[b]Constant initial profiles.
[c]Initial profile was in the vicinity of the profile II (see Fig. 4-6).

Example 4.6

A special form of the differential equation describing axial mixing in tubular reactors is

$$\frac{d^2 y}{dx^2} - \frac{dy}{dx} + \exp\left(-\frac{1}{y}\right) = 0 \qquad (4.105)$$

subject to

$$x = 0: \quad \frac{dy}{dx} = y - 1$$
$$x = 1: \quad \frac{dy}{dx} = 0 \qquad (4.106)$$

The transformation $z = y - 1$ yields

$$z'' - z' + \exp\left(\frac{-1}{1+z}\right) = 0 \qquad (4.107)$$

with the homogeneous boundary conditions

$$x = 0: \quad z' = z$$
$$x = 1: \quad z' = 0 \qquad (4.108)$$

The Green's function pertinent to the differential equation

$$z'' - z' = 0$$

subject to (4.108) is [see (4.79)]

$$G(x, t) = \begin{cases} -\exp(x - t) & x \leq t \\ -1 & x \geq t \end{cases} \tag{4.109}$$

The method of successive approximations yields

$$z_{k+1}(x) = -\int_0^1 G(x, t) \exp\left[\frac{-1}{1 + z_k(t)}\right] dt$$

$$= \int_0^x \exp\left[\frac{-1}{1 + z_k(t)}\right] dt + \exp(x) \int_x^1 \exp\left[-t - \frac{1}{1 + z_k(t)}\right] dt \tag{4.110}$$

It is obvious that $\max\limits_{x,t \in \langle 0,1 \rangle} |G(x, t)| = 1$ and thus referring to (4.85), the sufficient condition for convergence is

$$\max_{x \in \langle 0, 1 \rangle} \left| \frac{1}{(1 + z(x))^2} \exp\left[-\frac{1}{1 + z(x)}\right] \right| \leq \exp(-1) \doteq 0.37 < 1$$

supposing that only positive values of $z(x)$ are considered. The successive approximations converge and the solution is in the region $z(x) > 0$ unique. The resulting $z(x)$ is drawn in Fig. 4-7.

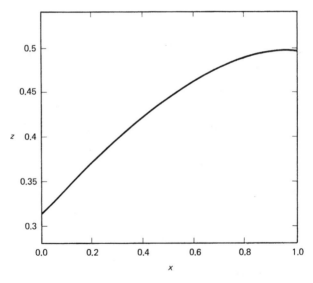

Figure 4-7 Solution of the NBVP given by (4.105) and (4.106).

4.2.4 Discussion

It has been pointed out that the iteration process using the Green's function (inverse operator) is analytically identical with that iterating the nonlinear boundary value problem in its differential form. However, the numerical realization of both procedures can be different. Unlike the method of successive approximations used for the problem in differential form, the method of Green's function makes it possible to estimate the region of convergence, uniqueness of solution, and so on.

Referring to (4.50), the matrix A can be viewed as a discretized form of a linear differential operator; (4.51) then corresponds to the method of successive approximations for the original nontransformed problem. After inversion of A an inverse operator of the discretized linear differential operator can be evaluated. It is a discretized counterpart of the Green's function. Equation (4.52) corresponds in a qualitative way to the successive approximations method with an inverse operator (i.e., to a "method of discretized Green's function").

Frequently, it is not complicated to evaluate the Green's function for a single second-order differential equation; however, for differential equations with variable coefficients the construction is not straightforward. Of course, the Green's function can be constructed for $y'' = 0$ and the linear terms y' and y can be incorporated in the nonlinear part; however, the convergence properties of the successive approximation method may be poor. For a set of differential equations a more complicated situation is to be expected. The method of Green's function can be used to solve the differential equations

$$\mathfrak{L}y_i = f_i(x, y_1, \ldots, y_n, y'_1, \ldots, y'_n) \qquad i = 1, 2, \ldots, n \qquad (4.111)$$

with boundary conditions R_i that are dependent on y_i only. Generally speaking, the method using the Green's functions is not apt for solving a general set of differential equations. The same conclusion is valid for implicit differential equations

$$F(x, y, y', \ldots, y^{(n)}) = 0 \qquad (4.112)$$

where the highest derivative cannot be expressed explicitly

PROBLEMS

1. Isothermal diffusion in a porous plate and simultaneously occurring chemical reaction of second order can be described by the dimensionless differential equation

$$\frac{d^2y}{dx^2} = \phi^2 y^2$$

subject to

$$x = 0: \qquad \frac{dy}{dx} = 0$$

$$x = 1: \qquad y = 1$$

Using the method of Green's function, calculate by successive approximations the profile $y(x)$ for $\phi = 0.05$. Making use of (4.85), estimate the region of ϕ where convergence may be expected.

$$\left[\phi < \frac{\sqrt{2}}{2}\right]$$

2. In electrodynamics the equilibrium of the neighboring drops at different potentials is given by the differential equation

$$\frac{d^2y}{dx^2} + \frac{1}{x}\frac{dy}{dx} = \alpha + \frac{\beta}{y^2}$$

subject to

$$x = 0: \quad \frac{dy}{dx} = 0$$

$$x = 1: \quad y = 1$$

By virtue of the Green's function approach, devise an algorithm for calculation of $y(x)$. Estimate the region of convergence.

BIBLIOGRAPHY

The theory of Green's function is presented in textbooks on differential equations, e.g.:

INCE, E. L.: *Ordinary Differential Equations.* Dover Publications, New York, 1956.
CODDINGTON, E. A., AND LEVINSON, N.: *Theory of Ordinary Differential Equations.* McGraw-Hill, New York, 1955.

For experience with application of the Green's function approach to the solution of nonlinear boundary value problems, see

SCHILSON, R. E., AND AMUNDSON, N. R.: *Chem. Eng. Sci. 13*, 226, 237 (1961).
BEUSCH, H.: Solution of boundary value problems in heterogeneous catalysis by the Green's function approach. Ph.D. thesis, Münster, West Germany, 1970 (accessible only in German).

4.3   Newton–Kantorovich Method and the Quasi-Linearization Technique

Among the techniques currently available, the Newton–Kantorovich method, sometimes also called the quasi-linearization technique, is worthwhile. The quasi-linearization technique was developed by Bellman and Kalaba (1965) in connection with the dynamic programming approach. Actually, it is not necessary to employ the dynamic programming approach because the generalized Newton method for operator equations yields the same results. Kantorovich and McGill and Kenneth (1963) studied the convergence properties of the generalized Newton–Raphson method, and their results can be used to develop an algorithm for the numerical solution of nonlinear boundary value problems. Because the quasi-linearization results in the same equations as the Newton–Kantorovich method, both terms are used in the literature at present.†

---

†In 1957, Fox proposed to linearize the nonlinear differential equations by expanding the particular nonlinearity in a Taylor's series up through first-order terms. He called this procedure the η-method. It seems that the importance of this paper has been completely overlooked in the literature. [See L. Fox, *The Numerical Solution of Two-Point Boundary Problems in Ordinary Differential Equations* (New York, Oxford University Press (1957).]

Our discussion in this chapter takes advantage of the abstract Newton method. Emphasis is placed on computational aspects of the Newton–Kantorovich procedure. The process with the second- as well as third-order rate of convergence will be discussed. Two of the most frequently used methods for solving the resulting set of linear ordinary differential equations—the superposition principle and the finite-difference method—are dealt with. To assure convergence of the iteration process, the Miele concept of the performance index will be presented.

4.3.1   Newton–Kantorovich Method

As mentioned in Chapter 2, the Newton–Kantorovich method is the Newton method for an operator equation $F(y) = 0$ in Banach space. The development of this method will be sketched for a single nonlinear second-order differential equation

$$F(y) = \frac{d^2y}{dx^2} + f\left(x, y, \frac{dy}{dx}\right) = 0 \qquad x \in \langle a, b \rangle \qquad (4.113)$$

subject to the linear homogeneous two-point boundary conditions

$$\alpha_0 y(a) + \beta_0 y'(a) = 0$$
$$\alpha_1 y(b) + \beta_1 y'(b) = 0 \qquad (4.114)$$

The function f is supposed to be continuously differentiable with respect to both y and y'. Here F is the mapping from the Banach space Y_1:

$$Y_1 = \{y \,|\, y \in Y, y \text{ fulfills } (4.114)\}$$
$$Y = \{y \,|\, y \in C^2_{\langle a, b\rangle}, \|y\| = \max\left(\max_{\langle a, b\rangle} |y(x)|, \max_{\langle a, b\rangle} |y'(x)|, \max_{\langle a, b\rangle} |y''(x)|\right)\}$$

into the Banach space of continuous functions C.

After using the Newton method on the operator equation $F(y) = 0$, we have

$$F'_y(y_k)\delta_y = -F(y_k) \qquad (4.115)$$

where

$$\delta y = y_{k+1} - y_k$$

Here $F'_y(y)$ is the Fréchet derivative of F:

$$F'_y(y) = \frac{d^2}{dx^2} + \frac{\partial f(x, y, y')}{\partial y'}\frac{d}{dx} + \frac{\partial f(x, y, y')}{\partial y}$$

Now we can rewrite (4.115) as

$$(\delta y)'' + \frac{\partial f}{\partial y'}(\delta y)' + \frac{\partial}{\partial y}\delta y = -y''_k - f(x, y_k, y'_k) \qquad (4.116)$$

Since the linear operator $F'_y(y)$ is considered in the Banach space Y_1, the boundary conditions are ($\delta y \in Y_1, y_k \in Y_1$)

$$\alpha_0 \delta y(a) + \beta_0 \delta y'(a) = 0, \qquad \alpha_1 \delta y(b) + \beta_1 \delta y'(b) = 0 \qquad (4.117a)$$
$$\alpha_0 y_{k+1}(a) + \beta_0 y'_{k+1}(a) = 0, \qquad \alpha_1 y_{k+1}(b) + \beta_1 y'_{k+1}(b) = 0 \qquad (4.117b)$$

Obviously, instead of solving the nonlinear boundary value problem, a sequence of linear boundary problems is to be solved. Another, more obvious development is based on an application of Taylor series. After expanding the function $f(x, y, y')$ in a Taylor series up through first-order terms around the nominal solution y_k, we have

$$f(x, y_{k+1}, y'_{k+1}) \doteq f(x, y_k, y'_k) + \frac{\partial f(x, y_k, y'_k)}{\partial y}(y_{k+1} - y_k)$$

$$+ \frac{\partial f(x, y_k, y'_k)}{\partial y}(y'_{k+1} - y'_k) \tag{4.118}$$

Insertion into (4.113) yields approximation:

$$y''_{k+1} + \frac{\partial f(x, y_k, y'_k)}{\partial y}(y_{k+1} - y_k) + \frac{\partial f(x, y_k, y'_k)}{\partial y'}(y'_{k+1} - y'_k) = -f(y, y_k, y'_k) \tag{4.119}$$

Equation (4.119) is identical with (4.116); that is, both method of development result in the same linear differential equations.

According to the Kantorovich theorem, the sequence $\{y_k\}$ converges monotonously with quadratic rate of convergence to the exact solution of the nonlinear boundary value problem. Of course, the convergence of the iteration process depends on the initial guess y_0.

We can use the same procedure for boundary value problems with nonlinear boundary conditions. For the boundary conditions

$$g_0[y(a), y'(a)] = 0, \qquad g_1[y(b), y'(b)] = 0 \tag{4.120}$$

an expansion of the left-hand side of (4.120) in a Taylor series around a nominal solution y_k yields

$$\frac{\partial g_0[y_k(a), y'_k(a)]}{\partial y}[y_{k+1}(a) - y_k(a)] + \frac{\partial g_0[y_k(a), y'_k(a)]}{\partial y'}[y'_{k+1}(a) - y'_k(a)]$$

$$= -g_0[y_k(a), y'_k(a)]$$

$$\tag{4.121}$$

$$\frac{\partial g_1[y_k(b), y'_k(b)]}{\partial y}[y_{k+1}(b) - y_k(b)] + \frac{\partial g_1[y_k(b), y'_k(b)]}{\partial y'}[y'_{k+1}(b) - y'_k(b)]$$

$$= -g_1[y_k(b), y'_k(b)]$$

Clearly, for nonlinear boundary conditions (4.120), inhomogeneous linear boundary conditions result.

Since the nonlinear boundary value problems can be approximated by the procedure described above by a sequence of linear boundary value problems, the methods for linear boundary value problems developed in Chapter 3 can be readily used.

To recapitulate the material presented, a flowchart for the Newton–Kantorovich method is shown in Fig. 4-8.

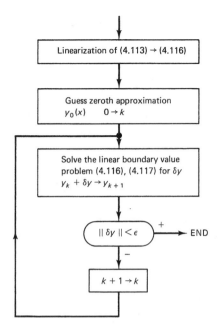

Figure 4-8 Schematic flowchart of the Newton–Kantorovich method for (4.113) and (4.114).

The method can be extended to a set of nonlinear first-order ordinary differential equations:

$$\frac{dy_i}{dx} = f_i(x, y_1, \ldots, y_N) \qquad i = 1, 2, \ldots, N \qquad (4.122)$$

subject to two-point linear boundary conditions

$$\sum_{j=1}^{N} a_{ij} y_j(a) + b_{ij} y_j(b) = c_i \qquad i = 1, 2, \ldots, N \qquad (4.123)$$

Obviously, the relevant linear differential equations are

$$\frac{dy_i^{k+1}}{dx} = f_i(x, y_1^k, \ldots, y_N^k) + \sum_{j=1}^{N} \frac{\partial f_i(x, y_1^k, \ldots, y_N^k)}{\partial y_j}(y_j^{k+1} - y_j^k)$$
$$i = 1, 2, \ldots, N \qquad (4.124)$$

subject to the two-point linear boundary conditions

$$\sum_{j=1}^{N} a_{ij} y_j^{k+1}(a) + b_{ij} y_j^{k+1}(b) = c_i \qquad i = 1, 2, \ldots, N \qquad (4.125)$$

It seems reasonable to use the superposition principle (see Chapter 3) to solve the set of linear differential equations rather than to construct the finite-difference methods. To solve (4.124) and (4.125), a set of differential equations for a

113

matrix Y^{k+1} of the type $N \times N$ must be integrated:

$$\frac{dY^{k+1}}{dx} = \Gamma(y^k)Y^{k+1}, \qquad Y^{k+1}(a) = E \tag{4.126}$$

along with the particular equations for a vector $\bar{y}^{k+1}$

$$\frac{d\bar{y}^{k+1}}{dx} = f(x, y^k) + \Gamma(y^k)(\bar{y}^{k+1} - y^k)$$

$$\bar{y}^{k+1}(a) = 0 \tag{4.127}$$

Here we have denoted the Jacobian matrix of the right-hand sides of (4.122) by $\Gamma = \{\partial f_i/\partial y_j\}$. A new approximation y^{k+1} is given by

$$y^{k+1} = \bar{y}^{k+1} + Y^{k+1}u^{k+1}$$

where u^{k+1} is a vector calculated after inserting y^{k+1} into (4.125). Obviously, it is

$$y^{k+1}(a) = u^{k+1} \tag{4.128}$$

The results of integration of (4.126) and (4.127) can be stored only for given mesh points. For the values of the right-hand sides of (4.126) and (4.127) it is necessary to interpolate when using common integration routines (with the exception of the Euler method and trapezoidal rule). In turn, on approximating the partial derivatives in $\Gamma(y^k(x))$ by a function piecewise constant or linear, the Newton method can be considered to be the "Newton-like" method. This procedure obviously has the same effect as if the values of the derivatives in the Newton method would be subjected to large errors. For the old approximation nN values must be stored; here n is the number of mesh points where $y_i^k(x)$ are stored. For large values of N and parametric-sensitive problems, the computer memory requirements can be the limiting factor. To overcome this difficulty, Bellman and Kalaba (1965) suggested an algorithm where the computer memory is trading for computer time. The procedure consists of the following steps:

1. Guess $y^0(x)$; frequently, simple functions (e.g., linear) are used.
2. Integration of (4.126) and (4.127) for $k = 0$ to evaluate u^1.
3. Simultaneous integration of (4.126) and (4.127) for $k = 1$ together with a set of differential equations

$$\frac{dy^1}{dx} = f(x, y^0) + \Gamma(y^0)(y^1 - y^0)$$

$$y^1(a) = u^1 \tag{4.129}$$

Obviously, the profile $y^1(x)$ need not be stored. A set of $N(N + 1) + N$ differential equations is integrated to calculate u^2 and $y^2(x)$.

⋮

$R + 2$. Simultaneous integration of (4.126) and (4.127) for $k = R$ together with the sets of equations

$$\frac{dy^i}{dx} = f(x, y^{i-1}) + \Gamma(y^{i-1})(y^i - y^{i-1})$$

$$y^i(a) = u^i \qquad i = 1, 2, \ldots, R \tag{4.130}$$

Hence the profile $y^R(x)$ need not be stored; it is sufficient to integrate simultaneously $N(N + 1) + RN$ differential equations to evaluate u^R and $y^R(x)$, as long as we have for the differences $\| y^{k+1} - y^k \| < \epsilon$. Since the Newton–Kantorovich procedure requires only a low number of iterations, the additional computer time expenditure due to repeated integrations of (4.130) is not excessive.

The number of iterations depends on the quality of the zeroth iteration. A poor initial guess may result in a divergent numerical process. There are several ways to establish reasonable initial profiles. Usually, some simple physical approximation of the model may lead to profiles that are not too far from the solution. Sometimes simple straight lines or parabola-like functions are satisfactory. Another technique to guess nominal profiles is to estimate the missing initial conditions and to integrate the set of differential equations across.

Apparently, the Newton–Kantorovich method is not suitable for problems where the function f is difficult to differentiate. Another disadvantage of this technique is the necessity of evaluation of derivates analytically, which for large systems of differential equations can be a formidable task. On the other hand, the type of boundary conditions does not essentially complicate the problem.

Ortega and Rheinboldt† proved that when using the same type of finite-difference approximations, the Newton–Kantorovich technique is identical with the Newton–Raphson method applied to the linearized original finite-difference approximations. Obviously, the sequence of both operations linearization and discretization is interchangable.

To illustrate the Newton–Kantorovich method, some examples are presented.

Example 4.7 ‡

Consider the second-order nonlinear ordinary differential equation

$$y'' - 2y^3 = 0 \tag{4.131}$$

subject to the nonlinear boundary conditions

$$y(1) = 1, \qquad y'(2) + [y(2)]^2 = 0 \tag{4.132}$$

On expanding the left-hand side in a Taylor's series, we have

$$\delta y'' - 6y_k^2\, \delta y = -y_k'' + 2y_k^3 \tag{4.133}$$

subject to inhomogeneous linear boundary conditions

$$\begin{aligned}
\delta(1) &= 1 - y_k(1) \\
[\delta y(2)]' + 2y_k(2)\, \delta y(2) &= -y_k'(2) - y_k^2(2)
\end{aligned} \tag{4.134}$$

†J. M. Ortega and W. C. Rheinboldt, *SIAM J. Num. Anal. 3*, 143 (1966).

‡R. E. Bellman and R. E. Kalaba, *Quasilinearization and Boundary Value Problems* (New York: American Elsevier, 1965).

Assume the nominal function $y_0 \equiv 1$; (4.133) and (4.134) are in the form $(\delta = y_1 - y_0)$

$$\delta'' - 6\delta = 2$$

$$\delta(1) = 0$$

$$\delta'(2) + 2\delta(2) = -1$$

The three-point finite-difference approximations yield for $h = 0.2$ $[\delta_i = \delta(x_i) = \delta(1 + ih)]$

$$\delta_{i-1} + \delta_{i+1} - 2.24\delta_i = 0.08 \qquad i = 1, 2, \ldots, 5$$

The boundary conditions after discretization are

$$\delta_0 = 0$$

$$\frac{\delta_6 - \delta_4}{2h} + 2\delta_5 = -1$$

The resulting set of six linear algebraic equations can be easily solved; the results are reported in Table 4-17. The last column in the table contains the exact solution $y = 1/x$.

TABLE 4-17
COURSE OF ITERATION OF THE
NEWTON–KANTOROVICH TECHNIQUE

x	y_0	y_1	y_2	y_3	$y = 1/x$
1	1.000	1.000	1.000	1.000	1.000
1.2	1.000	0.866	0.837	0.835	0.833
1.4	1.000	0.779	0.721	0.716	0.714
1.6	1.000	0.719	0.634	0.627	0.625
1.8	1.000	0.672	0.567	0.558	0.555
2.0	1.000	0.626	0.513	0.502	0.500
2.2	1.000	0.571	0.467	0.457	0.455

For some problems it is very difficult to select an appropriate initial profile so that divergence may occur. To circumvent this disadvantage, Miele and Iyer (1971) devised a way to reduce the calculated correction:

$$y_{k+1} = y_k + \alpha\delta y_k \qquad (4.135)$$

where $\alpha \in (0, 1 >$ must be chosen so as to minimize the performance index

$$J(\alpha) = \int_a^b [y'' + f(x, y, y')]^2 \, dx + \{g_0[y(a), y'(a)]\}^2$$
$$+ \{g_1[y(b), y'(b)]\}^2 \qquad (4.136)$$

considering the problem (4.113), (4.120) and putting $y = y_{k+1}$.

Since it is time consuming to find the value of α that leads to a minimum of $J(\alpha)$, it is sufficient to evaluate the value of α that results in reduction of J. We first assign the value $\alpha = 1$ to the step size and calculate the value of the performance index. If the reduction of J is recorded, y_{k+1} is calculated. Otherwise, a bisection algorithm is used; that is $\alpha = 1/2$ is tried, and so on. Once the value of α is known, the solution y_{k+1} is calculated. This procedure may force to convergence those problems which for a given nominal profile would diverge. An example is presented to illustrate this procedure.

Example 4.8†

Consider the nonlinear differential equations

$$y''' = -\tfrac{1}{6}(y'')^2 uu' \qquad (4.137)$$
$$u'' = -\tfrac{1}{2}y'(u')^3$$

subject to the boundary conditions

$$
\begin{aligned}
y(0) = u(0) = 1, \qquad y(1) = 16 \\
u'(0) = -1, \qquad u(1) = 0.5
\end{aligned}
\qquad (4.138)
$$

To transform (4.137) into a set of first-order equations, let

$$y = y_1, \qquad y' = y_2, \qquad y'' = y_3, \qquad u = y_4, \qquad u' = y_5$$

Then

$$
\begin{aligned}
y_1' = y_2, \qquad y_2' = y_3, \qquad y_3' = -\tfrac{1}{6}y_3^2 y_4 y_5, \\
y_4' = y_5, \qquad y_5' = -\tfrac{1}{2}y_2 y_3^3
\end{aligned}
\qquad (4.139)
$$

The following zeroth iterations have been selected:

$$y_1^0(x) = 1 + 15x, \qquad y_2^0(x) = 0, \qquad y_3^0(x) = 0, \qquad y_4^0(x) = 1 - 0.5x,$$
$$y_5^0(x) = -1$$

which satisfy the boundary conditions (4.138).

Starting the iteration with these initial profiles, the standard Newton–Kantorovich method fails, but the modified method does guarantee convergence. The results of the computation are reported in Tables 4-18 and 4-19.

TABLE 4-18
CONVERGED SOLUTION OF
(4.137) AND (4.138)

x	y	u
0.0	1.000	1.000
0.2	2.073	0.8333
0.4	3.841	0.7142
0.6	6.553	0.6250
0.8	10.49	0.5555
1.0	16.00	0.5000

†Taken from Miele and Iyer (1971).

k	α	J
0	—	0.2×10^3
1	$\frac{1}{16}$	0.2×10^3
2	$\frac{1}{8}$	0.1×10^3
3	$\frac{1}{4}$	0.1×10^3
4	$\frac{1}{2}$	0.4×10^2
5	$\frac{1}{2}$	0.2×10^2
6	1	0.1×10^1
7	$\frac{1}{2}$	0.3×10^0
8	$\frac{1}{2}$	0.1×10^0
9	1	0.2×10^{-1}
10	1	0.2×10^{-8}
11	1	0.1×10^{-21}

Example 4.9

Consider Example 4.3. With regard to (4.113), we have

$$f(x, y, y') = -\phi^2 y \exp\left[\frac{\gamma\beta(1 - y)}{1 + \beta(1 - y)}\right] + \frac{a}{x}y' \tag{4.140}$$

After performing the appropriate differentation, we have

$$\frac{\partial f}{\partial y} = -\phi^2\left[1 - \frac{\gamma\beta y}{[1 + \beta(1 - y)]^2}\right]\exp\left[\frac{\gamma\beta(1 - y)}{1 + \beta(1 - y)}\right]$$
$$\frac{\partial f}{\partial y'} = \frac{a}{x} \tag{4.141}$$

On replacing the second and first derivatives in (4.116) by standard three-point formulas having $O(h^2)$ and using the nonsymmetric formula

$$-3y_{k+1}(0) + 4y_{k+1}(h) - y_{k+1}(2h) = 0 \tag{4.142}$$

to approximate the boundary condition

$$y'_{k+1}(0) = 0$$

a set of linear algebraic equations with a tridiagonal structure results. The nominal profile has been assumed: $y(x) \equiv 1$. The course of calculation for the Newton–Kantorovich method is reported in Table 4-20, where a sequence of concentrations $y_k(0)$ in the center of the particle is presented. We can note that due to the poor nominal profile, convergence is relatively slow.

k	$y_k(0)$
0	1.0000
1	−5.0000
2	−3.7347
3	−2.6449
4	−1.7205
5	−0.9636
6	−0.3974
7	−0.0793
8	−0.0011
9	0.0023
10	0.0023

4.3.2 Third-Order Newton–Kantorovich Method

In this section we address ourselves to the third-order Newton–Kantorovich method. Obviously, the procedure makes use of an approximation of the operator equation $F(y) = 0$ by the first and second terms in the Taylor's expansion. The resulting set of nonlinear differential equations is either solved by the standard Newton–Kantorovich procedure or approximated by the method of tangent hyperbolas.

With the background of Section 4.3.1, we can develop the third-order method. Let us denote by δ_k the difference

$$\delta_k = y_{k+1} - y_k \tag{4.143}$$

Equation (4.116) can be written in an abridged form:

$$\delta_k'' = -y_k'' - f - f_y \delta_k - f_{y'} \delta_k' \tag{4.144}$$

where the functions $f, f_y,$ and $f_{y'}$ are to be evaluated for $y = y_k$. For y_{k+1} we have

$$y_{k+1} = y_k + \delta_k \tag{4.145}$$

For the nonhomogenous boundary conditions (3.2) the equation (4.117a) may be rewritten

$$\alpha_0 \delta_k(a) + \beta_0 \delta_k'(a) = \gamma_0 - \alpha_0 y_k(a) - \beta_0 y_k'(a)$$
$$\alpha_1 \delta_k(b) + \beta_1 \delta_k'(b) = \gamma_1 - \alpha_1 y_k(b) - \beta_1 y_k'(b) \tag{4.146}$$

Now let us expand (4.147):

$$y''_{k+1} = -f(x, y_{k+1}, y'_{k+1}) = -f(x, y_k + \delta_k, y'_k + \delta'_k) \qquad (4.147)$$

in a Taylor's series up through second-order terms around a nominal solution y_k:

$$\delta''_k = -y''_k - f - f_y\delta_k - f_{y'}\delta'_k - \tfrac{1}{2}f_{yy}\delta_k^2 - f_{yy'}\delta_k\delta'_k - \tfrac{1}{2}f_{y'y'}(\delta'_k)^2 \qquad (4.148)$$

Here $f, f_y, f_{y'}$, and so on, have to be evaluated for $y = y_k$. We may note that (4.148) is a nonlinear differential equation for δ_k; however, the nonlinearities are of the quadratic type. Let us rewrite (4.148) in an abridged form

$$\delta''_k = F(x, \delta_k, \delta'_k) \qquad (4.149)$$

subject to boundary conditions (4.146). To calculate the solution of (4.149), the classical Newton–Kantorovich approach can be used. Accordingly, we construct a sequence of functions $\delta_{k,i}$ which are solutions of the sequence of linear equations, (4.150):

$$\Delta''_i = -\delta''_{k,i} + F + F_{\delta_k}\Delta_i + F_{\delta'_k}\Delta'_i \qquad (4.150)$$

where the functions

$$\delta_{k,i+1} = \delta_{k,i} + \Delta_i \qquad (4.151)$$

meet boundary conditions (4.146). The functions $F, F_{\delta_k}, F_{\delta'_k}$ are to be evaluated for $\delta_k = \delta_{k,i}$. The limit value of this sequence represents the solution of (4.149). The partial derivatives F_{δ_k} and $F_{\delta'_k}$ are rather simple, owing to the power function in (4.148). The functions $f, f_y, f_{y'}$, and so on, have to be calculated in the local iteration cycle only in the first trial. The application of the Newton–Kantorovich method to (4.149) and a finite-difference approach leads to a set of linear equations with a tridiagonal or multidiagonal matrix.

We can expect a lower overall computing time for problems where the enumeration of the functions f, f_y, and $f_{y'}$ is time consuming and the evaluation of the terms $f_{yy}, f_{yy'}$, and $f_{y'y'}$ does not substantially increase the computing time. Let us note that for the initial guess $\delta_{k,o} = 0$, the result of the first local iteration $\delta_{k,1}$ is identical with the δ_k resulting from the original Newton–Kantorovich method given by (4.144). To recapitulate, the third-order Newton–Kantorovich procedure consists of the following steps, which are displayed in a flowchart in Fig. 4-9.

Apparently, the method of "tangent hyperbolas" can be adapted to solve (4.148). Let us denote by $\bar{\delta}_k$ the correction calculated by the classical Newton–Kantorovich method. Now supposing the tangent hyperbola approach, we can rewrite (4.148) as follows:

$$\delta''_k = -y''_k - f - f_y\delta_k - f_{y'}\delta'_k - \tfrac{1}{2}f_{yy}\bar{\delta}_k\delta_k - \tfrac{1}{2}f_{yy'}\bar{\delta}_k\delta'_k$$
$$- \tfrac{1}{2}f_{yy'}\delta_k\bar{\delta}'_k - \tfrac{1}{2}f_{y'y'}\bar{\delta}'_k\delta'_k \qquad (4.152)$$

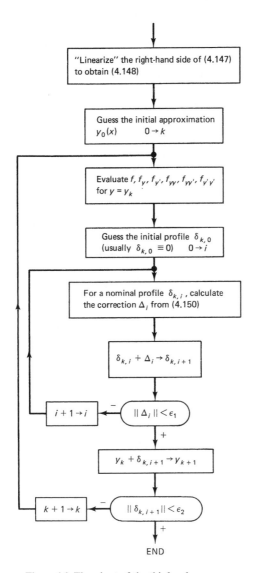

Figure 4-9 Flowchart of the third-order process.

Equation (4.152) constitutes a linear boundary value problem subject to (4.146). We again solve this equation to calculate the correction δ_k either by superposition or by finite-difference methods. To recapitulate, the development of flowchart of the algorithm proposed is shown in Fig. 4-10.

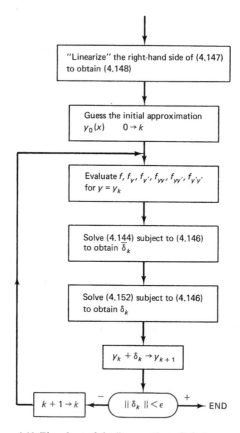

Figure 4-10 Flowchart of the "tangent hyperbolas" approach.

Example 4.10

Consider the nonlinear boundary value problem solved in Example 4.9 by the standard Newton–Kantorovich method. To use the third-order Newton–Kantorovich method, the derivatives f_{yy}, $f_{yy'}$, $f_{y'y'}$ must be evaluated:

$$f_{yy} = \phi^2 \left\{ \frac{2\gamma\beta[(1 + \beta)^2 - \beta y(1 + \beta + \gamma/2)]}{[1 + \beta(1 - y)]^4} \right\} \exp\left[\frac{\gamma\beta(1 - y)}{1 + \beta(1 - y)} \right]$$

$$f_{yy'} = f_{y'y} = 0$$

The course of iterations using the algorithm presented in Fig. 4-9 is reported in Table 4-21. A comparison with the results from Table 4-20 shows that this method is faster than the classical Newton–Kantorovich technique.

TABLE 4-21
COURSE OF ITERATIONS FOR THE
THIRD-ORDER CONVERGENCE METHOD:
$\gamma = 20$, $\beta = 0.05$, $\phi = 6$,
$a = 2$, $h = 0.0125$

k	$y_k(0)$	i	$\delta_{k,i}(0)$
0	1.0000	0	0.0000
		1	−6.0000
		2	−3.0780
		3	−1.7755
		4	−1.3722
		5	−1.3301
		6	−1.3297
1	−0.3297	7	−1.3297
		0	0.0000
		1	0.2592
		2	0.3408
		3	0.3499
		4	0.3500
2	0.0203	5	0.3500
		0	0.0000
		1	−0.0183
		2	−0.0181
3	0.0023	3	−0.0181
		0	0.0000
4	0.0023	1	0.0000

4.3.3 Discussion

The Newton–Kantorovich approach replaces a nonlinear boundary value problem by a sequence of linear boundary value problems which in general converges rapidly to the solution of the original nonlinear problem. Frequently, rough initial guesses for the unknown function lead to the solution of the original problem. We have shown how to guess initial profiles; however, the physical intuition and experiences are crucial to estimate an appropriate zeroth approximation. The success of the method, of course, depends essentially on the quality of the starting profile. In fact, for an overwhelming majority of physical and engineering problems a physically sound initial guess leads to a convergent sequence.

If there is no stability problem connected with the marching integration of relevant differential equations, the superposition principle technique may be used. However, the ill-conditioning can sometimes spoil the results of integration. This ill-conditioning problem is connected with the solution of algebraic equations which are solved to evaluate the integration constants. Another disadvantage of the superposition principle approach is the necessity of storing old profiles, which for large problems can be a tedious task. Bellman's approach of simultaneous solution of different iterations can overcome the storage problem, but since the numerical integration of the resulting large set of differential equations is highly time consuming, we are essentially trading the computer expenditure for the computer memory. It should be mentioned that the results obtained by marching integration techniques are more accurate than those calculated by the finite-difference approach. For numerically sensitive boundary value problems the finite-difference approach is superior to superposition. An appropriate ordering of finite-difference equations gives rise to a multidiagonal matrix which may be handled easily.

For a set of second-order differential equations, if using the finite-difference approximations, it is necessary to group appropriately the variables and finite-difference equations in order to obtain the multidiagonal set of linear algebraic equations. A detailed discussion of this problem is presented elsewhere (see Section 4.1) and the conclusions are also valid for the Newton–Kantorovich method.

Sometimes we wish to work only with tridiagonal matrices and thus a "relaxation type" of the Newton–Kantorovich method can be devised. Let us examine how this procedure can be carried out. For a set of two second-order differential equations

$$u'' + f(x, u, u', v, v') = 0$$
$$v'' + g(x, u, u', v, v') = 0 \tag{4.153}$$

the Newton–Kantorovich procedure yields an iteration formula:

$$u''_{k+1} + f + \frac{\partial f}{\partial u}(u_{k+1} - u_k) + \frac{\partial f}{\partial u'}(u'_{k+1} - u'_k) + \frac{\partial f}{\partial v}(v_{k+1} - v_k)$$
$$+ \frac{\partial f}{\partial v'}(v'_{k+1} - v'_k) = 0 \tag{4.154a}$$

$$v''_{k+1} + g + \frac{\partial g}{\partial u}(u_{k+1} - u_k) + \frac{\partial g}{\partial u'}(u'_{k+1} - u'_k) + \frac{\partial g}{\partial v}(v_{k+1} - v_k)$$
$$+ \frac{\partial g}{\partial v'}(v'_{k+1} - v'_k) = 0 \tag{4.154b}$$

where f, g, $\partial f/\partial u$, $\partial g/\partial v$, and so on, are evaluated for u_k and v_k. If we approximate v_{k+1} and v'_{k+1} in (4.154a) by v_k and v'_k, respectively, and on using the three-point finite-difference formulas to replace u''_{k+1} and u'_{k+1}, a tridiagonal set of linear algebraic equations results. Equation (4.154a) yields

$$u''_{k+1} + f + \frac{\partial f}{\partial u}(u_{k+1} - u_k) + \frac{\partial f}{\partial u'}(u'_{k+1} - u'_k) = 0 \tag{4.155a}$$

The same approximations can be constructed for (4.154b):

$$v''_{k+1} + g + \frac{\partial g}{\partial v}(v_{k+1} - v_k) + \frac{\partial g}{\partial v'}(v'_{k+1} - v'_k) = 0 \qquad (4.155b)$$

Obviously, (4.155a) and (4.155b) become independent and can be solved easily by means of the Thomas algorithm. However, we can make use of the profiles calculated [e.g., from (4.155a)] to approximate the terms

$$\frac{\partial g}{\partial u}(u_{k+1} - u_k) \qquad \text{and} \qquad \frac{\partial g}{\partial u'}(u'_{k+1} - u'_k)$$

in (4.154b). Sometimes, this forcing procedure enhances the rate of convergence. However, convergence of these modifications is not assured.

The Newton–Kantorovich procedure can also be used to solve nonlinear differential equations of higher order; after linearization, a linear boundary value problem of higher order results. The finite-difference approach gives rise to a multidiagonal matrix. The second possibility for solving the relevant linear boundary value problem is to reduce the linear differential equation to a set of first-order differential equations which are handled by the superposition principle approach.

Another disadvantage of the Newton–Kantorovich procedure is the necessity of evaluating the partial derivatives in the linearized equations. Of course, these derivatives should be determined analytically. For strongly nonlinear boundary value problems, the hand evaluation can be a tedious task. We wish to stress again that for most engineering problems the order of discretization and linearization is immaterial; that is, we can discretize the problem and then perform a linearization (Newton–Raphson method) or we may linearize the particular nonlinear boundary value problem and then replace the linear equations by the finite-difference analogy (Newton–Kantorovich approach). Both distinct paths lead in practice to equivalent approximations.

If the enumeration of partial derivatives is time consuming, it is possible, especially in the vicinity of the solution, to use the modified Newton–Kantorovich method. For this approach the derivatives $\partial f/\partial y$ and $\partial f/\partial y'$ are evaluated only once for $y = y_0(x)$. The iteration formula (4.124) becomes

$$\frac{dy_i^{k+1}}{dx} = f_i(x, y_1^k, \ldots, y_N^k) + \sum_{j=1}^{N} \frac{\partial f_i(x, y_1^0, \ldots, y_N^0)}{\partial y_j}(y_j^{k+1} - y_j^k)$$
$$i = 1, 2, \ldots, N \qquad (4.156)$$

The set of homogeneous equations [see (4.126)]

$$\frac{dY^{k+1}}{dx} = \Gamma(y^0)Y^{k+1} \qquad Y^{k+1}(a) = E \qquad (4.157)$$

does not depend on the particular iteration and can be precalculated. Hence for each iteration we must integrate only the set of particular equations (4.156) (supposing that the superposition principle approach is used). This modification reduces the computer time; moreover, in the vicinity of the solution, the modified Newton method exhibits also a high rate of convergence.

Let us discuss the benefits and shortcomings of both the superposition principle and the finite-difference approach for difficult linear boundary value problems. Generally speaking, for stable initial value problems the superposition principle may be adapted especially for a set of first-order differential equations. For unstable problems, however, the superposition principle cannot be easily adapted and the finite-difference approach is superior. Since the finite-difference approach is essentially an implicit procedure, it does not exhibit stability problems connected with marching techniques. Moreover, the refinement of the mesh size which is important in the vicinity of the solution can be easily accomplished (see Section 4.1). In addition, the accuracy of the approximation can be essentially improved, taking into consideration more grid points to approximate the particular derivatives. However, sometimes for finite-difference methods the storage requirements can be a limiting factor and thus the relaxation technique must be used; see (4.155).

The numerical solution of a set of linear algebraic equations resulting from the superposition principle may be a difficult problem because ill-conditioned sets arise. To overcome this problem the orthogonalization procedure was devised (see Chapter 3).

Despite all the disadvantages discussed, the Newton–Kantorovich technique appears to be a very powerful tool for solving nonlinear boundary value problems.

PROBLEMS

1. The infinite slab of reacting explosive material each side of which is held on the prescribed temperature is described by

$$\frac{d^2\Theta}{dx^2} = -\delta e^{\Theta}$$
$$\Theta(1) = 0$$
$$\Theta(0) = 0$$

Find the temperature profile by making use of the Newton–Kantorovich approach in the slab for $\delta = 1$.

x	0	0.2	0.4	0.6	0.8	1.0
Θ	0	0.0892	0.1347	0.1347	0.0892	0

2. The profile that gives rise to the shortest time of descent of a sphere moving be-

tween two fixed points is described by the Euler equation:

$$yy'' + (y')^2 + 1 = 0$$
$$y(1) = 1$$
$$y(2) = 2$$

Find the shape of the curve. Use the Newton–Kantorovich procedure. Show on this example that the order of discretization and linearization is immaterial.

x	1.00	1.25	1.50	1.75	2.00
y	1.0000	1.3919	1.6583	1.8541	2.0000

3. The restricted three-body problems in orbital mechanics is described by

$$\dot{x}_1 = x_3$$
$$\dot{x}_2 = x_4$$
$$\dot{x}_3 = 2x_4 + F_1(x_1, x_2)$$
$$\dot{x}_4 = -2x_3 + F_2(x_1, x_2)$$

where $F_1 = \partial f/\partial x_1$, $F_2 = \partial f/\partial x_2$, and

$$f = \frac{1}{2}(x_1^2 + x_2^2) + \frac{1}{r}(1 - \mu) + \frac{\mu}{\rho} + \frac{1}{2}\mu(1 - \mu)$$
$$r = \sqrt{(x_1 - \mu)^2 + x_2^2}, \qquad \rho = \sqrt{(x_1 + 1 - \mu)^2 + x_2^2}$$

subject to the boundary conditions

$$x_1(0) = -0.2, \qquad x_1(1) = -1.2$$
$$x_2(0) = -0.1, \qquad x_2(1) = 0.0$$

Solve this problem by the superposition principle for $\mu = 0.012$ and find x_i for $t = 0.5$:

$$\begin{bmatrix} x_1(0.5) = -0.8195 & x_2(0.5) = -0.3599 \\ x_3(0.5) = -1.042 & x_4(0.5) = 0.2497 \end{bmatrix}$$

4. The problem of isothermal axial mixing in tubular reactors for a reaction with second-order kinetics is described by a differential equation of second order:

$$\frac{1}{Pe}\frac{d^2y}{dx^2} - \frac{dy}{dx} - Ry^2 = 0$$

subject to Danckwert's boundary conditions

$$x = 0: \qquad 1 = y - \frac{1}{Pe}\frac{dy}{dx}$$

$$x = 1: \qquad \frac{dy}{dx} = 0$$

Calculate the profile $y = y(x)$ for $Pe = 6$ and $R = 2$ by the Newton–Kantorovich technique. Show that the linearized differential equations can be integrated by marching techniques only from $x = 1$ to $x = 0$. Use the Runge–Kutta method of fourth order with the step length $h = 0.1$. Compare these results with those obtained on the basis of finite-difference approach where 50 mesh points are used:

x	0	0.2	0.4	0.6	0.8	1.0
y	0.8313	0.6656	0.5517	0.4702	0.4131	0.3873

BIBLIOGRAPHY

Two monographs have been devoted to the Newton–Kantorovich method:

LEE, E. S.: *Quasilinearization and Invariant Imbedding.* Academic Press, New York, 1968.

BELLMAN, R. E., AND KALABA, R. E.: *Quasilinearization and Nonlinear Boundary Value Problems.* American Elsevier, New York, 1965.

For the fundamental ideas of the linearization of functional equations, see:

KANTOROVICH, L. V., AND AKILOV, G. P.: *Functional Analysis in Normed Spaces.* English translation, Pergamon Press, Elmsford, N.Y., 1964.

A number of new findings together with some fundamental theory of the Newton–Kantorovich method is presented in a very readable book:

ROBERTS, S. M., AND SHIPMAN, J. S.: *Two-Point Boundary Value Problems: Shooting Methods.* American Elsevier, New York, 1972.

Practical computational aspects of the Newton–Kantorovich approach in engineering problems have been presented in the following papers:

LEE, E. S.: Quasilinearization, difference approximation and nonlinear boundary value problems. *AICHE J. 14,* 490 (1968).

LEE, E. S.: Quasilinearization, nonlinear boundary value problems and optimization. *Chem. Eng. Sci. 21,* 183 (1966).

LEE, E. S., AND FAN, L. T.: Quasilinearization technique for solution of boundary layer equations. *Can. J. Chem. Eng. 46,* 200 (1968).

For a proof of convergence of the Newton–Kantorovich process, see:

MCGILL, R., AND KENNETH, P.: A convergence theorem on the iterative solution of nonlinear two-point boundary value systems. *Proc. 14th. Int. Astronaut. Congr., Paris IV. 12,* 173 (1963).

For use of the performance index for enhancing the rate of convergence, see:

MIELE, A., AND IYER, R. R.: Modified quasilinearization method for solving nonlinear two-point boundary value problems. *J. Math. Anal. Appl. 36,* 674 (1971).

For a development of the third-order Newton–Kantorovich method, see:

KUBÍČEK, M., HLAVÁČEK, V., AND MOKROŠ, J.: Construction of a high order convergence method based on Newton–Kantorovich algorithm. *Chem. Eng. Sci. 26,* 2113 (1971).

4.4 Method of False Transient for the Solution of Nonlinear Boundary Value Problems

4.4.1 Introduction

The method described herein results logically from the numerical solution of parabolic partial differential equations. The steady-state solution may be found by direct solving the steady-state equations or by solving the transient equations. Integrating the relevant parabolic equation, we proceed through time until the solution ceases to change significantly. Obviously, we are interested only in the final steady state and not in the shape of the transient profile. If the steady-state solution to parabolic partial differential equations exists and is unique, the numerical integration of the parabolic equations which were obtained by introduction of false transient terms leads to the final steady state, which is also the solution of the original nonlinear boundary value problem. This is the basis of the technique called the method of false transient. Clearly, the false transient term (also accumulation or fictitious term) leads to a set of parabolic differential equations which can be solved by marching integration through the distorted time. The rate of convergence may be essentially enhanced by making use of the numerous finite-difference schemes. The strongly implicit finite-difference method makes it possible to use a very long time step. Of course, the true transient solution is lost, but at large times the transient terms decay and the steady-state solution is approached rapidly.

Although the idea of the false transient method is very simple and the algorithm is straightforward, this technique was almost completely overlooked in the mathematical as well as engineering literature in the West. On the other hand, this method is very popular in the U.S.S.R., and sometimes it is recommended as a universal approach to solving nonlinear boundary value problems of diffusional type.

In the next section the numerical aspects of the false transient method will be dealt with. The aim is to discuss the typical features on the basis of the nonlinear diffusion equation. The false transient equations will be developed; however, there seems little point in investigating this numerical procedure without first having a fast efficient scheme for the integration of parabolic partial differential equations. A semi-implicit finite-difference method will be proposed to integrate easily the nonlinear false transient equations. This method allows reasonable stability and accuracy of the integration. A method will be suggested which permits us to use only a few mesh points at the early stage of integration. Later, the number increases as the steady-state solution is approached. The false transient technique will be applied to solve four problems arising in the chemical kinetics and boundary layer theory.

4.4.2 False Transient Equations

We have shown that by adding the fictitious accumulation terms the nonlinear boundary value problem becomes a set of parabolic partial differential equations. Let us formulate this problem for an autonomous differential equation

$$\frac{dy}{dt} = F(y) \tag{4.158}$$

in the Banach space. An element y^* for which

$$F(y^*) = 0 \tag{4.159}$$

is called the steady-state solution of (4.158). For a stable steady state, we have

$$y^* = \lim_{t \to \infty} y(t) \tag{4.160}$$

where $y(t)$ is a solution of the transient equation (4.158) subject to the initial condition

$$y(0) = y^0 \tag{4.161}$$

It is not simple to determine the stability of the particular solution and, generally speaking, it is more difficult than to solve the equation $F(y) = 0$. The local stability (the element y^0 is in the "vicinity" of y^*) can be tested by making use of the first Liapunov method. To make this calculation it is necessary to know the steady state solution y^*. The global stability (the element y^0 belongs to a particular region or to the whole space) can sometimes be evaluated by taking advantage of the second (also direct) Liapunov method. However, construction of the Liapunov functions can be a difficult task.

The effectiveness of the calculation of the steady state value y^* may be estimated according to the rate of decay of the transient term, that is, the time variable t for which

$$\| y(t) - y^* \| < \epsilon \tag{4.162}$$

or

$$\| F(y(t)) \| < E \tag{4.163}$$

is satisfied can be considered as a criterion of effectiveness. Here $y(t)$ is a solution of the differential equation (4.158) with the initial condition (4.161). The value of the independent variable t for which (4.162) is satisfied depends on the selection of the initial condition y^0. Generally speaking, for an initial value y^0 which is close to y^*, the "transition" time t will be short. Furthermore, since only the steady-state solution is desired, the values of the false transient term can be chosen arbitrarily; however, the rate of convergence to the steady-state solution can be enhanced for an appropriate value of the accumulation term. Hence (4.158) can be rewritten

$$A \frac{dy}{dt} = F(y) \tag{4.164}$$

where A is a time-independent operator. Of course, an inverse form of (4.164) can be considered:

$$\frac{dy}{dt} = A^{-1}F(y) \tag{4.165}$$

It is obvious that the steady-state solution of (4.164), y^*, is also the solution of (4.159). An appropriate choice of the operator A can enhance the effectiveness of the method and, in addition, may assure the stability of y^*. For instance, the steady-state solution, y^*, can be unstable with respect to (4.158), however, its stability is guaranteed for (4.164). Clearly, for $A = -F'(y^*)$, a continuous version of the modified Newton method results. Here the convergence is guaranteed for a sufficiently small norm $\| y^0 - y^* \|$.

Moreover, the operator A can be dependent on $y(t)$; (4.164) is now

$$A(y)\frac{dy}{dt} = F(y) \tag{4.166}$$

This relation can be considered to be a continuous version of the quasi-Newton method if the term $-A(y)$ is an approximation of the Frèchet derivative of the operator F.

Consider the operator F in the form

$$F(y) = \frac{d^2y}{dx^2} - f\left[x, y(x), \frac{dy}{dx}\right] = 0 \tag{4.167}$$

subject to homogeneous boundary conditions

$$\begin{aligned} x &= 0: \quad y = 0 \\ x &= 1: \quad y = 0 \end{aligned} \tag{4.168}$$

After adding the fictitious accumulation term, the ordinary differential equation (4.167) becomes a parabolic partial differential equation

$$\frac{\partial y(x, t)}{\partial t} = \frac{\partial^2 y(x, t)}{\partial x^2} - f\left[x, y(x, t), \frac{\partial y(x, t)}{\partial x}\right] \tag{4.169}$$

subject to the boundary conditions (4.168). The initial condition can be chosen arbitrarily, however, it should be physically consistent with the particular parabolic equation:

$$y(x, 0) = y^0(x) \tag{4.170}$$

4.4.3 Numerical Solution
of False Transient Equations

Consider for simplicity a set of second-order ordinary differential equations

$$\frac{d^2y_i}{dx^2} - f_i\left(x, y, \frac{dy}{dx}\right) = 0 \qquad i = 1, 2, \ldots, N \tag{4.171}$$

subject to boundary conditions

$$
\begin{array}{ll}
x = 0: & y_i = \alpha_i \\
x = 1: & y_i = \beta_i
\end{array} \qquad i = 1, 2, \ldots, N \qquad (4.172)
$$

Here we have denoted $y = y(x) = (y_1(x), y_2(x), \ldots, y_N(x))$. The pertinent parabolic differential equations are

$$
\frac{\partial y_i}{\partial t} = \frac{\partial^2 y_i}{\partial x^2} - f_i\left(x, y, \frac{\partial y}{\partial x}\right) \qquad (4.173)
$$

The initial condition is

$$
y(x, 0) = y^0(x) = \varphi(x) \qquad (4.174)
$$

The boundary conditions are given by (4.172). Because of the similarity of the false transient equations to the real transient processes, the parameter t will be referred to as the time parameter.

There are a number of numerical methods for finding the solution of parabolic partial differential equations (4.173). The reader who is interested in a detailed study of methods for numerical integration of parabolic equations is referred to the references presented in the Bibliography.

The object of this section is to show two numerical procedures, explicit and implicit finite-difference schemes, to provide solutions of the false transient equations.

4.4.3.1 Simple explicit finite-difference method

Let (x_m, t_j) denote a mesh, $(x_m, t_j) = (mh, jk)$, where h and k are the increments in the x and t directions, respectively. On replacing the differential equation (4.173) by the finite-difference analogy, we have

$$
\frac{y_{i,m}^{j+1} - y_{i,m}^j}{k} = \frac{y_{i,m-1}^j - 2y_{i,m}^j + y_{i,m+1}^j}{h^2} - f_i\left[x_m, y_m^j, \frac{y_{m+1}^j - y_{m-1}^j}{2h}\right]
$$

$$
i = 1, 2, \ldots, N; \quad m = 1, 2, \ldots, n-1; \quad j = 0, 1, \ldots \qquad (4.175)
$$

Here we have denoted $y_i(x_m, t_j) = y_{i,m}^j$, $n = 1/h$. Looking at the finite-difference scheme we can note that each particular point $y_{i,m}^{j+1}$ at the new profile $j + 1$ can be calculated easily from the known values at the profile j. At the boundaries the boundary conditions in the form

$$
y_{i,0}^{j+1} = \alpha_i, \qquad y_{i,n}^{j+1} = \beta_i \qquad (4.176)
$$

must be combined with (4.175). Obviously, to move from the profile j to the profile $j + 1$ it is necessary to perform $N(n - 1)$ evaluations of $y_{i,m}^{j+1}$ from (4.175). It is very simple to work with the explicit finite-difference scheme; the need for computer care is very low. To evaluate the economy of the method, one should bear in mind that the time increment, k, is subjected to the inequality (4.177):

$$
\frac{k}{h^2} \leq \frac{1}{2} \qquad (4.177)
$$

which governs its maximum permissible value. For higher values of k the finite-difference scheme becomes unstable. If the steady-state space profile $y(x)$ must be calculated with high precision, a short space step, h, has to be used and hence the relationship (4.177) gives rise to a very short time step k. Clearly, the computing time requirements are proportional to h^3. To improve the economy of the calculation, one must turn to the implicit finite-difference methods.

4.4.3.2 Simple implicit finite-difference method

After replacing the second derivative, $\partial^2 y_i/\partial x^2$, at the profile $j + 1$, an implicit finite-difference scheme results:

$$\frac{y_{i,m}^{j+1} - y_{i,m}^j}{k} = \frac{y_{i,m-1}^{j+1} - 2y_{i,m}^{j+1} + y_{i,m+1}^{j+1}}{h^2} - f_i\left(x_m, y_m^j, \frac{y_{m+1}^j - y_{m-1}^j}{2h}\right)$$

$$m = 1, 2, \ldots, n - 1; \quad i = 1, 2, \ldots, N \qquad (4.178)$$

Note that the nonlinear functions f_i have been approximated at the "old" profile j. The finite-difference approximation (4.178) is now implicit; that is, a set of linear algebraic equations, given by (4.178), must be solved simultaneously to evaluate a new time profile $j + 1$. Accordingly, using this approximation, N separated sets of linear algebraic equations result which can be solved independently. The calculation is very simple because only linear equations with tridiagonal matrices are to be solved. The Thomas algorithm, described, for example, by Lapidus,[†] can be adopted easily. An approximation of f_i at the profile $j + 1$ leads to a set of nonlinear algebraic equations, which are, generally speaking, not separable. While the pure implicit schemes where all derivatives and nonlinearities are approximated at the new profile are usually unconditionally stable, there is no guarantee of the unconditional stability for mixed approximations of the type given by (4.178). However, as has been shown by Richtmyer,[‡] the stability-determining term is the approximation to the second derivative. Hence it can be expected that the stability of the finite-difference approximation is satisfactory. The approximation given by (4.178) thus seems to offer distinct advantages over the pure implicit scheme.

4.4.3.3 Improving steady-state solutions

Since the initial condition (4.174) is sometimes very far from the steady-state profiles at the early stage of integration of false transient equations, only a few mesh points in the space direction are required to bring the inappropriate initial profile to the vicinity of the steady-state solution. As the steady-state solution is approached, the number of mesh points increases gradually. An algorithm that can be adopted to improve the steady-state solution during the computation is presented in Fig. 4-11. For problems with very slow decay it is necessary to choose an appropriate value of the tolerance because a large value results in the inaccurate steady-state profiles, while a too small value may give rise to excessive computer time.

[†] L. Lapidus, *Digital Computation for Chemical Engineers* (New York: McGraw-Hill, 1962).

[‡] R. D. Richtmyer, *Difference Methods for Initial-value Problems* (New York: Interscience 1955).

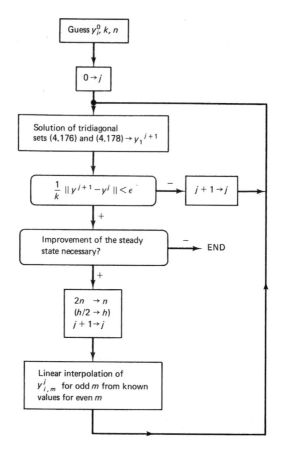

Figure 4-11 Flowchart showing gradual improvement in the steady-state profiles.

The time t is only an auxiliary parameter; the evaluation of the new "time" profile $j + 1$ can be considered as one iteration step which is characterized by the parameter k:

$$y_i^{j+1} = \phi_i(y_1^j, y_2^j, \ldots, y_N^j, k) \tag{4.179}$$

To speed up the convergence of the method a modification of the nonlinear block Gauss–Seidel method can be used:

$$y_i^{j+1} = \phi_i(y_1^{j+1}, y_2^{j+1}, \ldots, y_{i-1}^{j+1}, y_i^j, \ldots, y_N^j, k) \tag{4.180}$$

Here, obviously, the function ϕ_i [and also f_i in (4.178)] takes use of the computed value at the $j + 1$ profile. This is a logical result of the fact that the sets of linear algebraic equations are separated for each value of i.

4.4.4 Examples

The relative merits of the method of false transient are illustrated by examples of differential equations arising in the chemical kinetics, hydrodynamics, and optimum control. The first example shows the method of approximation of highly nonlinear diffusion equations and the problems of location of solutions if multiplicity can be expected.

Example 4.11

Axial heat and mass transfer with an exothermic chemical reaction taking into consideration the temperature dependence of the physicochemical variables is in the form†

$$\frac{1}{\text{Pe}}\left(\frac{d^2y}{dx^2} + \frac{1}{T}\frac{dT}{dx}\frac{dy}{dx}\right) - \frac{dy}{dx} - \frac{y}{T}\frac{dT}{dx} - \text{Da}\frac{y}{T}\exp\left[\frac{\gamma(T-1)}{T}\right] = 0 \qquad (4.181)$$

$$\frac{1}{\text{Pe}}\frac{d^2T}{dx^2} - \frac{dT}{dx} + \beta\,\text{Da}\,\frac{y}{T}\exp\left[\frac{\gamma(T-1)}{T}\right] = 0 \qquad (4.182)$$

subject to the nonlinear boundary conditions

$$y(0) = \frac{1}{T(0)} + \frac{1}{\text{Pe}}\frac{dy(0)}{dx}$$
$$T(0) = 1 + \frac{1}{\text{Pe}}\frac{dT(0)}{dx} \qquad (4.183)$$

$$\frac{dT(1)}{dx} = \frac{dy(1)}{dx} = 0 \qquad (4.184)$$

The false transient equations can be written in the form‡

$$w_1\frac{\partial y}{\partial t} = \frac{1}{\text{Pe}}\left(\frac{\partial^2 y}{\partial x^2} + \frac{1}{T}\frac{\partial T}{\partial x}\frac{\partial y}{\partial x}\right) - \frac{\partial y}{\partial x} - \frac{y}{T}\frac{\partial T}{\partial x} - \text{Da}\frac{y}{T}\exp\left[\frac{\gamma(T-1)}{T}\right] \qquad (4.185)$$

$$w_2\frac{\partial T}{\partial t} = \frac{1}{\text{Pe}}\frac{\partial^2 T}{\partial x^2} - \frac{\partial T}{\partial x} + \beta\,\text{Da}\,\frac{y}{T}\exp\left[\frac{\gamma(T-1)}{T}\right] \qquad (4.186)$$

subject to the boundary conditions $(t > 0)$

$$y(0, t) = \frac{1}{T(0, t)} + \frac{1}{\text{Pe}}\frac{\partial y(0, t)}{\partial x} \qquad (4.187a)$$

$$T(0, t) = 1 + \frac{1}{\text{Pe}}\frac{\partial T(0, t)}{\partial x} \qquad (4.187b)$$

$$\frac{\partial T(1, t)}{\partial x} = \frac{\partial y(1, t)}{\partial x} = 0 \qquad (4.188)$$

The initial conditions are $(0 < x < 1)$

$$y(x, 0) = \varphi(x), \qquad T(x, 0) = \psi(x) \qquad (4.189)$$

†From J. J. Carrberry, and M. H. Wendel, *AICHE Jour.* 9, 129 (1963).
‡M. Kubíček et al, *Chem. Eng. Sci.* 31, 727 (1976).

For the weights $w_1 = w_2 = 1$, the false transient equations describe the true transient behavior of the physical system.

Analogously to (4.178), the temperature false transient equation [i.e., (4.186)] can be approximated by the following finite-difference analogy:

$$\frac{w_2}{k}(T_m^{j+1} - T_m^j) = \frac{1}{\text{Pe }h^2}(T_{m-1}^{j+1} - 2T_m^{j+1} + T_{m+1}^{j+1}) - \frac{1}{2h}(T_{m+1}^{j+1} - T_{m-1}^{j+1})$$
$$+ \beta \text{ Da } \frac{y_m^j}{T_m^j} \exp\left[\frac{\gamma(T_m^j - 1)}{T_m^j}\right] \qquad m = 1, 2, \ldots, n-1 \qquad (4.190)$$

The boundary conditions (4.187b) and (4.188) are replaced by the nonsymmetrical finite-difference formulas

$$T_0^{j+1} - \frac{1}{2\text{ Pe }h}(-3T_0^{j+1} + 4T_1^{j+1} - T_2^{j+1}) = 1 \qquad (4.191a)$$
$$3T_n^{j+1} - 4T_{n-1}^{j+1} + T_{n-2}^{j+1} = 0 \qquad (4.191b)$$

Here $h = 1/n$.

After simple algebraic rearrangements, a set of linear algebraic equations yields a set with the tridiagonal matrix which can be easily solved to give the values T_m^{j+1}, $m = 0, 1, \ldots, n$. Clearly, we have implicitly approximated the space partial derivatives while the nonlinearities have been replaced at the "old" profile j. With regard to the fact that the temperature profile at $j+1$ has been evaluated, a number of finite-difference approximations can be devised to determine the concentration profile described by (4.185). The four suggested approximations are presented in Table 4-22. For instance, for method 3 the finite-difference approximation is

TABLE 4-22
POSSIBILITIES OF THE FINITE-DIFFERENCE APPROXIMATIONS
OF THE CONCENTRATION FALSE TRANSIENT EQUATION

Method	$\dfrac{1}{\text{Pe}}\dfrac{\partial^2 y}{\partial x^2} - \dfrac{\partial y}{\partial x}$	$\dfrac{1}{T}\dfrac{\partial T}{\partial x}\left(\dfrac{1}{\text{Pe}}\dfrac{\partial y}{\partial x} - y\right)$	$\text{Da }\dfrac{y}{T}\exp\left[\dfrac{\gamma(T-1)}{T}\right]$
1	$j+1$	$j+1$	j
2	$j+1$	$j+1$	$j+1$
3	$j+1$	$j+1/j^a$	j
4	$j+1$	j	j

^{a}Approximated by $\dfrac{1}{T^j}\dfrac{\partial T^j}{\partial x}\left[-y^{j+1} + \dfrac{1}{\text{Pe}}\dfrac{\partial y^{j+1}}{\partial x}\right]$

$$\frac{w_1}{k}(y_m^{j+1} - y_m^j) = \frac{1}{\text{Pe }h^2}(y_{m-1}^{j+1} - 2y_m^{j+1} + y_{m+1}^{j+1}) - \frac{1}{2h}(y_{m+1}^{j+1} - y_{m-1}^{j+1})$$
$$+ \frac{1}{2hT_m^j}\left[\frac{1}{2\text{ Pe }h}(y_{m+1}^{j+1} - y_{m-1}^{j+1}) - y_m^{j+1}\right](T_{m+1}^j - T_{m-1}^j) \qquad (4.192)$$
$$- \text{Da }\frac{y_m^j}{T_m^j}\exp\left[\frac{\gamma(T_m^j - 1)}{T_m^j}\right] \qquad m = 1, 2, \ldots, n-1$$

It is worth noting that for method 2, the exponential function for the calculation of the new profile $j+1$ must be evaluated $2(n-1)$ times, while for the other three methods only $(n - 1)$ exponential function evaluations are necessary. The boundary condition (4.187a) can be replaced, for instance, by the finite-difference formula

$$y_0^{j+1} - \frac{1}{2 \, Pe \, h}(-3y_0^{j+1} + 4y_1^{j+1} - y_2^{j+1}) = \frac{1}{T_0^{j+1}} \qquad (4.193a)$$

Here the value T_0^{j+1} is known because the temperature profile has been evaluated. The finite-difference analogy to the second boundary condition is

$$3y_n^{j+1} - 4y_{n-1}^{j+1} + y_{n-2}^{j+1} = 0 \qquad (4.193b)$$

After some simple algebraic substitutions, a set of linear algebraic equations results with a tridiagonal matrix so that the "new" concentration profile, y_0^{j+1}, $y_1^{j+1}, \ldots, y_n^{j+1}$, can be readily evaluated. For the calculation the algorithm taking advantage of the gradual improvement of the steady states has been adapted. The tolerance has been determined by

$$\frac{1}{k} \sum_{s=0}^{10} (|T_{sr}^{j+1} - T_{sr}^{j}| + |y_{sr}^{j+1} - y_{sr}^{j}|) < \epsilon, \qquad r = \frac{n}{10} \qquad (4.194)$$

Four factors affect the rate of convergence:

1. The selection of the weights w_1 and w_2.
2. The way of calculation of the new concentration profile.
3. The length of the time increment, k.
4. The guess of the initial conditions φ and ψ.

Accordingly, tests were conducted on related type of methods and the effect of the time step k was investigated; see Table 4-23. Here the number of time steps necessary to reach the steady-state solution T^* and y^* is presented when

TABLE 4-23

NUMBER OF TIME STEPS TO SATISFY (4.195):
$Pe = 5$, $Da = 0.15$, $\beta = 0.2$, $\gamma = 40$, $w_1 = w_2 = 1$, $\varphi \equiv 0$, $\psi \equiv 1$,
$y^*(x = 1) = 0.000006$, $T^*(x = 1) = 1.17548$, $n = 20$[a]

E	k	Method 1	Method 2	Method 3	Method 4
0.1	0.025	104	106	104	103
	0.05	55	56	55	55
	0.1	d(?)	32	d	d
	0.5	d	12	d	d
	1.0	d	9	d	d
	2.0	d	8	d	d
	∞	d	6	d	d
0.005	0.025	151	158	154	190
	0.05	o	84	o	o
	0.1	d(?)	46	d	d
	0.5	d	17	d	d
	1.0	d	13	d	d
	2.0	d	11	d	d
	∞	d	8	d	d

[a]d, divergence; o, oscillations.

the following conditions are satisfied:

$$E_1 = |T(1, t) - T^*(x = 1)| + |y(1, t) - y^*(x = 1)| < E \qquad (4.195)$$

The correct values T^* and y^* were obtained by the shooting method. Experience has shown that for higher values of the parameter Pe the shooting method suffers from the stiffness of the relevant initial value differential equations and computer time can be prohibitive. On the other hand, the false transient method results frequently in a convergent sequence. For instance, for Pe = 200 and Da = 0.1 [$\gamma = 40$, $\beta = 0.2$, $\psi \equiv 1$, $\varphi \equiv 0.5$, $T^*(1) = 1.03215$, $y^*(1) = 0.81058$], all four methods yield the steady-state solution in a wide range of parameters w_1 and w_2, $w_1/k \in \langle 0, 1 \rangle$, $w_2/k \in \langle 0, 1 \rangle$. The dependence of E_1 in (4.195) on the number of iterations (for $w_1 = w_2 = 0$) is presented in Table 4-24. For higher

TABLE 4-24
DEPENDENCE OF THE ERROR E_1 IN (4.195) ON THE NUMBER
OF ITERATIONS FOR $w_1 = w_2 = 0$, $n = 20$: Pe = 200, $\gamma = 40$,
$\beta = 0.2$, Da = 0.1, $T^*(x = 1) = 1.03215$, $y^*(1) = 0.81058$,
$\varphi \equiv 0.5$, $\psi \equiv 1$

Iteration	Method 1	Method 2	Method 3	Method 4
0	0.3427	0.3427	0.3427	0.3427
1	0.1527	0.0894	0.1622	0.1622
2	0.0598	0.0363	0.0713	0.0710
3	0.0215	0.0145	0.0247	0.0244
4	0.0086	0.0070	0.0091	0.0090
5	0.0052	0.0049	0.0052	0.0052
6	0.0044	0.0044	0.0044	0.0044
7	0.0043	0.0043	0.0043	0.0043
8	0.0043[a]	0.0043[a]	0.0043[a]	0.0043[a]

[a]Nonzero asymptotic value caused by the discretization error in the x-direction.

values of the parameter Da, Da = 0.2, methods 1, 3, and 4 became unstable for higher values of k.

We have chosen these equations as an illustration of the false transient method because for some values of governing parameters the steady-state equations can exhibit multiplicity. Of course, for this event the "time" trajectory depends on the guess of the initial profile. The particular steady state attracts only those trajectories that originate from a given region of initial conditions. Hence to calculate all stable steady states a number of different initial profiles must be guessed. For following values of parameters, $\gamma = 40$, $\beta = 0.3$, Da = 0.03, Pe = 2, three steady states exist—the lower $T^*(1) = 1.0125$, $y^*(1) = 0.946$; the middle $T^*(1) = 1.111$, $y^*(1) = 0.506$; and the upper $T^*(1) = 1.246$, $y^*(1) = 0.00002$. It can be shown that the middle steady state is unstable.

In an attempt to calculate the middle profile, the following false transient equations have been investigated:[†]

$$w_{11}\frac{\partial y}{\partial t} + w_{12}\frac{\partial T}{\partial t} = \frac{1}{Pe}\left(\frac{\partial^2 y}{\partial x^2} + \frac{1}{T}\frac{\partial T}{\partial x}\frac{\partial y}{\partial x}\right) - \frac{\partial y}{\partial x} - \frac{y}{T}\frac{\partial T}{\partial x}$$

$$- Da\frac{y}{T}\exp\left[\frac{\gamma(T-1)}{T}\right] \qquad (4.196a)$$

$$w_{21}\frac{\partial y}{\partial t} + w_{22}\frac{\partial T}{\partial t} = \frac{1}{Pe}\frac{\partial^2 T}{\partial x^2} - \frac{\partial T}{\partial x} + \beta\, Da\frac{y}{T}\exp\left[\frac{\gamma(T-1)}{T}\right] \qquad (4.196b)$$

subject to boundary conditions (4.187) and (4.188) and initial conditions (4.189). The matrix

$$W = \begin{pmatrix} w_{11} & w_{12} \\ w_{21} & w_{22} \end{pmatrix}$$

is assumed to be nonsingular. The partial differential equations have been solved by a simple explicit method with the nonsymmetrical finite-difference approximations at the boundary. As the initial profiles, the following relations have been used:

$$\varphi(x) = 0.536 + 0.294(x - 1)^2, \qquad \psi(x) = 1.111 - 0.078(x - 1)^2$$

which approximates fairly well the middle profile calculated by the shooting method. Table 4-25 presents some numerical experiments performed for different values of the coefficients w_{ij}. The table reveals that this procedure is also not capable of calculating the middle steady-state profile. The standard steady-state solution was obtained by the shooting procedure combined with the

TABLE 4-25
ATTEMPT TO REACH THE UNSTABLE (MIDDLE)
SOLUTION FOR A CASE OF MULTIPLE
SOLUTIONS: TRANSIENT EQUATION (4.196)[a]

w_{11}	w_{12}	w_{21}	w_{22}	Solution obtained
1	1	0	1	Upper
1	−1	0	1	Upper
1	0	−1	1	Upper
1	0	0	1	Upper
0	1	−1	1	Upper
1	0.4	−0.2	1	Upper
1	1	−1	1	Upper
1	1	−1	0	Upper
−1	1	−1	0	Upper
0	1	−1	0	Upper

[a]All other possible combinations of 1, 0, −1 for $w_{11}, w_{12}, w_{21}, w_{22}$ lead to divergence.

[†]M. Kubíček et al, *Chem. Eng. Sci. 31,* 727 (1976).

root-finding secant method (see Warner's method, Section 4.6). Table 4-23 reveals that method 2 is the most convenient because even if $k = \infty$ (i.e., $w_1 = w_2 = 0$), the steady-state solutions can be obtained. Essentially, this type of iteration was dealt with in Section 4.1. For these conditions, approximations of the particular nonlinear terms can be considered as a linearization of the original differential equations.

The effect of the ratio of weights w_1 and w_2 can be followed for method 2 in Table 4-26. It appears from these results that the number of iterations can be affected substantially.

TABLE 4-26

EFFECT OF VALUES OF w_1 AND w_2, METHOD 2, $k = 1$:
$Pe = 5$, $Da = 0.15$, $\beta = 0.2$, $\gamma = 40$, $\varphi \equiv 0$, $\psi \equiv 1$, $n = 20$[a]

	w_2			
w_1	0	1	2	5
0	6-7-8-17	9-11-13-20		
1	7-12-14-19	9-10-13-22	11-14-15-23	18-29-40-57
2		9-16-19-25		
5		12-20-21-28		

[a]The number of time steps to satisfy (4.195) is presented, sequentially for $E = 0.1 - 0.01 - 0.005 - 0.001$.

The effect of the initial (constant) profiles can be inferred for all four methods from Table 4-27. Based on these results it seems that method 2 is superior. In

TABLE 4-27

NUMBER OF TIME STEPS TO SATISFY (4.195), $E = 0.1$:
$\gamma = 40$, $\beta = 0.2$, $Da = 0.15$, $Pe = 5$, $w_1 = w_2 = 40$,
$k = 1$, $n = 20$[a,b]

ψ	φ	Method 1	Method 2	Method 3	Method 4
1	0	104	106	104	103
1	1	69	70	69	69
1.15	1	d	43	d	d
1.15	0.5	d	23	d	d
1	0.5	91	92	90	90

[a]Different initial (constant) profiles.
[b]d, divergence.

addition, as we have shown, method 2 is also safe against divergence, even for low values of w_1 and w_2 (i.e., for high values of k).

The second and third examples explore further the situation of multiplicity on difficult diffusion problems arising in chemical kinetics.

Example 4.12

Nonisothermal nonadiabatic axial mixing in a tubular reactor is described by the dimensionless differential equations

$$\frac{1}{Pe}\frac{d^2\theta}{dx^2} - \frac{d\theta}{dx} - \beta(\theta - \theta_c) + B\,Da\,(1-y)\exp\left(\frac{\theta}{1+\theta/\gamma}\right) = 0 \qquad (4.197)$$

$$\frac{1}{Pe}\frac{d^2y}{dx^2} - \frac{dy}{dx} + Da\,(1-y)\exp\left(\frac{\theta}{1+\theta/\gamma}\right) = 0 \qquad (4.198)$$

subject to the boundary conditions

$$x = 0: \qquad Pe\,\theta = \frac{d\theta}{dx}, \qquad Pe\,y = \frac{dy}{dx} \qquad (4.199)$$

$$x = 1: \qquad \frac{d\theta}{dx} = \frac{dy}{dx} = 0 \qquad (4.200)$$

The false transient equations are

$$w_1\frac{\partial\theta}{\partial t} = \frac{1}{Pe}\frac{\partial^2\theta}{\partial x^2} - \frac{\partial\theta}{\partial x} - \beta(\theta - \theta_c) + B\,Da(1-y)\exp\left(\frac{\theta}{1+\theta/\gamma}\right) \qquad (4.201)$$

$$w_2\frac{\partial y}{\partial t} = \frac{1}{Pe}\frac{\partial^2 y}{\partial x^2} - \frac{\partial y}{\partial x} + Da\,(1-y)\exp\left(\frac{\theta}{1+\theta/\gamma}\right) \qquad (4.202)$$

The initial conditions are in the form

$$\theta(x,0) = \psi(x), \qquad y(x,0) = \varphi(x) \qquad (4.203)$$

It will be shown in Example 4.32 that these equations for particular values of the parameters β, B, Da, γ, and Pe possess multiple steady states. For instance, for $Pe = 2$, $\beta = 2$, $B = 12$, $Da = 0.12$, and $\gamma \rightarrow \infty$, five steady states exist. However, some of these steady states do not attract the transient trajectories (i.e., these profiles behave as unstable solutions). For example, for $w_1 = 1$ and $w_2 = 1$, only steady state 1 is stable (see Fig. 4-12). Steady states 2 through 5 are unstable. Around

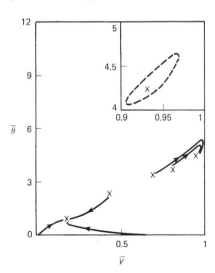

Figure 4-12 Example 4.12 trajectories: $Pe = 2$, $\beta = 2$, $B = 12$, $Da = 0.12$, $\gamma = \infty$, $w_1 = w_2 = 1$.

steady state 5 a limit cycle exists (i.e., undamped oscillations occur). To illustrate this situation, Fig. 4-12 is shown. Here we have denoted by $\bar{\theta}$ and $\bar{y}$ the integral average values of temperature and conversion, respectively. The cross denotes the steady state.

This example shows that for problems with multiple steady states a careful examination of all possible steady states must be performed if the false transient method is used.

Example 4.13

Heat and mass transfer accompanied by an exothermic chemical reaction occurring in a porous catalyst is described by two nonlinear differential equations of the boundary value type:

$$\frac{d^2y}{dx^2} + \frac{a}{x}\frac{dy}{dx} - \frac{\delta}{\gamma\beta}y^n \exp\left(\frac{\theta}{1+\theta/\gamma}\right) = 0 \tag{4.204}$$

$$\frac{d^2\theta}{dx^2} + \frac{a}{x}\frac{d\theta}{dx} + \delta y^n \exp\left(\frac{\theta}{1+\theta/\gamma}\right) = 0 \tag{4.205}$$

subject to the boundary conditions

$$y(1) = 1, \qquad \theta(1) = 0 \tag{4.206}$$

$$\frac{dy(0)}{dx} = 0, \qquad \frac{d\theta(0)}{dx} = 0 \tag{4.207}$$

The false transient equations are

$$w_1\frac{\partial y}{\partial t} = \frac{\partial^2 y}{\partial x^2} + \frac{a}{x}\frac{\partial y}{\partial x} - \frac{\delta}{\gamma\beta}y^n \exp\left(\frac{\theta}{1+\theta/\gamma}\right) \tag{4.208}$$

$$w_2\frac{\partial \theta}{\partial t} = \frac{\partial^2 \theta}{\partial x^2} + \frac{a}{x}\frac{\partial \theta}{\partial x} + \delta y^n \exp\left(\frac{\theta}{1+\theta/\gamma}\right) \tag{4.209}$$

The boundary conditions are given by (4.206) and (4.207), and the initial conditions are

$$y(x, 0) = \varphi(x), \qquad \theta(x, 0) = \psi(x) \tag{4.210}$$

For the parameter values $\gamma = 20$, $\beta = 0.2$, $\delta = 2.56$, $a = 0$, and $n = 1$, only one steady state exists. For the weights $w_1 = 1$ and $w_2 = 1$, this steady state can be calculated easily by the false transient method. A careful examination of the stability shows that for $w_1 \in \langle 1.0, 2.5 \rangle$ and $w_2 = 1$, the false transient method always converges to the steady state. For other values of w_1 and w_2, the situation can be more complicated. For example, for $w_1 = 4$ and $w_2 = 1$, the false transient method does not approach the steady state, but undamped oscillations are produced (see Fig. 4-13).

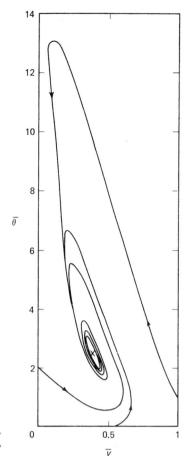

Figure 4-13 Example 4.13 trajectories: $\gamma = 20$, $\beta = 0.2$, $\delta = 2.56$, $a = 0$, $n = 1$, $w_1 = 4$, $w_2 = 1$.

The following example is devoted to problems connected with the appropriate choice of false transient equations. While for nonlinear diffusion equations the construction of false transient equations is a trivial problem and leads to nonlinear parabolic equations, the situation for other differential equations, arising, for example, in the boundary layer theory, is more complicated. Sometimes it is difficult to construct a convergent false transient procedure.

Example 4.14

The Blasius equation describing the boundary layer of a non-Newtonian fluid over a plate for a steady-state problem is in the form

$$\frac{d^3y}{dx^3} + y\left(\frac{d^2y}{dx^2}\right)^{2-\alpha} = 0 \tag{4.211}$$

subject to the boundary conditions

$$y(0) = 0, \qquad \frac{dy(0)}{dx} = 0, \qquad \frac{dy(\infty)}{dx} = 1 \tag{4.212}$$

143

Before considering the false transient equations it is worthwhile to look at the alternatives to (4.211). There are four possibilities which can be handled by the false transient method:

$$(1) \quad y''' + y(y'')^{2-\alpha} = 0 \tag{4.213}$$

$$(2) \quad y' - z = 0, \; z'' + y(z')^{2-\alpha} = 0 \tag{4.214}$$

$$(3) \quad y'' - z = 0, \; z' + yz^{2-\alpha} = 0 \tag{4.215}$$

$$(4) \quad y' - z = 0, \; z' - u = 0, \; u' + yu^{2-\alpha} = 0 \tag{4.216}$$

The boundary conditions for a particular alternative are obvious. The following false transient equations will be considered:

$$(1) \quad w\frac{\partial y}{\partial t} = \frac{\partial^3 y}{\partial x^3} + y\left(\frac{\partial^2 y}{\partial x^2}\right)^{2-\alpha} \tag{4.217}$$

$$(2) \quad w_{11}\frac{\partial y}{\partial t} + w_{12}\frac{\partial z}{\partial t} = \frac{\partial y}{\partial x} - z \tag{4.218a}$$

$$w_{21}\frac{\partial y}{\partial t} + w_{22}\frac{\partial z}{\partial t} = \frac{\partial^2 z}{\partial x^2} + y\left(\frac{\partial z}{\partial x}\right)^{2-\alpha} \tag{4.218b}$$

$$(3) \quad w_{11}\frac{\partial y}{\partial t} + w_{12}\frac{\partial z}{\partial t} = \frac{\partial^2 y}{\partial x^2} - z \tag{4.219a}$$

$$w_{21}\frac{\partial y}{\partial t} + w_{22}\frac{\partial z}{\partial t} = \frac{\partial z}{\partial x} + yz^{2-\alpha} \tag{4.219b}$$

$$(4) \quad w_{11}\frac{\partial y}{\partial t} + w_{12}\frac{\partial z}{\partial t} + w_{13}\frac{\partial u}{\partial t} = \frac{\partial y}{\partial x} - z \tag{4.220a}$$

$$w_{21}\frac{\partial y}{\partial t} + w_{22}\frac{\partial z}{\partial t} + w_{23}\frac{\partial u}{\partial t} = \frac{\partial z}{\partial x} - u \tag{4.220b}$$

$$w_{31}\frac{\partial y}{\partial t} + w_{32}\frac{\partial z}{\partial t} + w_{33}\frac{\partial u}{\partial t} = \frac{\partial u}{\partial x} + yu^{2-\alpha} \tag{4.220c}$$

Unlike the diffusion problems which have been considered in Examples 4.11 through 4.13, it is difficult to assess which type of false transient method will converge. To test the convergence properties of the false transient equations, a simple explicit finite-difference analogy has been constructed.

For the false transient equation given by (4.217), the following finite-difference approximation has been used:

$$\frac{w}{k}(y_i^{j+1} - y_i^j) = \frac{1}{2h^3}(-y_{i-2}^j + 2y_{i-1}^j - 2y_{i+1}^j + y_{i+2}^j)$$
$$+ y_i^j\left(\frac{y_{i-1}^j - 2y_i^j + y_{i+1}^j}{h^2}\right)^{2-\alpha} \qquad i = 2, 3, \ldots, n-2 \tag{4.221}$$

The boundary conditions are constructed in a semi-implicit way:

$$y_0^{j+1} = 0$$

$$\frac{w}{k}(y_1^{j+1} - y_1^j) = \frac{1}{2h^3}(-y_{-1}^{j+1} + 2y_0^{j+1} - 2y_2^{j+1} + y_3^{j+1})$$
$$+ y_1^{j+1}\left(\frac{y_0^j - 2y_1^j + y_2^j}{h^2}\right)^{2-\alpha} \tag{4.222}$$

Because of $y_0^j = y_0^{j+1}$ and because of symmetry $y_{-1}^{j+1} = y_1^{j+1}$, the variable y_1^{j+1} can be easily calculated from (4.222) after some simple algebraic manipulations.

For practical calculations the infinite length of the interval will be approximated by the finite value $x = x_f$. The implicit approximation of (4.217) at the mesh points

$n - 1$ and n yields

$$\frac{w}{k}(y_{n-1}^{j+1} - y_{n-1}^j) = \frac{1}{2h^3}(-y_{n-3}^{j+1} + 2y_{n-2}^{j+1} - 2y_n^{j+1} + y_{n+1}^{j+1})$$
$$+ y_{n-1}^{j+1}\left(\frac{y_{n-2}^j - 2y_{n-1}^j + y_n^j}{h^2}\right)^{2-\alpha} \quad (4.223)$$

$$\frac{w}{k}(y_n^{j+1} - y_n^j) = \frac{1}{2h^3}(3y_{n-4}^{j+1} - 14y_{n-3}^{j+1} + 24y_{n-2}^{j+1} - 18y_{n-1}^{j+1} + 5y_n^{j+1})$$
$$+ y_n^{j+1}\left(\frac{y_{n-1}^j - 2y_n^j + y_{n+1}^j}{h^2}\right)^{2-\alpha} \quad (4.224)$$

The approximation of the boundary condition at $x = x_f$ results in

$$y_{n+1}^{j+1} = y_{n-1}^{j+1} + 2h \quad (4.225)$$

The linear algebraic equations (4.223)–(4.225) must be solved simultaneously to calculate y_{n-1}^{j+1}, y_n^{j+1}, and y_{n+1}^{j+1}. The interior mesh points at $j + 1$ are calculated recurrently from the explicit scheme.

The false transient equation was capable of determining the steady-state profiles. An experimentation with method 2 through method 4 has shown that these iteration procedures do not converge toward the required steady-state profile. For instance, for a diagonal matrix $W(w_{ij} = 0$ for $i \neq j)$, a modified Wendroff[†] approximation has been used and eight combinations with $w_{ii} = \pm 1$, $i = 1, 2, 3$ have been examined. However, all experiments have failed. Thus only the results attained for method 1 will be discussed. The effectiveness of the false transient method has been investigated for a Newtonian fluid (i.e., for $\alpha = 1$ and $x_f = 10$). The initial condition has been chosen in the form

$$y(x, 0) = 8\left[1 + \sin\left(\frac{\pi}{2}\left(3 + \frac{x}{x_f}\right)\right)\right] \quad (4.226)$$

The course of calculations by method 1 is presented in Tables 4-28 and 4-29. Table

TABLE 4-28
SOLUTION OF FALSE TRANSIENT EQUATION (4.217)
FOR THE BLASIUS EQUATION[a]

t	$x = 2$	$x = 4$	$x = 6$	$x = 8$	$x = 10$
0.00	0.391	1.528	3.298	5.528	8.000
0.04	0.394	1.537	3.312	5.538	7.923
1.0	0.458	1.736	3.525	5.512	7.521
2.0	0.527	1.897	3.655	5.571	7.555
5.0	0.735	2.246	4.012	5.903	7.876
10.0	1.024	2.693	4.526	6.444	8.423
20.0	1.327	3.191	5.131	7.101	9.093
30.0	1.428	3.362	5.343	7.335	9.333
40.0	1.458	3.413	5.407	7.405	9.406
50.0	1.466	3.427	5.425	7.426	9.427
60.0	1.469	3.432	5.431	7.432	9.433
70.0	1.469	3.433	5.432	7.433	9.434
80.0	1.470	3.433	5.432	7.434	9.435

[a]Boundary conditions in the form of (4.223)–(4.225),
$w = 1$, $n = 5$, $k = 0.04$, $x_f = 10$

[†]A. R. Mitchell, *Computational Methods in Partial Differential Equations* (London: Wiley, 1969).

4-29 reveals that from a relatively good initial profile a great number of time steps must be done to reach the steady-state solution. Apparently, the excessive number of the time steps is the result of the explicit scheme used. Practical tests indicated

TABLE 4-29
SOLUTION OF FALSE TRANSIENT
EQUATION (4.217) FOR THE
BLASIUS EQUATION[a]

t	$y(x_f, t)$	j
0	8.000	0
1	7.503	25
2	7.530	50
5	7.834	125
10	8.285	250
20	8.689	500
30	8.799	750
40	8.827	1000
50	8.834	1250
60	8.836	1500
70	8.836	1750

[a]Boundary conditions replaced by the approximation given by (4.223)–(4.225), $w = 1$, $n = 10$, $k = 0.04$, $x_f = 10$

TABLE 4-30
DEPENDENCE OF THE STATIONARY SOLUTION OF (4.217) ON x_f AND n[a]

x_f	n	h	$x = 2$	$x = 4$	$x = 5$	$x = 8$	$x = 10$	$x = 16$	$x = 20$	k	t
5	5	1	0.943	2.894	3.915					0.1	30
	10	0.5	0.901	2.810	3.812					0.01	30
	20	0.25	1.420	4.149	5.616					0.002	15
	5[b]	1	0.933	2.872	3.872					0.1	25
	10[b]	0.5	0.900	2.806	3.806					0.01	25
	20[b]	0.25	0.891	2.791	3.790					0.002	25
10	5	2	1.470	3.433	—	7.434	9.435			0.04	70
	10	1	0.921	2.838	3.837	6.836	8.836			0.04	70
	20	0.5	0.898	2.802	3.801	3.801	8.801			0.01	70
	10[b]	1	0.921	2.838	3.837	6.836	8.836			0.04	60
20	5	4		3.746		7.742		15.741	19.741	0.1	150
	10	2	1.468	3.431		7.429	9.429	15.428	19.428	0.1	150
	20	1	0.920	2.838	3.836	6.836	8.835	14.835	18.835	0.02	150
	10[b]	2	1.468	3.431		7.429	9.429	15.429	19.429	0.1	150

[a]Boundary conditions in the form (4.223)–(4.225). In the column t, the time necessary to reach a stationary solution satisfying $\| \partial y(x, t)/\partial t \| < 10^{-5}$ is presented.
[b]Boundary conditions in the form $y_{n-1}^{j+1} = y_{n-2}^{j+1} + h$, $y_n^{j+1} = y_{n-2}^{j+1} + 2h$ instead of (4.223)–(4.225).

that the computer time expenditure can be lowered by an order of magnitude if an implicit finite-difference scheme would be used.

Tests have been conducted to estimate the accuracy of the steady-state profile as a function of the mesh points used and the value of x_f as well. The results reported in Table 4-30 show that for $x_f = 5$ and $x_f = 10$ (for $h = 0.5$), comparable profiles can be calculated. Clearly, the accuracy of the results in not essentially affected for $x_f > 5$. On the other hand, the mesh dimension, h, exhibits a profound effect on the accuracy.

The false transient method can be adapted to convective nonlinear problems with split boundary conditions (heat exchangers, tubular reactors with counter-current cooling, etc.). The following example illustrates this situation.

Example 4.15

The model of a tubular recycle reactor with piston flow is described by two non-linear differential equations of the boundary value type:

$$\frac{dy}{dx} = \text{Da}\,(1 - y)\exp\left(\frac{\theta}{1 + \theta/\gamma}\right) = R_1(y, \theta) \tag{4.227}$$

$$\frac{d\theta}{dx} = \text{Da}\,B(1 - y)\exp\left(\frac{\theta}{1 + \theta/\gamma}\right) - \beta(\theta - \theta_c) = R_2(y, \theta) \tag{4.228}$$

subject to the mixed boundary conditions

$$(1 - \lambda)\theta(1) = \theta(0), \qquad (1 - \lambda)y(1) = y(0) \tag{4.229}$$

Here Da, γ, B, β, θ_c, and λ are the governing parameters. The false transient equations are

$$w_1 \frac{\partial y}{\partial t} + \frac{\partial y}{\partial x} = R_1(y, \theta) \tag{4.230}$$

$$w_2 \frac{\partial \theta}{\partial t} + \frac{\partial \theta}{\partial x} = R_2(y, \theta) \tag{4.231}$$

the boundary conditions are given by (4.229). The false transient equations which are of hyperbolic type can be solved easily by the finite-difference method. The type of approximation of, for example, (4.230), is evident from (4.232):

$$\frac{w_1}{2k}(y_{i+1}^{j+1} + y_i^{j+1} - y_{i+1}^j - y_i^j) + \frac{y_{i+1}^{j+1} - y_i^{j+1}}{h} = R_1\left(\frac{y_{i+1}^j + y_i^j}{2}, \frac{\theta_{i+1}^j + \theta_i^j}{2}\right)$$

$$i = 0, 1, \ldots, n - 1 \tag{4.232}$$

$$y_0^{j+1} = (1 - \lambda)y_n^{j+1}$$

Clearly, the approximation given by (4.232) is implicit. The bidiagonal set of linear algebraic equations with one off-diagonal element can be solved easily by modified Gaussian elimination.

To illustrate the applicability of the false transient method, three figures are presented in which different situations are displayed. Here the integral average values $\bar{y}$ and $\bar{\theta}$ are drawn. The integral average value of, for example, y is

$$\bar{y}(t) = \int_0^1 y(x, t)\, dx \tag{4.233}$$

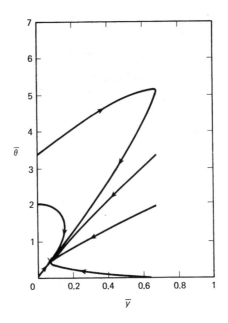

Figure 4-14 Example 4.15 trajectories for $w_1 = w_2 = 1$: $\gamma = 20$, $B = 6$, Da $= 0.05$, $\lambda = 0.65$, $\beta = 0$.

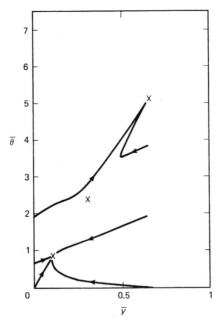

Figure 4-15 Example 4-15 trajectories for $w_1 = w_2 = 1$: $\gamma = 20$, $B = 8$, Da $= 0.0025$, $\beta = 0$, $\lambda = 0.718$.

In Fig. 4-14 the false transient trajectories are drawn for $w_1 = w_2 = 1$. Obviously, the steady state is stable and the false transient method is capable of calculating the steady-state solution. However, for three steady states (see Fig. 4-15) and $w_1 = w_2 = 1$, only two solutions are approached by the false transient method, while the third solution is unstable and does not attract the false transient trajectories. Finally, Fig. 4-16 shows the situation where one unstable steady state exists and the undamped oscillations around the steady-state profile exist.

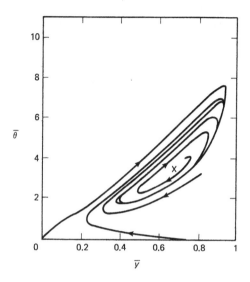

Figure 4-16 Example 4-15 trajectories for $w_1 = w_2 = 1$: $\gamma = 20$, $B = 13$, Da $= 0.059$, $\beta = 0.8$, $\lambda = 0.4$, $\theta_c = 0$.

The false transient method can be used successfully to solve problems arising in the optimal control theory. The way of introduction of false transient term for optimal control problems is discussed in the example presented below.

Example 4.16

For a consecutive isothermal first-order chemical reaction,

$$A \xrightarrow{k_1} B \xrightarrow{k_2} C$$

the concentration profiles of the components A and B are described by

$$\frac{dy_1}{dx} = -k_1(T)y_1 \tag{4.234}$$

$$\frac{dy_2}{dx} = k_1(T)y_1 - k_2(T)y_2 \tag{4.235}$$

The initial conditions are

$$y_1(0) = y_1^0, \qquad y_2(0) = y_2^0 \tag{4.236}$$

The reaction-rate constants depend exponentially on temperature

$$k_i = k_{i0} \exp\left(-\frac{E}{RT}\right) \qquad i = 1, 2$$

It is necessary to find the temperature profile $T(x)$ which maximizes the exit concentration of B [i.e., $y_2(x_f)$ should be maximum]. On using the maximum principle, a set of adjoint equations results:

$$\frac{d\psi_1}{dx} = k_1(T)(\psi_1 - \psi_2), \qquad \frac{d\psi_2}{dx} = k_2(T)\psi_2 \qquad (4.237)$$

with the terminal condition

$$\psi_1(x_f) = 0, \qquad \psi_2(x_f) = 1 \qquad (4.238)$$

The maximization of the Hamiltonian function

$$H = k_1(T)(\psi_2 - \psi_1)y_1 - k_2(T)\psi_2 y_2 \qquad (4.239)$$

with respect to T for fixed values of y and ψ gives rise to $(dH/dT = 0)$:

$$T = \frac{E_2 - E_1}{R \ell n \dfrac{k_{20}E_2 y_2 \psi_2}{k_{10}E_1 y_1(\psi_2 - \psi_1)}} \qquad (4.240)$$

Obviously, to find the optimum temperature profile it is necessary to solve (4.234) and (4.235) together with (4.237) subject to boundary conditions (4.236) and (4.238) and the explicit relation for the control variable (4.240). The false transient equations are

$$\frac{\partial y_1}{\partial t} = -\frac{\partial y_1}{\partial x} - k_1(T)y_1 \qquad (4.241a)$$

$$\frac{\partial y_2}{\partial t} = -\frac{\partial y_2}{\partial x} + k_1(T)y_1 - k_2(T)y_2 \qquad (4.241b)$$

$$\frac{\partial \psi_1}{\partial t} = \frac{\partial \psi_1}{\partial x} + k_1(T)(\psi_2 - \psi_1) \qquad (4.241c)$$

$$\frac{\partial \psi_2}{\partial t} = \frac{\partial \psi_2}{\partial x} - k_2(T)\psi_2 \qquad (4.241d)$$

subject to the boundary conditions

$$x = 0: \qquad y_1 = y_1^0, \qquad y_2 = y_2^0 \qquad (4.242a)$$

$$x = x_f: \qquad \psi_1 = 0, \qquad \psi_2 = 1 \qquad (4.242b)$$

The initial conditions are, for example,

$$t = 0: \qquad y_1 \equiv y_1^0, \qquad y_2 \equiv y_2^0, \qquad \psi_1 \equiv 0, \qquad \psi_2 \equiv 1 \qquad (4.243)$$

The first-order partial differential equations can be solved easily, for instance, by the explicit method, which gives rise to simple recurrent calculations. For 20 to 100 mesh points in the x-direction and for the time increment $t = 0.06$, the steady-state solution has been obtained after calculation of 50 to 100 time profiles (i.e., for $t = 3 - 6$).

4.4.5 Discussion

The false transient method is a very powerful technique if real transient equations can be constructed. For instance, for diffusion nonlinear boundary value problems it is very simple to construct the false transient equations. On the other hand, for nonlinear differential equations which have no direct physical meaning as, for example, the differential equations describing the hydrodynamic boundary layer, it is, generally speaking, difficult to construct a reliable and successful false transient equation. The difficulties associated with the boundary layer equations solved by the false transient method have been illustrated in Example 4.14. It appears that for such type of problems the false transient method is not convenient to find the solution rapidly. However, if a proper false transient alternative was found, the method can be used to provide solution effectively.

We are faced with the problem of the estimation of the weights w_i. Unfortunately, no general simple law can be developed and numerical experimentation with the particular problem is necessary.

For diffusion-like problems the effectiveness of the false transient technique is high and the method can be easily implemented. The technique seems to offer distinct advantages over the traditional shooting methods:

1. Stiff and parametrically sensitive boundary value problems can be readily handled.
2. Strongly nonlinear problems can be dealt with without calculating derivatives.
3. The sensitivity to a poor initial guess is very low and for a stable profile, convergence is usually assured.
4. The dimension of the boundary value problem is immaterial.

Apart from these advantages the method of false transients has two shortcomings:

1. If multiplicity occurs, it is difficult to locate all solutions because of the instability of some steady states.
2. For differential equations without direct physical meaning, it is not easy to construct the false transient method which converges.

The false transient method may also be adapted to solve elliptic partial differential equations. After approximating the particular elliptic equation by the "method of lines," the resulting set of ordinary differential equations can be solved by the false transient method.

We have also shown that the false transient method can be used to solve the problems of optimal control. Unfortunately, there is insufficient information in the literature on the solution of optimal control problems by the false transient method.

PROBLEMS

1. Heat and mass transfer in a porous spherical catalyst may be described by two nonlinear ordinary differential equations:

$$\frac{d^2Y}{d\xi^2} + \frac{2}{\xi}\frac{dY}{d\xi} = \phi^2 Y \exp\left(\frac{\theta}{1+\theta/\gamma}\right) \tag{4.244}$$

$$\frac{d^2\theta}{d\xi^2} + \frac{2}{\xi}\frac{d\theta}{d\xi} = -\gamma\beta\phi^2 Y \exp\left(\frac{\theta}{1+\theta/\gamma}\right) \tag{4.245}$$

$$\xi = 0: \qquad \frac{dY}{d\xi} = \frac{d\theta}{d\xi} = 0 \tag{4.246}$$

$$\xi = 1: \qquad Y = 1, \qquad \theta = 0 \tag{4.247}$$

Solve these equations by the false transient method for $\gamma = 40$, $\beta = 0.125$, and $\phi = 0.337$. Show that it is more convenient to solve the equation

$$\frac{d^2Y}{d\xi^2} + \frac{2}{\xi}\frac{dY}{d\xi} = \phi^2 Y \exp\left[\frac{\gamma\beta(1-Y)}{1+\beta(1-Y)}\right] \tag{4.248}$$

where $\theta = \gamma\beta(1-Y)$, by the false transient method.

2. For Problem 1, where the resistances against heat and mass transfer at the catalyst surface exist, the governing equation is

$$\frac{d^2Y}{d\xi^2} + \frac{2}{\xi}\frac{dY}{d\xi} = \phi^2 Y \exp\left[\frac{\gamma\beta(\chi-Y)}{1+\beta(\chi-Y)}\right] \tag{4.249}$$

The boundary conditions are

$$\xi = 1: \qquad -\frac{dY}{d\xi} = \text{Sh}\,(Y-1) \tag{4.250}$$

$$\xi = 0: \qquad \frac{dY}{d\xi} = 0 \tag{4.251}$$

Here we have denoted

$$\chi = \left(1 - \frac{\text{Sh}}{\text{Nu}}\right)Y(1) + \frac{\text{Sh}}{\text{Nu}}$$

Solve (4.249) by the false transient method. Construct a semi-implicit scheme which gives rise to a tridiagonal matrix.

3. Solve the nonlinear boundary value problem from Example 4.12 [i.e., (4.197) and (4.198)] subject to the boundary conditions

$$x = 0: \qquad \text{Pe}\,y = \frac{dy}{dx} \tag{4.252}$$

$$x = 1: \qquad \frac{d\theta}{dx} = \frac{dy}{dx} = 0 \tag{4.253}$$

and

$$\beta\int_0^1 (\theta - \theta_c)\,dx = C \tag{4.254}$$

Calculate the profiles of y and θ for parameters $\mathrm{Da} = 0.1$, $B = 1$, $\beta = 0.05$, $\gamma = 40$, $\theta_c = 0$, $\mathrm{Pe} = 400$, and $C = 0.005$. Use the semi-implicit finite-difference scheme. Compare the explicit and semi-implicit approximations.

4. Solve the nonlinear elliptic equation describing heat transfer in a two-dimensional body:

$$\frac{\partial^2\theta}{\partial x^2} + \frac{\partial^2\theta}{\partial y^2} = -\delta e^\theta \qquad (4.255)$$

with boundary conditions

$$\theta = 0 \text{ on the lines } x = 0,\ x = 1,\ y = 0,\ y = 1 \qquad (4.256)$$

Use the "method of lines" to approximate this equation. The resulting set of ordinary differential equations is solved by the false transient method. Use five grid points for discretization in the x-direction. Solve for $\delta = 0.2$.

BIBLIOGRAPHY

Application of the false transient method to the solution of nonlinear boundary value problems is presented in:

SHEAN-LIN-LIU: Numerical solution of two-point boundary value problems in simultaneous second order nonlinear ordinary differential equations. *Chem. Eng. Sci. 22*, 871 (1967).

MALLISON, G. D., AND DE VAHL, D. G.: The method of false transient for the solution of coupled elliptic equations. *J. Comput. Phys. 12*, 435 (1973).

FILIPPOV, C. C.: Numerical methods of solution of some nonlinear boundary value problems arising in the heat and mass transfer theory. *Eng. Phys. J. 34*, 756 (1975) (in Russian).

KUBÍČEK, M., HLAVÁČEK, V., HOLODNIOK, M.: Solution of nonlinear boundary value problems-X. The false transient method. *Chem. Eng. Sci. 31*, 727 (1976).

DOSS, S., AND MILLER, K.: Dynamic ADI Methods for Elliptic Equations, SIAM *J. Numer. Anal 16*, 837 (1979).

For a solution of optimal control problems by the false transient method, see:

BYKOV, V. J., KUZIN, V. A., AND FEDOTOV A. V.: A numerical method of solution of optimization problems in chemical technology. *Control Syst. 1*, 83 (1968) (in Russian).

Information on numerical integration of nonlinear parabolic equations can be found in:

SAULYEV, V. K.: *Integration of Equations of Parabolic Type by the Method of Nets.* Macmillan, New York, 1969.

Description of multilevel adaptive solutions to boundary value problems is presented in the paper:

BRANDT, A.: Multi-level Adaptive Solution to Boundary Value Problems, *Math Comput. 31*, 333 (1977).

4.5 One-Parameter Imbedding Technique

4.5.1 Introduction

In an attempt to improve the convergence properties of iterative methods for solving the nonlinear boundary value problems for ordinary differential equations, numerous methods involving solution of an associated system of differential equations have been suggested recently in the literature. Particularly promising is the application of the method devised by Davidenko to solve the set of nonlinear differential equations subject to boundary conditions. Davidenko's algorithm appears to be fairly powerful to solve functional equations and, obviously, can be applied to wide classes of nonlinear boundary value problems.

The one-parameter imbedding method (also the continuation method or method of artificially introduced parameter) has been used for the proofs of existence and uniqueness of numerous problems; however, so far, only insufficient attention has been devoted to the numerical application of this algorithm to solve nonlinear boundary value problems.

This section is oriented toward description of principal features of the one-parameter imbedding method for a general operator equation. The one-loop and multiloop one-parameter imbedding procedures are introduced. The general discussion is focused on the numerical aspects of application of both groups of methods to nonlinear boundary value problems. The similarity of the methods described in Sections 4.1, 4.3, 4.4, and 4.5 is discussed. A number of interesting examples illustrating this method are presented.

4.5.2 Details of One-Parameter Imbedding Technique

To illustrate the basic idea of the Davidenko approach, a general operator equation

$$F(y) = 0 \qquad (4.257)$$

will be considered. Here F is a suitably restricted map of a Banach space D into itself. Suppose that we can find a one-parameter family of operators $H(t, y)$ depending on a real parameter t in the following way:

$$H(t_0, y_0) = 0 \qquad (4.258)$$

where t_0 and y_0 are fixed;

$$H(t_1, y) = 0 \qquad (4.259)$$

which for a given t_1 has the same solution as (4.257), for example,

$$H(t_1, y) \equiv F(y) \qquad (4.260)$$

The idea of one-parameter imbedding is based on evaluation of the dependence

$$y = y(t), \qquad y_0 = y(t_0) \qquad (4.261)$$

which for $t \in \langle t_0, t_1 \rangle$ satisfies (4.262):

$$H(t, y(t)) \equiv 0 \qquad (4.262)$$

The element y_0 is known or can be easily calculated. However, from the dependence (4.261) only the terminal value

$$y^* = y(t_1) \tag{4.263}$$

can be utilized. Clearly, if $y(t)$, $t \in \langle t_0, t_1 \rangle$, is found, then y^* is the solution of the original equation (4.257).

4.5.2.1 Some general forms of H

In this section we present some convenient forms of $H(t, y)$ which can be adapted for the one-parameter imbedding method and which are not dependent on the type of (4.257). A wide group of one-parameter imbedding procedures can be written in the form

$$H(t, y) = a(t)G(y) + b(t)F(y) \tag{4.264}$$

where $a(t)$ and $b(t)$ are continuously differentiable for $t \in \langle t_0, t_1 \rangle$. The functions $a(t)$ and $b(t)$ are chosen such that (4.265)

$$H(t_0, y) = a(t_0)G(y) + b(t_0)F(y) = 0 \tag{4.265}$$

possesses the solution $y = y_0$. On the other hand, we suppose that for $t = t_1$, $H(t_1, y) \equiv F(y)$, that is,

$$a(t_1) = 0, \qquad b(t_1) = 1 \tag{4.266}$$

The most frequently used one-parameter imbedding formulas are

$$H(t, y) = F(y) + (t - 1)F(y_0) \tag{4.267}$$

and

$$H(t, y) = (1 - t)(y - y_0) + \omega t F(y) \tag{4.268}$$

where $t_0 = 0$ and $t_1 = 1$. The parameter ω can be chosen as either $+1$ or -1.

4.5.2.2 Solution of the family of operator equations

Two different approaches can be used to find the dependence $y(t)$ on the interval $t \in \langle t_0, t_1 \rangle$. The first method makes use of the Newton–Kantorovich method applied to (4.262) for a mesh

$$t = t_0 + ih, \qquad i = 1, 2, \ldots, n; \qquad h = \frac{t_1 - t_0}{n};$$

$$y^{k+1}(t_0 + ih) = y^k(t_0 + ih) - [H'_y(t_0 + ih, y^k(t_0 + ih))]^{-1} \tag{4.269}$$
$$\cdot H(t_0 + ih, y^k(t_0 + ih)) \qquad k = 0, 1, \ldots$$

The initial approximation for the iteration process can be chosen; for example,

$$y^0(t_0 + ih) = y(t_0 + (i - 1)h) \qquad i = 2, 3, \ldots, n \tag{4.270}$$

For $i = 1$ we have a guess

$$y^0(t_0 + h) = y_0 \tag{4.271}$$

Of course, any root-finding method can be used instead of the Newton method. Since the dependence $y(t)$ on $t \in \langle t_0, t_1 \rangle$ is not required, it is not necessary to calculate this dependence with a high precision and usually one to two iterations are sufficient. For $i = n$ (i.e., at the terminal point where $t = t_1$) the solution must be established accurately to obey the predetermined tolerance.

The second approach takes advantage of the implicit function theorem applied to (4.262). We can write supposing that H is continuously differentiable with respect to y and t,

$$\frac{dy}{dt} = -[H'_y(t, y)]^{-1} H'_t(t, y) \tag{4.272}$$

with the initial condition

$$y(t_0) = y_0 \tag{4.273}$$

Obviously, (4.272) results from the relation $dH/dt = 0$. Equation (4.272) can be considered as an initial value problem and integrated from $t = t_0$ to $t = t_1$. For $t = t_1$ the integration yields the desired solution $y(t_1) \equiv y^*$. The accuracy attained depends both on the length of the integration step and on the order of a particular marching integration technique. If, as a result of approximation errors, the approximation of y^* by $y(t_1)$ is insufficient, we can easily improve the value of $y(t_1)$ by using the Newton–Kantorovich method.

4.5.2.3 One-loop and multiloop one-parameter imbedding

Linear one-parameter imbedding (4.264) can be basically divided into two different groups: one-loop and multiloop algorithms.

The one-loop algorithm requires:

1. Iterative solving (4.265) to calculate y_0, that is, the operator $H(t, y)$; for example,

$$H(t, y) = a(t)G(y, y_0) + b(t)F(y) = 0 \tag{4.274}$$

cannot be explicitly solved with respect to y for $t = t_0$; or

2. For $t = t_0$ only one solution $y = y_0$ of (4.265) [or (4.258)] exists.

The multiloop algorithm is represented, for example, by the one-parameter imbedding given by (4.267) and (4.268). It is not necessary to solve (4.265) iteratively in order to determine y_0. Finally, let us note that the one-parameter imbedding given by (4.267) and (4.268) is a special case of (4.274) which can be solved explicitly for $t = t_0$.

We have distinguished the one-parameter imbedding methods to one- and multiloop algorithms because of their close relation to the resulting iteration process. For one-loop one-parameter imbedding the operator equation is solved in a straightforward way and the resulting solution cannot be improved by a repeated continuation process. On the other hand, multiloop imbedding makes it possible to enhance (by an iteration process) the accuracy of the solution gradually.

To illustrate the construction of the iteration algorithm for multiloop imbedding, the following form of the operator equation will be considered:

$$H(t, y, y_0) = (1 - t)(y - y_0) + \omega t F(y) \tag{4.275}$$

By guessing y_0^0 we can construct the imbedding $H(t, y^0, y_0^0)$ in the form

$$\frac{dy^0}{dt} = -[H'_y(t, y^0, y_0^0)]^{-1} H'_t(t, y^0, y_0^0)$$

$$y^0(0) = y_0^0 \tag{4.276}$$

The differential equation (4.276) can be integrated step by step until we finally reach $t = 1$. The calculated value $y^0(1)$ approximates the exact solution y^*. This value is used to construct a new imbedding $H(t, y^1, y_0^1)$ according to (4.275) which can be handled analogously as $H(t, y^0, y_0^0)$; that is, the differential equation (4.277) is integrated:

$$\frac{dy^1}{dt} = -[H_y'(t, y^1, y_0^1)]^{-1} H_t'(t, y^1, y_0^1)$$

$$y^1(0) = y_0^1 = y^0(1)$$

(4.277)

This algorithm is performed as long as the preassigned tolerance of y^* is not reached. The flowchart of the iteration algorithm is displayed in Fig. 4-17. It

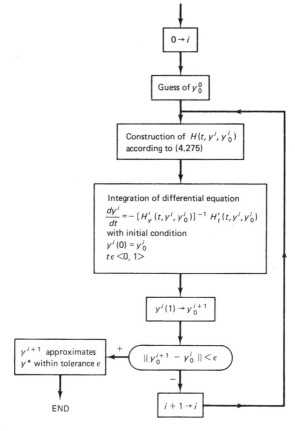

Figure 4-17 Flowchart of the iteration process according to imbedding given by (4.275).

is worth noting that the algorithm can fail if the initial guess y_0^0 does not belong to the "domain of attraction"; however, for the initial guesses which are physically sound for a particular engineering problem, the method is usually always convergent. To illustrate the methods mentioned above, we present a simple example.

Example 4.17

Consider a simple nonlinear boundary value problem

$$\frac{d^2y}{dx^2} = \alpha e^y \tag{4.278}$$

subject to the boundary conditions

$$y(0) = 0, \qquad y(1) = 0 \tag{4.279}$$

which describes the steady-state temperature distribution in a homogeneous rod of unit thickness where heat is generated at a rate αe^y per unit time per unit volume.

For this problem the Banach space D is a subset of the space $C^2_{\langle 0,1\rangle}$ for the functions y which meet the homogeneous boundary conditions (4.279). Comparing with (4.257), the differential equation (4.278) can be rewritten

$$F(y) = \frac{d^2y}{dx^2} - \alpha e^y = 0 \tag{4.280}$$

The one-parameter imbedding (4.267) yields

$$H(t, y) = \frac{d^2y}{dx^2} - \alpha e^y + (t - 1)\left[\frac{d^2y_0}{dx^2} - \alpha e^{y_0}\right] \tag{4.281}$$

For the imbedding formula (4.268), we have ($\omega = 1$ or $\omega = -1$)

$$H(t, y) = (1 - t)(y - y_0) + \omega t\left(\frac{d^2y}{dx^2} - \alpha e^y\right) \tag{4.282}$$

For both types of this imbedding the solution of the operator equation

$$H(t, y) = 0 \tag{4.283}$$

is known for $t = 0$ since it is the guessed "parameter" y_0. For $t = 1$ the solution to (4.283) is also the solution of the nonlinear boundary value problem given by (4.278) and (4.279). Obviously, the imbedding mentioned above is of the multiloop type.

The one-loop imbedding is, for instance,

$$H(t, y) = \frac{d^2y}{dx^2} - t\alpha e^y \tag{4.284}$$

Clearly, for $t = 0$ (4.283) possesses the solution $y \equiv 0$ because of boundary conditions (4.279). For $t = 1$ the solution of (4.278) and (4.279) results. It is evident that for this type of imbedding the solution $y(1)$ cannot be iteratively improved. For the sake of simplicity, the dependence of the element $y \in D$ on t will be denoted $y(x, t)$ in order to stress that the function $y(x)$ depends on the parameter t. On using this type of notation, (4.284) [considering (4.283)] can be rewritten

$$H(t, y) = \frac{\partial^2 y(x, t)}{\partial x^2} - t\alpha e^{y(x,t)} = 0 \tag{4.285}$$

subject to the boundary conditions

$$y(0, t) = 0, \qquad y(1, t) = 0 \tag{4.286}$$

Equations (4.281) and (4.282) can be rewritten in an analogous way. All three imbedding formulas employed in this section will be rewritten in terms of $y(x, t)$:
Case 1 [(4.281)]:

$$\frac{\partial^3 y(x, t)}{\partial x^2 \partial t} - \alpha e^{y(x,t)} \frac{\partial y(x, t)}{\partial t} + \frac{d^2 y_0(x)}{dx^2} - \alpha e^{y_0(x)} = 0 \qquad (4.287)$$

$$y(0, t) = 0, \qquad y(1, t) = 0 \qquad (4.288)$$

$$y(x, 0) = y_0(x) \qquad (4.289)$$

Case 2 [(4.282)]:

$$(1 - t)\frac{\partial y(x, t)}{\partial t} - [y(x, t) - y_0(x)] + \omega \left\{ \frac{\partial^2 y(x, t)}{\partial x^2} - \alpha e^{y(x,t)} \right. \qquad (4.290)$$

$$\left. + t\left[\frac{\partial^3 y(x, t)}{\partial x^2 \partial t} - \alpha e^{y(x,t)} \frac{\partial y(x, t)}{\partial t} \right] \right\} = 0$$

The boundary and initial conditions are given by (4.288) and (4.289), respectively.
Case 3 [(4.284)]:

$$\frac{\partial^3 y(x, t)}{\partial x^2 \partial t} - \alpha e^{y(x,t)} \left[1 + t\frac{\partial y(x, t)}{\partial t} \right] = 0 \qquad (4.291)$$

The boundary conditions are given by (4.288); the initial condition is

$$y(x, 0) \equiv 0 \qquad (4.292)$$

The function $y(x, t)$ evaluated from (4.287), (4.290), and (4.291) is for $y_0(x) \equiv 0$ and $\alpha = 1$ presented in Figs. 4-18 through 4-21. The method of integration of imbedding equations is described below. Whereas for the events presented in Figs. 4-18, 4-20, and 4-21 the solution can be found, imbedding of the type (4.290) and $\omega = 1$ is not capable of calculating the solution because there does not exist a continuous dependence $y(x, t)$ on t.

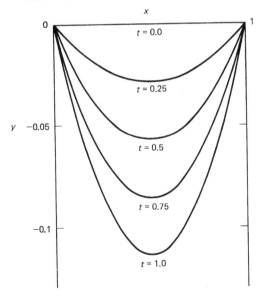

Figure 4-18 Course of $y(x, t)$ evaluated from (4.287): $\alpha = 1$.

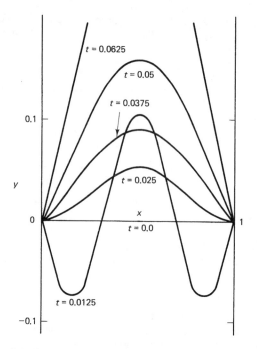

Figure 4-19 Course of $y(x, t)$ evaluated from (4.290): $\omega = 1$, $\alpha = 1$.

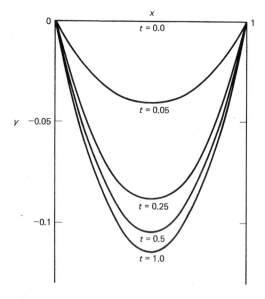

Figure 4-20 Course of $y(x, t)$ evaluated from (4.290): $\omega = -1$, $\alpha = 1$.

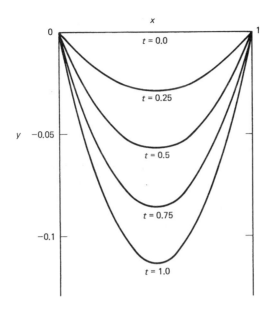

Figure 4-21 Course of $y(x, t)$ evaluated from Eq. (4.291): $\alpha = 1$.

4.5.3 Continuous Version of the Newton–Kantorovich Method for NBVP

In this section we address ourselves in detail to imbedding according to (4.267); that is,

$$H(t, y) = F(y) + (t - 1)F(y_0) \tag{4.267}$$

where $t \in \langle 0, 1 \rangle$. We introduce the variable λ,

$$1 - t = e^{-\lambda} \tag{4.293}$$

Thus y becomes a function of a continuous parameter λ on the interval $\lambda \in \langle 0, \infty \rangle$. Equation (4.267) can be rewritten

$$H(\lambda, y) = F(y) - e^{-\lambda}F(y_0) \tag{4.294}$$

The dependence $y(\lambda)$ which fulfills the equation $H[\lambda, y(\lambda)] = 0$ is governed obviously by

$$F(y) = e^{-\lambda}F(y_0) \tag{4.295}$$

The differential equation (4.272) after using the imbedding (4.294) is

$$\frac{dy}{d\lambda} = -[F'_y(y)]^{-1}e^{-\lambda}F(y_0) = -(F'_y(y))^{-1}F(y) \tag{4.296}$$

161

Here we have utilized the relation (4.295). Equation (4.296) represents a continuous analogy of the Newton–Kantorovich method. An alternative form of (4.296) is provided by noting that

$$F'_y(y)\frac{dy}{d\lambda} = -F(y) \tag{4.297}$$

with the initial condition

$$\lambda = 0: \qquad y = y_0 \tag{4.298}$$

If the dependence $y(\lambda)$ with given properties exists (it can be proven for y_0, which is in the vicinity of y^*), the value of $\|F(y(\lambda))\| = e^{-\lambda}\|F(y_0)\|$ for $\lambda \to \infty$ estimates the rate of convergence. The relation between the continuous and the discrete form of the Newton–Kantorovich method is evident if the continuous version is integrated by the Euler method with the step $h = 1$. Accordingly,

$$F'_y(y_k)(y_{k+1} - y_k) = -F(y_k) \tag{4.299}$$

where we have denoted $y_k = y(kh) = y(k)$.

Let us return to imbedding given by (4.267). The dependence $y(t)$ is described by a differential equation

$$\frac{dy}{dt} = -[F'_y(y(t))]^{-1}F(y_0)$$

$$y(0) = y_0 \tag{4.300}$$

This equation can be integrated by the Euler method with step length $h = 1$, and the value $y_1 = y(1)$ is obtained. Of course, because on the interval $\langle 0, 1 \rangle$ only one step of the Euler method has been used, the solution is not accurate enough. However, because of the multiloop character of the imbedding (4.267), y_1 may be used instead of y_0, integration by the Euler method is repeated, and so on. It is evident that this algorithm is formally identical with the Newton–Kantorovich method (4.299). To sum up, the mutiloop imbedding (4.267), together with Euler integration with the step $h = 1$, gives rise to the classical Newton–Kantorovich method.

Example 4.18

The nonlinear boundary value problem governed by (4.278) and (4.279) will be solved by the one-parameter imbedding (4.267) [i.e., (4.281) is considered]. For this particular case (4.300) results in (4.287) subject to boundary conditions (4.288). The initial condition is (4.289), where y_0 is a guessed function. The nonlinear third-order partial differential equation can be solved by taking advantage of the finite difference approximations (a different approach to solving this equation will be presented in Section 4.5.6). Let us denote $Y_i^j \sim y(ih, jk)$, $h = 1/n$, and $k = 1/m$. On replacing the derivatives by finite-differences, we have

$$
\frac{1}{kh^2}(Y_{i-1}^{j+1} - 2Y_i^{j+1} + Y_{i+1}^{j+1} - Y_{i-1}^j + 2Y_i^j - Y_{i+1}^j)
$$

$$
- \frac{\alpha}{k}(Y_i^{j+1} - Y_i^j)\exp(Y_i^j) \tag{4.301}
$$

$$
+ \frac{1}{h^2}(Y_{i-1}^0 - 2Y_i^0 + Y_{i+1}^0) - \alpha\exp(Y_i^0) = 0
$$

$$
i = 1, 2, \ldots, n - 1
$$

$$
Y_0^{j+1} = 0, \qquad Y_n^{j+1} = 0 \tag{4.302}
$$

The initial condition is approximated by

$$
Y_i^0 = y_0(ih) \qquad i = 0, 1, \ldots, n \tag{4.303}
$$

Equations (4.301) and (4.302) represent a set of $n + 1$ linear algebraic equations with a tridiagonal matrix for $n + 1$ unknowns $Y_0^{j+1}, Y_1^{j+1}, \ldots, Y_n^{j+1}$. These equations can be solved by means of the Thomas algorithm. Three strategies of computation can be adapted:

Algorithm A: Integration of (4.301) and (4.302) using a short integration step k. A relatively exact approximation of solution of the original NBVP is established.

Algorithm B: Integration of (4.301) and (4.302) using a long integration step k (e.g., $k = 1$). The accuracy of this solution is gradually improved by a successive imbedding (i.e., the first approximation is used to replace y_0, etc.).

Algorithm C: The continuous analogy of the Newton–Kantorovich method is used and the Euler integration method with a short integration step is employed. According to (4.301) and (4.302), the values of Y_i^1, $i = 0, 1, \ldots, n$, are calculated and in the second loop used instead of the values Y_i^0, $i = 0, 1, \ldots, n$, and so on. Evidently, for $k = 1$ algorithms B and C are identical with the classical Newton–Kantorovich method.

To demonstrate the advantages and shortcomings of each particular algorithm, a comparison of them will be presented:

1. Algorithm A requires a very fine mesh (i.e., a short step k which may result in an excessive computer time).
2. Algorithm C using a step $k < 1$ is analogous to the Miele modification of the Newton–Kantorovich method, which does not take advantage of the performance index.
3. For a shorter step k it is necessary to solve a number of tridiagonal systems within one integration from $t = 0$ to $t = 1$; however, the number of loops is lowered, and vice versa. Accordingly, an optimum value of the integration step k exists for both algorithms B and C since we are not interested in having the accurate dependence $y(x, t)$ on t but in the number of steps necessary to reach the solution y^* within the predetermined tolerance.

These conclusions are demonstrated in Tables 4-31 through 4-34. A comparison of Tables 4-31 and 4-32 reveals that algorithm A is not effective since the number of steps for algorithm A is by a factor of 10 higher than that for the algorithm B using

TABLE 4-31
DEPENDENCE OF SOLUTION ON t:
$n = 10$, $\alpha = 4$, $y_0(x) \equiv 0$

	$y(0.5, t)$	
t	$k = 0.1$	$k = 0.2$
0.0	0.0000	0.0000
0.2	−0.0706	−0.0702
0.4	−0.1424	−0.1418
0.6	−0.2156	−0.2146
0.8	−0.2900	−0.2887
1.0	−0.3656	−0.3640
0.0	−0.3656	−0.3640
0.2	−0.3660	−0.3646
0.4	−0.3663	−0.3653
0.6	−0.3666	−0.3660
0.8	−0.3670	−0.3666
1.0	−0.3673	−0.3673

$k = 1$. This is a general property of imbedding of the type A. The behavior of the continuous analogy of the Newton–Kantorovich method (i.e., algorithm C) is presented in Table 4-33. Obviously, this procedure shows that an optimal length of integration step exists ($k \sim 1.0$). The case where the initial guess $y_0(x)$ is far from the solution is presented in Table 4-34. From this table it can also be inferred that the optimum step k is approximately $k \sim 1.0$. The behavior of the Newton–Kantorovich method for the event with multiple solutions can be followed in Table 4-35. It is interesting to note that for a poor initial guess of the solution with higher values of y, the method using $k = 1$ (i.e., the classical Newton–Kantorovich) results in convergence, while for a shorter step $k = 0.1$, the algorithm has failed. The reason is, perhaps, that the dependence of $y(x, t)$ on t is not continuous.

TABLE 4-32
RESULTS FOR THE NEWTON–
KANTOROVICH METHOD:
$n = 10$, $\alpha = 4$, $y_0(x) \equiv 0$,
$k = 1.0$

t	$y(0.5, t)$
0.0	0.0000
1.0	−0.3511
0.0	−0.3511
1.0	−0.3673
0.0	−0.3673
1.0	−0.3673

TABLE 4-33
RESULTS FOR CONTINUOUS VERSION OF THE NEWTON–KANTOROVICH
METHOD: $n = 10$, $\alpha = 4$, $y_0(x) \equiv 0$

	$y(0.5, t)$				
t	$k = 0.05$	$k = 0.1$	$k = 0.2$	$k = 0.5$	$k = 1.0$
0.0	0.0000	0.0000	0.0000	0.0000	0.0000
0.2	−0.0656	−0.0670	−0.0702	—	—
0.5	−0.1433	−0.1461	—	−0.1756	—
0.8	−0.2015	−0.2048	−0.2120	—	—
1.0	−0.2318	−0.2352	−0.2425	−0.2694	−0.3511
1.5	−0.2857	−0.2888	—	−0.3178	—
2.0	−0.3182	−0.3208	−0.3259	−0.3424	−0.3673
3.0	−0.3497	−0.3510	−0.3537	−0.3610	−0.3673
4.0	−0.3610	−0.3616	−0.3628	−0.3657	−0.3673

TABLE 4-34
CASE OF A POOR INITIAL GUESS:
$n = 20$, $\alpha = 4$, $y_0(x) = 8x(1 - x) + \sin 2\pi x$

	$y(0.5, t)$		
t	$k = 0.1$	$k = 0.5$	$k = 1$
0.0	2.0000	2.0000	2.0000
0.5	1.2768	1.4332	—
1.0	−0.1200	0.5018	0.8663
0.0	−0.1200	0.5018	0.8663
0.5	−0.2419	0.1239	—
1.0	−0.3672	−0.3080	−0.1039
0.0	−0.3672	−0.3080	−0.1039
1.0	−0.3679	−0.3677	−0.3602
0.0	−0.3679	−0.3677	−0.3602
1.0	−0.3679	−0.3679	−0.3679

TABLE 4-35
CASE OF MULTIPLE SOLUTIONS, NEWTON–KANTOROVICH METHOD:
$\alpha = -3.0$, $n = 20$[a]

	$y_0(x) \equiv 0$		$y_0(x) = 12x(1 - x)$		$y_0(x) = 20x(1 - x)$	
t	$k = 0.1$	$k = 1.0$	$k = 0.1$	$k = 1.0$	$k = 0.1$	$k = 1.0$
0.0	0.0000	0.0000	3.0000	3.0000	5.0000	5.0000
0.5	0.2858	—	2.7003	—	4.3287	—
1.0	0.6262	0.5441	2.1305	2.5089	Divergence	3.9875
2.0	0.6414	0.6373	1.9731	2.1359		3.5547
3.0	0.6414	0.6414	1.9702	1.9947		2.6473
4.0		0.6414	1.9703	1.9713		2.2746
5.0				1.9707		2.0363
6.0				1.9707		1.9750
7.0						1.9707

[a] The values $y(0.5, t)$ are presented.

4.5.4 Multiloop One-Parameter Imbedding for the NBVP

We dealt in the preceding section with one type of the multiloop imbedding procedure, based on (4.267), because of its close relation to the Newton–Kantorovich method. Here we shall consider the imbedding procedure of the type (4.268) as a typical representative of a convenient multiloop algorithm. For a general operator equation it is difficult to formulate a variety of imbedding equations (4.264) so that y_0 could be evaluated from (4.265) in a noniterative way. Owing to the fact that linear boundary value problems can be solved easily, unlike the general linear operator equations, we can take advantage of splitting the nonlinear boundary value problem into linear and nonlinear parts.

Assume (4.257) in the form

$$F(y) = Ly - f(y) = 0 \qquad (4.304)$$

where L is a linear operator. The one-parameter imbedding procedure can be performed, for example, according to formulas (4.305a) or (4.305b):

$$H(t, y, \bar{y}) = Ly - tf(y) - \omega(1 - t)f(\bar{y}) = 0 \qquad (4.305a)$$

$$H(t, y, \bar{y}) = Ly - tf(y) - \omega(1 - t)(y - \bar{y}) = 0 \qquad (4.305b)$$

The relevant partial differential equations are

$$\frac{\partial}{\partial t} Ly - f(y) - tf'(y)\frac{\partial y}{\partial t} + \omega f(\bar{y}) = 0 \qquad (4.306a)$$

or

$$\frac{\partial}{\partial t} Ly - f(y) - [tf'(y) + \omega(1 - t)]\frac{\partial y}{\partial t} + \omega(y - \bar{y}) = 0 \quad (4.306b)$$

The initial condition

$$t = 0: \qquad y = y_0 \qquad (4.307)$$

can be obtained by solving the linear operator equation

$$Ly_0 = \omega f(\bar{y}) \qquad (4.308a)$$

or

$$Ly_0 = \omega(y_0 - \bar{y}) \qquad (4.308b)$$

The multiloop algorithm of this type can be formulated in the following way:

1. Guess $\bar{y}$.
2. Calculate y_0 from (4.308).
3. Integrate (4.306) from $t = 0$ to $t = 1$.
4. Insert the solution calculated for $t = 1$ instead of $\bar{y}$ and go to step 2 if the solution is not established with a given tolerance.

The splitting of the operator equation (4.257) into linear and nonlinear parts can be accomplished according to (4.304); however, artificial splitting is possible. For instance, to solve $\varphi(y) = 0$ we choose in (4.304) $f(y) = Ly - \varphi(y)$, where L is an arbitrary linear operator. Of course, a broad range of imbedding is possible and a successful imbedding procedure is a matter of experience and the ingenuity of the analyst.

The advantage of the imbedding procedure discussed in this section is, for instance, the multiloop character (i.e., the possibility of using a coarse integration grid). Two different cases are schematically sketched in Fig. 4-22. In the first event (Fig. 4-22a) the accuracy obtained is limited by the accuracy of the integration routine; that is, this type of imbedding exhibits the features of the one-loop imbedding procedure because only the first iterations do improve the accuracy of the solution. On the other hand, the second event (Fig. 4-22b) allows us to find the solution with any arbitrarily low tolerance. While the former imbedding procedure is limited by the accuracy of the integration method, the precision of the latter one is given by the computer arithmetic alone.

To cast more light on this problem, the following consideration can be useful. Consider that for $\bar{y}$ an approximation close to the exact solution y^* is set. Since

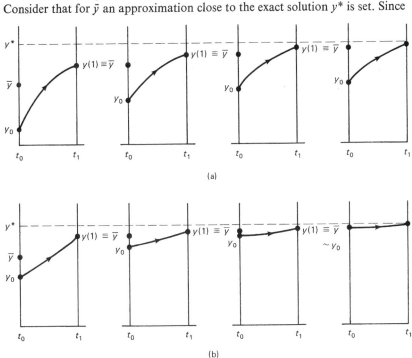

Figure 4-22 (a) schematic illustration of the behavior of an incorrect multiloop algorithm. (b) Schematic illustration of the behavior of a correct multiloop algorithm.

the value of y_0 is far from y^*, for $t = 1$ the integration errors cause the solution $y(1)$ to be less accurate than the original guess $\bar{y}$. Obviously, the properties of the multiloop algorithm are not favourable. To guarantee the favorable properties of the multiloop algorithm it is necessary that for a sequence of functions $\bar{y}_n$, where $\bar{y}_n \rightarrow y^*$, is $y_{0n} \rightarrow y^*$ [and also $y(1)_n \rightarrow y^*$]. Evidently, this consideration yields the relation

$$\bar{y} = y^* \Longrightarrow y_0 = y^* \quad \text{i.e.,} \quad H(0, y^*, y^*) = 0 \tag{4.309}$$

The imbedding procedure that obeys these relations will be referred to as the correct multiloop algorithm†. The reader is asked to verify that the imbedding formulas given by (4.267) and (4.268) yield the correct multiloop algorithms and, moreover, that $y_0 = \bar{y}$. Clearly, the imbedding procedures (4.305a) and (4.305b) comprise a correct multiloop algorithm only for $\omega = 1$ in the case (4.305a).

Some aspects of the multiloop procedures are demonstrated in the following example.

Example 4.19

The NBVP given by (4.278) and (4.279) will be solved by the one-parameter imbedding algorithm according to (4.268) [i.e., (4.282) is considered]. The relevant partial differential equation is given by (4.290) subject to boundary conditions (4.288) and initial conditions (4.289). Here $y_0(x)$ is a guessed function. The partial differential equation (4.290) will be replaced by a finite-difference analogy. We can write for $j = 0, 1, \ldots, m - 1$:

$$\frac{\omega \bar{t}}{kh^2}[\delta_x^2 Y_i^{j+1} - \delta_x^2 Y_i^j] + \frac{1}{k}(1 - \bar{t} - \bar{t}\omega\alpha \exp{(Y_i^j)})(Y_i^{j+1} - Y_i^j)$$

$$- (Y_i^{j+1} - Y_i^0) + \frac{\omega}{2h^2}[\delta_x^2 Y_i^{j+1} + \delta_x^2 Y_i^j] - \omega\alpha \exp{(Y_i^j)} = 0 \tag{4.310}$$

$$i = 1, 2, \ldots, n - 1$$

$$Y_0^{j+1} = 0, \qquad Y_n^{j+1} = 0 \tag{4.311}$$

where we have denoted $\bar{t} = (j + 1/2)k$ and

$$\delta_x^2 Y_i = Y_{i-1} - 2Y_i + Y_{i+1}$$

The initial condition is approximated by

$$Y_i^0 = y_0(ih) \qquad i = 0, 1, \ldots, n \tag{4.312}$$

Equations (4.310) and (4.311) represent a set of linear algebraic equations with a tridiagonal matrix which for a particular j can be solved by the Thomas algorithm.

The results calculated for $\alpha = 4$ and $y_0 \equiv 0$ are presented in Table 4-36. The dependence of $y(x, t)$ on t for a short integration step $k = 0.025$ is displayed in Fig. 4-23a.

---

†One parameter imbedding procedures satisfying $H(t, y^*, y^*) \equiv 0$ are called strongly correct multiloop algorithms in the paper by authors (1978).

TABLE 4-36
RESULTS FOR THE IMBEDDING ALGORITHM (4.282):
$\alpha = 4$, $n = 10$, $y_0(x) \equiv 0$

	$y(0.5, t)$				
t	$k = 0.05$	$k = 0.1$	$k = 0.25$	$k = 0.5$	$k = 1.0$
			$\omega = 1$		
0.0	0.0000	0.0000	0.0000	0.0000	0.0000
0.5	−0.3462	−0.3650	−0.4112	−0.4318	—
1.0	−0.3426	−0.3512	−0.3716	−0.3797	−0.3955
0.0	−0.3426	−0.3512	−0.3716	−0.3797	−0.3955
0.5	−0.3692	−0.3688	−0.3667	−0.3650	—
1.0	−0.3673	−0.3674	−0.3672	−0.3668	−0.3650
0.0	−0.3673	−0.3674	−0.3672	−0.3668	−0.3650
1.0	−0.3673	−0.3673	−0.3673	−0.3673	−0.3675
0.0		−0.3673	−0.3673	−0.3673	−0.3675
1.0		−0.3673	−0.3673	−0.3673	−0.3673
0.0					−0.3673
1.0					−0.3673
			$\omega = -1$		
0.0	0.0000	0.0000	0.0000	0.0000	0.0000
0.5	−0.3450	−0.3504	−0.3666	−0.3955	—
1.0	−0.3705	−0.3735	−0.3825	−0.3977	−0.4318
0.0	−0.3705	−0.3735	−0.3825	−0.3977	−0.4318
0.5	−0.3675	−0.3676	−0.3673	−0.3648	—
1.0	−0.3673	−0.3672	−0.3667	−0.3647	−0.3554
0.0	−0.3673	−0.3672	−0.3667	−0.3647	−0.3554
1.0	−0.3673	−0.3673	−0.3673	−0.3675	−0.3694
0.0		−0.3673	−0.3673	−0.3675	−0.3694
1.0		−0.3673	−0.3673	−0.3673	−0.3669
0.0				−0.3673	−0.3669
1.0				−0.3673	−0.3674

It is obvious that the dependence $y(x, t)$ on t is discontinuous. The discontinuity occurs approximately at $t = 0.025$. Apparently, following the solution of (4.290), $dH/dt = 0$ results. Evidently, the solution must satisfy this condition together with the condition $H = 0$. Figure 4-23b reveals that as long as the dependence $y(x, t)$ on t is continuous, the value of H is $H = 0$. At the singular point a sudden jump occurs and outside this point the function $y(x, t)$ is continuous and hence the value H is invariable again. Of course, the behavior of the dependence $y(x, t)$ on t near the singular point is given by the finite-difference approximation and is a random result of the finite number of arithmetic operations. The imbedding algorithm for $\omega = 1$ cannot be used for calculations; however, by chance the integration brings the initial guess close to the solution (see Fig. 4-23c). Here the trajectories are drawn for higher values of k. On the other hand, for $\omega = -1$ the imbedding procedure is capable of calculating the solution (see Fig. 4-24). This figure and Table 4-36 reveal the effect of the length of the integration step k. It can be inferred, in accordance

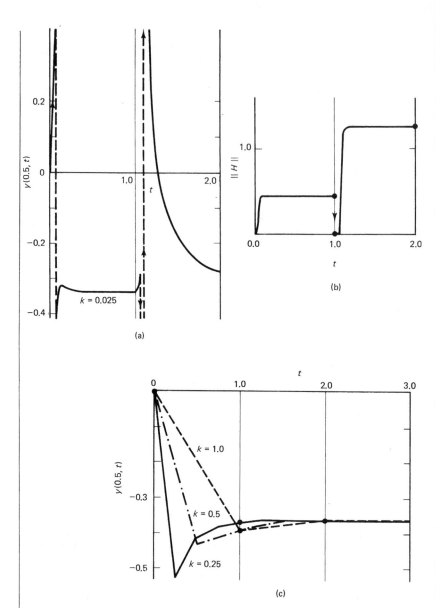

Figure 4-23 (a) Dependence of $y(0.5, t)$ on t, Example 4.19, (4.282): $h = 0.1$, $\omega = 1$, $y \equiv 0$, $\alpha = 4$. (b) Behavior of $\|H\|$, Example 4.19, (4.282), $h = 0.1$, $\omega = 1$, $y \equiv 0$, $\alpha = 4$. (c) Trajectories for higher k, Example 4.19, (4.282), $h = 0.1$, $\omega = 1$, $y \equiv 0$, $\alpha = 4$.

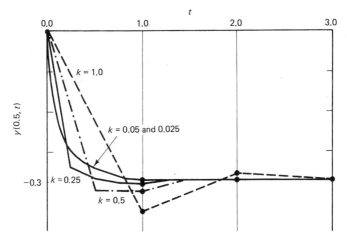

Figure 4-24 Trajectories for $\omega = -1$, Example 4.19, (4.282): $h = 0.1$, $y_0 \equiv 0$, $\alpha = 4$.

with the conclusions from Section 4.5.3, that an integration with a coarse grid is more convenient.

For the multiloop imbedding procedure according to (4.305), we have

$$H(t, y, \bar{y}) = \frac{d^2 y}{dx^2} - t\alpha e^y - \omega(1 - t)\alpha e^{\bar{y}} = 0 \qquad (4.313a)$$

or

$$H(t, y, \bar{y}) = \frac{d^2 y}{dx^2} - t\alpha e^y - \omega(1 - t)(y - \bar{y}) = 0 \qquad (4.313b)$$

The relevant partial differential equations for $y(x, t)$ are

$$\frac{\partial^3 y}{\partial x^2 \, \partial t} - \alpha e^y - t\alpha e^y \frac{\partial y}{\partial t} + \omega \alpha e^{\bar{y}} = 0 \qquad (4.314a)$$

$$\frac{\partial^3 y}{\partial x^2 \, \partial t} - \alpha e^y - [t\alpha e^y + \omega(1 - t)]\frac{\partial y}{\partial t} + \omega(y - \bar{y}) = 0 \qquad (4.314b)$$

The boundary conditions are (4.288), the initial condition (4.289), and $y_0(x)$ can be calculated from the linear boundary value problem

$$\frac{d^2 y_0}{dx^2} - \omega \alpha e^{\bar{y}} = 0$$

or $\qquad\qquad\qquad\qquad\qquad\qquad\qquad\qquad\qquad\qquad\qquad\qquad (4.315)$

$$\frac{d^2 y_0}{dx^2} - \omega(y_0 - \bar{y}) = 0$$

$$y_0(0) = 0, \qquad y_0(1) = 0$$

The function $\bar{y}$ should be an approximation to the solution of the original NBVP. The finite-difference approximation may be developed in an analogous way to (4.290) and (4.310); that is, for (4.314a) we have

$$\frac{1}{kh^2}[\delta_x^2 Y_i^{j+1} - \delta_x^2 Y_i^j] - \alpha \exp(Y_i^j) - \frac{t\alpha}{k}\exp(Y_i^j)(Y_i^{j+1} - Y_i^j) + \omega\alpha\exp(\bar{y}_i) = 0$$

$$(4.316)$$

The approximation of (4.315) can be accomplished analogously. Evidently, linear algebraic equations with a tridiagonal matrix result. The results calculated for the imbedding algorithm (4.313a) are presented in Table 4-37 for different values of the step length k. Obviously, for $\omega = 1$ the one-parameter imbedding procedure is of the correct multiloop type; on the other hand, for $\omega = -1$ the accuracy of the solution depends on the precision of integration (i.e., on the step length k). The differences between these two-parameter imbedding procedures can be followed in Fig. 4-25. For the imbedding (4.313b) the results are reported in Table 4-38 and Fig. 4-26. Clearly, for $\omega = 1$ as well as for $\omega = -1$, we cannot construct a correct multiloop algorithm.

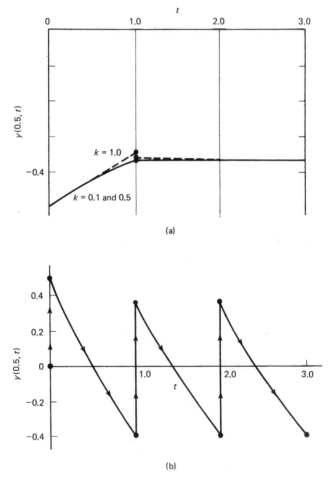

(a)

(b)

Figure 4-25 (a) Trajectories, Example 4.19, (4.313a): $\omega = 1$, $h = 0.1$, $\bar{y} \equiv 0$, $\alpha = 4$. (b) Trajectories, Example 4.19, (4.313a): $\omega = -1$, $h = 0.1$, $\bar{y} \equiv 0$, $k = 0.1$, $\alpha = 4$.

TABLE 4-37
RESULTS FOR THE IMBEDDING PROCEDURE (4.313a): $\alpha = 4$, $n = 10$, $\bar{y}(x) \equiv 0$

	$y(0.5, t)$					
t	$k = 0.025$	$k = 0.05$	$k = 0.1$	$k = 0.25$	$k = 0.5$	$k = 1.0$
			$\omega = 1$			
0.0	−0.5000	−0.5000	−0.5000	−0.5000	−0.5000	−0.5000
0.5	−0.4251	−0.4248	−0.4243	−0.4229	−0.4203	—
1.0	−0.3670	−0.3666	−0.3658	−0.3634	−0.3592	−0.3499
0.0	−0.3674	−0.3675	−0.3677	−0.3684	−0.3697	−0.3725
0.5	−0.3674	−0.3674	−0.3675	−0.3678	−0.3682	—
1.0	−0.3673	−0.3673	−0.3673	−0.3673	−0.3671	−0.3666
0.0	−0.3673	−0.3673	−0.3673	−0.3673	−0.3673	−0.3675
1.0	−0.3673	−0.3673	−0.3673	−0.3673	−0.3673	−0.3673
0.0						−0.3673
1.0						−0.3673
			$\omega = -1$			
0.0	0.5000	0.5000	0.5000	0.5000	0.5000	0.5000
0.5	−0.0022	−0.0045	0.0090	−0.0226	−0.0433	—
1.0	−0.3700	−0.3728	−0.3784	−0.3947	−0.4196	−0.4520
0.0	0.3665	0.3656	0.3640	0.3592	0.3520	0.3426
0.5	−0.0574	−0.0596	−0.0639	−0.0768	−0.0969	—
1.0	−0.3696	−0.3719	−0.3765	−0.3902	−0.4115	−0.4442
0.0	0.3666	0.3659	0.3645	0.3605	0.3543	0.3449
1.0	−0.3696	−0.3719	−0.3765	−0.3902	−0.4116	−0.4443
0.0			0.3645		0.3543	0.3449
1.0			−0.3765		−0.4116	−0.4443

TABLE 4-38
RESULTS FOR THE IMBEDDING PROCEDURE (4.313b): $\alpha = 4$, $n = 10$, $\bar{y}(x) \equiv 0$

	$y(0.5, t)$					
t	$k = 0.025$	$k = 0.05$	$k = 0.1$	$k = 0.25$	$k = 0.5$	$k = 1.0$
			$\omega = 1$			
0.0	0.0000	0.0000	0.0000	0.0000	0.0000	0.0000
0.5	−0.2016	−0.2025	−0.2044	−0.2104	−0.2209	—
1.0	−0.3689	−0.3703	−0.3733	−0.3825	−0.3985	−0.4318
0.0	−0.0354	−0.0355	−0.0358	−0.0367	−0.0382	−0.0413
0.5	−0.2175	−0.2184	−0.2202	−0.2259	−0.2359	—
1.0	−0.3688	−0.3700	−0.3727	−0.3810	−0.3954	−0.4253
0.0	−0.0354	−0.0355	−0.0357	−0.0365	−0.0379	−0.0407
1.0	−0.3688	−0.3700	−0.3727	−0.3810	−0.3954	−0.4254
0.0			−0.0357	−0.0365		−0.0407
1.0			−0.3727	−0.3810		−0.4254

TABLE 4-38 continued

t	k = 0.025	k = 0.05	k = 0.1	k = 0.25	k = 0.5	k = 1.0
			$y(0.5, t)$			
			$\omega = -1$			
0.0	0.0000	0.0000	0.0000	0.0000	0.0000	0.0000
0.5	−0.2202	−0.2208	−0.2220	−0.2257	−0.2317	—
1.0	−0.3681	−0.3688	−0.3703	−0.3748	−0.3822	−0.3955
0.0	0.0434	0.0435	0.0436	0.0442	0.0450	0.0466
0.5	−0.2028	−0.2035	−0.2048	−0.2088	−0.2153	—
1.0	−0.3682	−0.3690	−0.3707	−0.3757	−0.3840	−0.3985
0.0	0.0434	0.0414	0.0437	0.0443	0.0452	0.0469
1.0	−0.3682	−0.3690	−0.3707	−0.3757	−0.3840	−0.3985
0.0			0.0437	0.0443	0.0452	0.0469
1.0	—		−0.3707	−0.3757	−0.3840	−0.3985

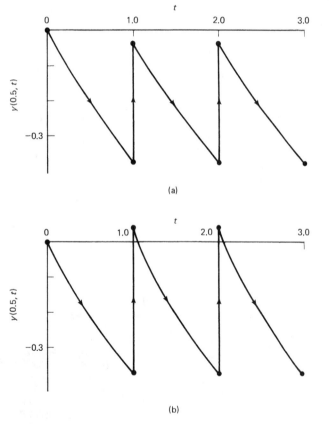

Figure 4-26 (a) Trajectories, Example 4.19, (4.313b): $\omega = 1$, $h = 0.1$, $\bar{y} \equiv 0$, $k = 0.1$, $\alpha = 4$. (b) Trajectories, Example 4.19, (4.313b): $\omega = -1$, $h = 0.1$, $\bar{y} \equiv 0$, $k = 0.1$, $\alpha = 4$.

4.5.5 One-Loop One-Parameter Imbedding Procedure for the NBVP

A typical feature of the one-loop one-parameter imbedding procedure is the existence of one y_0 which is dependent on the particular imbedding algorithm. Hence for this event the multiloop imbedding algorithm cannot be realized and for the imbedding procedure in question the accuracy of the solution calculated depends on the accuracy of the integration method only. From this point of view the one-loop one-parameter imbedding methods are apt for getting an appropriate initial guess for some iteration method. The two most frequently used types of the one-loop one-parameter imbedding procedure are discussed in the following text.

Consider a single second-order ordinary differential equation

$$y'' = f(x, y, y') \tag{4.317}$$

subject to the linear boundary conditions

$$\alpha_0 y(a) + \beta_0 y'(a) = \gamma_0, \qquad \alpha_1 y(b) + \beta_1 y'(b) = \gamma_1 \tag{4.318}$$

Now we introduce the one-loop one-parameter imbedding procedure as follows:

$$y'' = tf(x, y, y') \tag{4.319}$$

for $t \in \langle 0, 1 \rangle$ subject to the boundary conditions (4.318). Obviously, for $t = 0$ a linear homogeneous differential equation for y_0 results,

$$y_0'' = 0 \tag{4.320}$$

which satisfies (4.318). The partial differential equation for $y(x, t)$ is

$$\frac{\partial^3 y}{\partial x^2 \partial t} = t \left(\frac{\partial f}{\partial y} \frac{\partial y}{\partial t} + \frac{\partial f}{\partial y'} \frac{\partial^2 y}{\partial x \partial t} \right) + f \left(x, y, \frac{\partial y}{\partial x} \right) \tag{4.321}$$

with the initial condition

$$y(x, 0) = y_0(x) \tag{4.322}$$

and boundary conditions associated with (4.318). To clarify the method, an example will be discussed.

Example 4.20

Consider the NBVP described in Example 4.17, that is, (4.278) and (4.279) and an imbedding procedure of the type (4.319). The relevant partial differential equation is given by (4.291) with the initial condition

$$y_0(x) \equiv 0$$

because of (4.292). The finite-difference approximation is analogous to that described in Examples 4.18 and 4.19. Table 4-39 presents the results calculated for $\alpha = 4.0$, and Table 4-40 reveals the dependence of the accuracy on the number of the mesh points n and the length of the step k. Since the imbedding is of the one-loop type, the accuracy cannot be improved unless a refined discretization is used or a more accurate finite-difference approximation is adopted. Let us note that for $\alpha = -3.0$, two solutions of the NBVP exist; however, the one-parameter imbedding with the foregoing initial conditions is capable of determining only one solution. To overcome this dilemma we must make use of a new initial profile calculated, for instance, for $t = 0.8$. Of course, this solution is not trivial and the relevant NBVP must be solved by some iterative algorithm such as the Newton–Kantorovich method or shooting. Obviously, for this t two initial profiles $y_0(x)$ can be calculated and accordingly the continuation of the solution from $t = 0.8$ to $t = 1.0$ yields two solutions of the original NBVP. From the practical point of view the combination of the one-parameter imbedding algorithm with other methods (e.g., shooting) can be useful if the solution for $t = t_0$ may be readily calculated (by shooting) while the solution to the original NBVP is impossible and must be continued from $t = t_0$ to $t = 1$. Clearly, for some technical problems the solution for $t = 0$ need not exist and we must try to find the solution for $t = t_0$ in order to have an initial profile for the one-parameter imbedding technique.

TABLE 4-39

RESULTS FOR THE ONE-LOOP IMBEDDING ALGORITHM (4.284):
$\alpha = 4$, $n = 40$, $k = 0.00625$[a]

			t		
x	0.0	0.25	0.5	0.75	1.0
0.1	0.0000	−0.0415	−0.0774	−0.1092	−0.1380
0.2	0.0000	−0.0733	−0.1361	−0.1914	−0.2409
0.3	0.0000	−0.0959	−0.1774	−0.2488	−0.3124
0.4	0.0000	−0.1094	−0.2020	−0.2827	−0.3544
0.5	0.0000	−0.1138	−0.2101	−0.2939	−0.3683

[a]Double-precision arithmetic. The solution is symmetrical with respect to $x = 0.5$.

TABLE 4-40

ACCURACY OF $y(0.5, 1.0)$ AS A FUNCTION OF n AND k FOR $\alpha = 4$[a]

			n	
k	10	20	40	80
0.1	−0.3718	−0.3723	−0.3725	−0.3725
0.05	−0.3695	−0.3701	−0.3702	−0.3703
0.025	−0.3684	−0.3690	−0.3691	−0.3692
0.0125	−0.3678	−0.3684	−0.3686	−0.3686
0.00625	−0.3676	−0.3681	−0.3683	−0.3683

[a]Exact value $y^*(0.5) = -0.3680$. Double-precision arithmetic.

Consider now a set of first-order nonlinear differential equations

$$y_i' = f_i(x, y_1, \ldots, y_N) \qquad i = 1, 2, \ldots, N \qquad (4.323)$$

subject to the linear boundary conditions

$$y_i(a) = c_i \qquad i = 1, 2, \ldots, r \qquad (4.324a)$$

$$y_{i_m}(b) = d_m \qquad m = 1, 2, \ldots, N - r \qquad (4.324b)$$

A detailed analysis has shown that according to the specification of the boundary conditions, four groups of problems can be established. Of course, we try to introduce the parameter t in such a way in order to avoid the trial-and-error calculation of the solution for $t = 0$. Hence:

1. The left end for y_i is given only:

$$y_i' = f_i(x, y_1, \ldots, y_N) \qquad i = 1, 2, \ldots, r \quad \text{and} \quad i \neq i_m \qquad (4.325a)$$
$$m = 1, 2, \ldots, N - r$$

2. The right end for y_{i_m} is given only:

$$y_{i_m}' = t f_{i_m}(x, y_1, \ldots, y_N) \qquad \begin{array}{l} i_m > r \quad \text{and} \\ m = 1, 2, \ldots, N - r \end{array} \qquad (4.325b)$$

3. Both the left and right ends are given for y_{i_m}:

$$y_{i_m}' = t f_{i_m}(x, y_1, \ldots, y_N) + (1 - t)\frac{(d_m - c_{i_m})}{b - a}$$
$$m = 1, 2, \ldots, N - r \quad \text{and} \quad i_m \leq r \qquad (4.325c)$$

4. Neither the left nor the right ends for y_i are given:

$$y_i' = t f_i(x, y_1, \ldots, y_N) \qquad \begin{array}{l} i > r \quad \text{and} \quad i \neq i_m \\ m = 1, \ldots, N - r \end{array} \qquad (4.325d)$$

or

$$y_i' = f_i(x, y_1, \ldots, y_N) \qquad (4.325e)$$

The initial profiles are:

For (4.325b): $y_{i_m}^0(x) \equiv d_m$

For (4.325c): $y_{i_m}^0(x) = c_{i_m} + \dfrac{(d_m - c_{i_m})(x - a)}{b - a}$ $\qquad (4.326)$

For (4.325d): $y_i^0(x) \equiv$ arbitrary constant

For (4.325a): $y_i^0(x)$ is calculated by integration of the initial value problem (4.325a) with the initial conditions (4.324a) when for other y_i the already known values $y_i^0(x)$ are set

If we use (4.325e) instead of (4.325d) it is necessary to integrate (4.325a) with (4.325e). The values of $y_i^0(a)$ in (4.325e) are guessed as arbitrary constants. Differentiation of (4.325) with respect to t yields a partial differential equation of the hyperbolic type:

$$\frac{\partial^2 y_i}{\partial x\, \partial t} = F_i\left(x, t, y_1, \ldots, y_N, \frac{\partial y_1}{\partial t}, \ldots, \frac{\partial y_N}{\partial t}\right) \qquad (4.327)$$

subject to boundary conditions (4.324). The initial condition is

$$y_i(x, 0) = y_i^0(x) \qquad (4.328)$$

The application of the aforementioned procedure will be demonstrated in an example. The second example is presented in Section 4.5.7.

Example 4.21

Consider the NBVP described in Example 4.17 [i.e., (4.278) and (4.279)]. The problem can be considered as a set of two first-order ordinary differential equations:

$$y_1' = y_2, \qquad y_2' = \alpha e^{y_1} \tag{4.329}$$

subject to boundary conditions

$$y_1(0) = 0, \qquad y_1(1) = 0 \tag{4.330}$$

The one-parameter imbedding procedure given by (4.325) is

$$y_1' = t y_2 \qquad y_2' = t \alpha e^{y_1} \tag{4.331a}$$

or

$$y_1' = t y_2 \qquad y_2' = \alpha e^{y_1} \tag{4.331b}$$

Clearly, according to (4.326) the initial conditions are

$$y_1^0(x) \equiv 0 \qquad y_2^0(x) \equiv \gamma \tag{4.332a}$$

or

$$y_1^0(x) \equiv 0 \qquad y_2^0(x) = \alpha x + \gamma \tag{4.332b}$$

where γ is an arbitrary constant. The relevant partial differential equations are

$$\frac{\partial^2 y_1}{\partial x\, \partial t} = y_2 + t\frac{\partial y_2}{\partial t}, \qquad \frac{\partial^2 y_2}{\partial x\, \partial t} = \alpha e^{y_1}\left[t\frac{\partial y_1}{\partial t} + 1\right] \tag{4.333a}$$

or

$$\frac{\partial^2 y_1}{\partial x\, \partial t} = y_2 + t\frac{\partial y_2}{\partial t}, \qquad \frac{\partial^2 y_2}{\partial x\, \partial t} = \alpha e^{y_1}\frac{\partial y_1}{\partial t} \tag{4.333b}$$

The initial conditions are (4.328) and (4.332). After rewriting the boundary conditions (4.330), we have

$$y_1(0, t) = 0, \qquad y_1(1, t) = 0 \tag{4.334}$$

The simplest finite-difference approximation will be used to replace the hyperbolic equations (4.333):

$$\frac{\partial^2 u(x, t)}{\partial x\, \partial t} \doteq \frac{u(x + h, t + k) - u(x, t + k) - u(x + h, t) + u(x, t)}{kh}$$

$$\frac{\partial u(x, t)}{\partial t} \doteq \frac{u(x, t + k) + u(x + h, t + k) - u(x, t) - u(x + h, t)}{2k} \tag{4.335}$$

$$u\left(x + \frac{h}{2}, t + \frac{k}{2}\right) \doteq \frac{1}{4}[u(x, t) + u(x + h, t) + u(x, t + k) + u(x + h, t + k)]$$

It is obvious that the finite-difference analogy is centered at the point $(x + h/2, t + k/2)$. On using this finite-difference approximation to (4.333), a set of linear algebraic equations results for the values at the new profile $t + k$. An appropriate grouping of variables gives rise to a five-diagonal structural matrix (see, e.g., Section 4.1.3). The nonlinearity appearing in the NBVP is approximated at the old profile, namely at the point $(x + h/2, t)$. The numerical results for (4.331a) and (4.331b) are reported in Tables 4-41 and 4-42. It can be shown that the value of γ is not important; for $t > 0$ the value $y_2(0.5, t)$ approaches zero independent of h and k. These tables reveal the dependence of the accuracy of the solution on both steps h and k.

TABLE 4-41
RESULTS FOR THE IMBEDDING PROCEDURE (4.331a), $\alpha = 4$[a]

| h: | 0.05 | 0.05 | 0.05 | 0.025 | 0.025 | 0.025 |
k:	0.05	0.025	0.0125	0.05	0.025	0.0125
$t = 0.0$	0.0000	0.0000	0.0000	0.0000	0.0000	0.0000
0.2	−0.0197	−0.0197	−0.0197	−0.0197	−0.0197	−0.0197
0.4	−0.0754	−0.0753	−0.0752	−0.0754	−0.0753	−0.0752
0.6	−0.1586	−0.1583	−0.1581	−0.1586	−0.1582	−0.1580
0.8	−0.2596	−0.2590	−0.2586	−0.2595	−0.2588	−0.2585
1.0	−0.3703	−0.3694	−0.3689	−0.3700	−0.3691	−0.3686

[a]The values $y_1(0.5, t)$ are presented. γ is arbitrary (e.g. $\gamma = 0$).

TABLE 4-42
RESULTS FOR THE IMBEDDING PROCEDURE (4.331b), $\alpha = 4$[a]

| h: | 0.05 | 0.05 | 0.05 | 0.025 | 0.025 | 0.025 |
k:	0.05	0.025	0.0125	0.05	0.025	0.0125
$t = 0.0$	0.0000	0.0000	0.0000	0.0000	0.0000	0.0000
0.2	−0.0925	−0.0926	−0.0926	−0.0925	−0.0925	−0.0926
0.4	−0.1730	−0.1731	−0.1732	−0.1729	−0.1731	−0.1731
0.6	−0.2444	−0.2447	−0.2448	−0.2443	−0.2445	−0.2446
0.8	−0.3088	−0.3092	−0.3093	−0.3086	−0.3090	−0.3092
1.0	−0.3676	−0.3680	−0.3682	−0.3673	−0.3677	−0.3680

[a]The values $y_1(0.5, t)$ are presented. γ is arbitrary (e.g., $\gamma = 0$).

4.5.6 Numerical Aspects of Methods

In the preceding sections we have developed certain partial differential equations which govern the dependence of the solution on the parameter introduced. To solve these partial differential equations the finite-difference methods have been suggested. However, an entirely different approach to this problem exists. The gist of this approach consists of finding the solution of a certain linear boundary value problem for each t. To demonstrate the method, consider a set of first-order differential equations

$$\frac{dy_i}{dx} = f_i(x, y_1, \ldots, y_N) \qquad i = 1, 2, \ldots, N \qquad (4.323)$$

subject to the boundary conditions

$$y_i(a) = c_i \qquad i = 1, 2, \ldots, r \qquad (4.336a)$$

$$y_i(b) = c_i \qquad i = r + 1, \ldots, N \qquad (4.336b)$$

The simple boundary conditions have been chosen for the sake of simplicity. According to (4.325) the one-parameter imbedding procedure can be implemented in the following way:

$$\frac{dy_i}{dx} = f_i(x, y_1, \ldots, y_N) \qquad i = 1, 2, \ldots, r \qquad (4.337a)$$

$$\frac{dy_i}{dx} = tf_i(x, y_1, \ldots, y_N) \qquad i = r + 1, \ldots, N \qquad (4.337b)$$

The initial profiles y^0 for $t = 0$ can be guessed using the boundary conditions

$$y_i(x) \equiv c_i \qquad i = r + 1, \ldots, N \qquad (4.338)$$

while the first r differential equations [i.e., (4.337a)] can be integrated as an initial value problem using the initial conditions (4.336a).

Differentiation of (4.337) with respect to t and denoting

$$\Omega_i(x) = \frac{\partial y_i}{\partial t} \qquad (4.339)$$

yields a set of ordinary differential equations

$$\frac{d\Omega_i}{dx} = \sum_{j=1}^{N} \frac{\partial f_i(x, y_1, \ldots, y_N)}{\partial y_j} \Omega_j \qquad i = 1, 2, \ldots, r \qquad (4.340a)$$

$$\frac{d\Omega_i}{dx} = f_i(x, y_1, \ldots, y_N) + t \sum_{j=1}^{N} \frac{\partial f_i(x, y_1, \ldots, y_N)}{\partial y_j} \Omega_j \qquad i = r + 1, \ldots, N$$
$$(4.340b)$$

subject to the boundary conditions

$$\Omega_i(a) = 0 \qquad i = 1, 2, \ldots, r \qquad (4.341a)$$
$$\Omega_i(b) = 0 \qquad i = r + 1, \ldots, N \qquad (4.341b)$$

The auxiliary differential equations (4.340) with boundary conditions (4.341) represent a linear boundary value problem for Ω_i if the profiles $y_i(x)$ are known. For $t = 0$ we have already calculated $y_i(x)$ by integration of the initial value problem (4.337) with the initial conditions $y_i(a) = c_i$, $i = 1, 2, \ldots, N$. The initial conditions $y_i(a)$ satisfy the differential equation

$$\frac{dy_i(a)}{dt} = \Omega_i(a) \qquad i = 1, 2, \ldots, N \qquad (4.342)$$

Equation (4.342) can be rewritten

$$y_i(a, t) - y_i(a, 0) = \int_0^t \frac{dy_i(a)}{dt} dt \qquad (4.343)$$

To integrate the differential equation (4.342) standard integration procedures such as the Euler method or the Runge–Kutta method can be adapted. The right-hand sides for the numerical integration are calculated by solving the linear BVP for the variables Ω_i. In Chapter 5 we address ourselves to the auxiliary equations for Ω_i. The method will be illustrated next in an example.

Example 4.22

Equations describing axial mass transfer in tubular reactors are

$$\frac{dy}{dx} = \text{Pe}\,y + \text{Pe}\,h\!\left(z + \frac{1}{\text{Pe}}y\right)^2$$
$$\frac{dz}{dx} = -h\!\left(z + \frac{1}{\text{Pe}}y\right)^2 \tag{4.344}$$

subject to boundary conditions

$$z(0) = 1, \qquad y(1) = 0 \tag{4.345}$$

Since this problem cannot be integrated by marching techniques from $x = 0$ to $x = 1$, a backward integration must be performed. After using the one-parameter one-loop imbedding procedure, we have, for example,

$$\frac{dy}{dx} = \text{Pe}\,y + \text{Pe}\,h\!\left(z + \frac{1}{\text{Pe}}y\right)^2$$
$$\frac{dz}{dx} = -ht\!\left(z + \frac{1}{\text{Pe}}y\right)^2 \tag{4.344a}$$

The integration in the x- and t-directions has been performed by making use of the modified Euler method. For $z(0)$ the following results have been calculated ($\text{Pe} = 6,\ h = 2$):

	$\Delta t = 0.2$	$\Delta t = 0.1$	Correct value
$\Delta x = 0.01$	1.0225	1.0052	1.0000
$\Delta x = 0.1$	1.0224	1.0052	

It is obvious that the coarse grid yields results that can be used as a good starting profile if a solution with high accuracy is required. The refinement of the grid in the x-direction does not increase the accuracy substantially, whereas the effect of the step length Δt is worthwhile.

The inevitable question to be asked regarding the one-loop one-parameter imbedding procedures is the precision of the integration in the t-direction of the relevant initial value problem. The more accurate the integration performed, the higher is the computer time expenditure. As a waste by-product of the accurate integration is the dependence of the solution on the imbedded parameter t calculated with high precision. To enhance the economy of the calculation the one-loop imbedding procedure is frequently combined with other refining iterative techniques such as the Newton–Kantorovich method. This combination may have a marked effect on the accuracy of the final solution and the computer time expenditure as well. Sometimes this refinement can be used gradually during the calculation [i.e., at some points $t, t \in \langle 0, 1 \rangle$] to assure the high precision of the continuation algorithm. The reader is asked to compare this algorithm with the predictor–corrector GPM technique in Section 5.5.

For the multiloop algorithms the accuracy of the integration plays a secondary role. Generally speaking, an optimum precision of integration exists (i.e., an optimum length of the integration step), which gives rise to minimum computer time for the predetermined tolerance of the solution (compare the results of foregoing examples).

A very important question is the problem of imbedding the parameter t for large sets of differential equations. For a set of second-order differential equations the method of differentiation with respect to t may be used; the resulting set of third-order partial differential equations can be solved by the finite-difference methods. For an appropriate approximation of the nonlinear terms appearing in these equations, the large set of algebraic equations can be decomposed into small independent sets, usually with a tridiagonal matrix, which may be solved independently. For the multiloop imbedding procedure, the way of linearization is frequently immaterial; however, for the one-loop algorithm this simplification may spoil the accuracy of the final answer. For problems that are described by a set of first-order differential equations a set of auxiliary differential equations are usually to be solved. Moreover, in both cases the development of auxiliary equations can be a formidable task.

The following example demonstrates the "method of lines" used as an alternative to integrate the resulting partial differential equations.

Example 4.23

Heat and mass transfer and an exothermic chemical reaction in a flat plate yield under simplifying conditions a nonlinear boundary value problem:

$$F(y) = y'' + 4\delta(1 - y) \exp\left(\frac{\gamma\beta y}{1 + \beta y}\right) \tag{4.346}$$

$$y(0) = y(1) = 0 \tag{4.347}$$

The imbedding procedure of the type (4.267) and (4.268) will be used. The resulting partial differential equation is, for example, for the type (4.267):

$$\frac{\partial^3 y}{\partial x^2 \partial t} - g(y)\frac{\partial y}{\partial t} = r\left(y^0, \frac{d^2 y^0}{dx^2}\right) \tag{4.348}$$

where

$$g(y) = 4\delta\left[1 - (1 - y)\frac{\gamma\beta y}{(1 + \beta y)^2}\right] \exp\left(\frac{\gamma\beta y}{1 + \beta y}\right)$$

$$r\left(y^0, \frac{d^2 y^0}{dx^2}\right) = -\frac{d^2 y^0}{dx^2} - 4\delta(1 - y^0) \exp\left(\frac{\gamma\beta y^0}{1 + \beta y^0}\right)$$

The basic idea of the "method of lines" is to approximate the derivatives by finite differences in only one direction (e.g., in the x-direction). After replacing the derivatives with respect to x at the mesh points $x_i = ih$, $i = 1, 2, \ldots, N - 1$, and on denoting $y_i = y(x_i)$, $i = 0, 1, \ldots, N$, $x_0 = 0$, $x_N = 1$, we have for $y_i(t)$ a set of ordinary differential equations:

$$\frac{1}{h^2}\left(\frac{dy_{i-1}}{dt} - 2\frac{dy_i}{dt} + \frac{dy_{i+1}}{dt}\right) - g(y_i)\frac{dy_i}{dt} = r\left(y_i^0, \frac{y_{i-1}^0 - 2y_i^0 + y_{i+1}^0}{h^2}\right) \quad (4.349)$$

$$y_0(t) \equiv 0, \qquad y_N(t) \equiv 0 \qquad i = 1, 2, \ldots, N-1$$

with the initial condition

$$y_i(0) = y^0(x_i) \tag{4.350}$$

The set of equations (4.349) can be written in the matrix form

$$A(Y)\frac{dY}{dt} = R(Y) \tag{4.351}$$

where $Y = (y_0, y_1, \ldots, y_N)^T$ and A and R can be dependent on the $y^0(x)$.

A comparison of the Euler and modified Euler methods[†] used to integrate (4.351) is presented for different N in Table 4-43. The value of the exact solution at $x = 0.5$, $y^*(0.5) = 0.4479$, has been calculated by the shooting method. Let us note that the use of the Euler method is identical with the finite-difference method presented in this section. The advantage of adopting a better integration formula is important only in the first iteration; however, during the computation the differences are immaterial. For the imbedding procedure (4.268), $y^0 \equiv 0$, $\omega = 1$, and $\Delta t = 1.0, 0.5, 0.1$, the iteration process diverges. The same results

TABLE 4-43
COMPARISON OF INTEGRATION METHODS FOR DIFFERENT N:
$\gamma = 20$, $\beta = 0.05$, $\delta = 1$, $y^0 \equiv 0$, $\Delta t = 0.1$[a]

	N	Euler method	Modified Euler method
Results	4	0.00214	0.00277
in the	6	0.00364	0.00116
first	8	0.00412	0.00064
loop	10	0.00433	0.00041
for	20	0.00461	0.00011
$t = 1$	30	0.00466	0.00005
	40	0.00468	0.00004
	50	0.00469	0.00003
Results	4	0.00274	0.00274
in the	6	0.00113	0.00113
second	8	0.00061	0.00061
loop	10	0.00038	0.00038
for	20	0.00008	0.00008
$t = 1$	30	0.00003	0.00003
	40	0.00001	0.00001
	50	0.00000	0.00000

[a]The values of $|y(0.5, 1) - y^*(0.5)|$ are presented.

[†]The modified Euler procedure is

$$k_1 = hf(x, y), \qquad k_2 = hf(x + h/2, y + k_1/2), \qquad \bar{y} = y + k_2$$

TABLE 4-44
RESULTS FOR EXAMPLE 4.23 AND IMBEDDING PROCEDURE (4.268):
$\gamma = 20$, $\beta = 0.05$, $\delta = 1$, $N = 20$, $y^0 \equiv 0$, $\omega = -1$, $y^*(0.5) = 0.4479$[a,b]

t	$\Delta t = 0.1$		$\Delta t = 0.05$		$\Delta t = 0.01$	$\Delta t = 0.02$
	E	ME	E	ME	E	ME
1.0	0.0392	0.0081	0.0189	0.0016	0.0036	0.0004
2.0	0.0041	0.0046	0.0010	0.0001	0.0001	0.0001
3.0	0.0003	2.8580	0.0000		0.0001	
4.0	0.0001	!				

[a]The values of $|y(0.5, 1) - y^*(0.5)|$ are presented.
[b]E, Euler method; M, modified Euler method.

have been observed for $\omega = -1$ and $\Delta t = 1.0$ and 0.5; however, for $\Delta t = 0.1$ the problem can be solved. The dependence of the solution on the integration method and Δt is reported in Table 4-44. It can be inferred from this table that a more accurate integration routine does not improve the economy of the overall calculation (see Table 4-44). Results of application of 14 one-parameter imbedding algorithms can be found in the paper by Kubíček et al.[†]

4.5.7 Discussion

4.5.7.1 How to imbed the parameter

In the preceding text we have shown some types of one-parameter imbedding procedures. The goal of this section is to investigate the question of whether we are in a position to draw a priori conclusions on the capability of one-parameter imbedding techniques to solve particular physical problems. It has been pointed out that, generally speaking, the multiloop imbedding procedures are more convenient due to the fact that this algorithm does not require us to integrate the pertinent differential equations with high precision and a relatively coarse grid is sufficient. On the other hand, the one-loop algorithm depends strongly on the accuracy of the integration method. The same conclusions are valid for the imbedding procedures of the multiloop type, which are not correct.

Let us address ourselves to the problem of an appropriate initial guess. For multiloop imbedding procedures the initial conditions can be guessed quite arbitrarily, in contradistinction to the one-parameter imbedding methods, where the initial guess is fixed. Frequently, the imbedding procedures are implemented in such a way that the initial condition may be determined easily by means of analytical integration. Sometimes the parameter can be introduced so that the initial condition may be calculated by iteration procedures only. This strategy of calculation may be used under those circumstances where we can easily solve the NBVP for $t = t_0$ by certain iterative methods to determine

[†]M. Kubíček, M. Holodniok, and V. Hlaváček, Chem. Eng. Sci. 34, 645 (1979).

an appropriate initial profile. The one-parameter imbedding procedures continue the solution to such regions where the direct iteration methods fail because, for instance, of high parametric sensitivity.

The main drawback of the one-parameter imbedding method is the fact that a solution of the original problem can be found if the dependence of the solution of the associated family of imbedding equations on the parameter t is continuously differentiable. Perhaps this can be proved for the correct multiloop imbedding procedures supposing that the initial profile is close to the exact solution (analogously as for the Newton–Kantorovich method). It has been shown in the preceding examples that the solution of the NBVP can be found by the correct multiloop one-parameter imbedding method even if the dependence $y(t)$ is not continuous. These favorable features are caused by using a coarse grid. The solution improves in the first steps by the multiloop character while the effect of integration is not favorable.

The problem of the selection of an appropriate integration method is of great practical importance. For one-loop algorithms it is necessary to employ integration formulas having a high order of approximation, or the results obtained by use of a coarse grid must be improved by the Newton–Kantorovich method. For the correct multiloop imbedding procedures the most effective procedure appears to be the low-order integration formulas on a coarse grid with the iterative improvements. Roughly speaking, integration formulas of the first and second order (e.g., Euler methods) are sufficient. The imbedding formulas presented in this text may be adopted to the overwhelming majority of physical problems. Of course, for particular problems special imbedding formulas exist; however, this type of imbedding procedure will not be discussed.

Sometimes the development of the imbedding differential equations is a formidable task owing to the arduous algebraic operations; however, the one-parameter imbedding procedures, for various reasons, seem to be beneficial. It is convenient to solve this problem for a sequence of values of t by some iteration technique (e.g., by the Newton–Kantorovich method).

To illustrate various features of the one-parameter imbedding methods a number of examples are presented below.

Example 4.24[†]

The nonlinear boundary value problem arising from boundary layer theory[‡] is

$$\begin{aligned}
y_1' &= y_2 \\
y_2' &= y_3 \\
y_3' &= -cy_1y_3 - ny_2^2 + 1 - y_4^2 + sy_2 \\
y_4' &= y_5 \\
y_5' &= -cy_1y_2 - (n-1)y_2y_4 + s(y_4 - 1)
\end{aligned} \qquad (4.352)$$

[†]S. M. Roberts and J. S. Shipman, *J. Optimiz. Theory Appl. 12*, 136 (1973).
[‡]J. F. Holt, *Commun. ACM 7*, 366 (1964).

subject to the boundary conditions

$$y_1(0) = y_2(0) = y_4(0) = 0 \qquad (4.353)$$
$$y_2(x_f) = 0, \qquad y_4(x_f) = 1$$

Here $c = (3 - n)/2$, and the values of the governing parameters are

$$n = -0.1, \qquad s = 0.2, \qquad x_f = 3.5 \qquad (4.354)$$

The one-parameter imbedding procedure can be performed in the following way:

$$\begin{bmatrix} y_1' \\ y_2' \\ y_3' \\ y_4' \\ y_5' \end{bmatrix} = (1 - t) \begin{bmatrix} 0 & 1 & 0 & 0 & 0 \\ 0 & 0 & 1 & 0 & 0 \\ 0 & s & 0 & 0 & 0 \\ 0 & 0 & 0 & 0 & 1 \\ 0 & 0 & 0 & s & 0 \end{bmatrix} \begin{bmatrix} y_1 \\ y_2 \\ y_3 \\ y_4 \\ y_5 \end{bmatrix}$$

$$+ t \begin{bmatrix} y_2 \\ y_3 \\ -cy_1 y_3 - ny_2^2 + 1 - y_4^2 + sy_2 \\ y_5 \\ -cy_1 y_5 - (n - 1)y_2 y_4 + s(y_4 - 1) \end{bmatrix} \qquad (4.355)$$

Obviously, the differential equations are partitioned into linear and nonlinear parts. Roberts and Shipman call this imbedding procedure the linear partition scheme. For $t = 0$ the linear differential equations associated with (4.355) can be solved analytically:

$$y_1 \equiv y_2 \equiv y_3 \equiv 0$$
$$y_4 = \frac{\sinh x\sqrt{s}}{\sinh x_f\sqrt{s}} \qquad (4.355')$$
$$y_5 = \frac{\sqrt{s} \, \cosh x\sqrt{s}}{\sinh x_f\sqrt{s}}$$

The imbedding procedure used is of the one-loop type. To calculate the dependence $y(t)$, Roberts and Shipman made use of the variational equations for determination of $\Omega_i = \partial y_i/\partial t$ [see (4.340)], Euler predictor method and the Newton–Kantorovich method for correcting the values $y(x, t + k)$, Roberts and Shipman also employed a correct multiloop imbedding procedure which is referred to as the quasilinear partition scheme. Using the notation of Section 4.5.4, we can construct for the functional equation

$$F(y) = 0 \qquad (4.257)$$

an imbedding function $H(t, y)$,

$$H(t, y, \bar{y}) = (1 - t)[F(\bar{y}) + F_y'(\bar{y})(y - \bar{y})] + tF(y) = 0 \qquad (4.356)$$

Here $\bar{y}$ is an appropriate element (e.g., the preceding iteration). The initial condition y_0 can be calculated by solving (4.356) for $t = 0$:

$$y_0 = \bar{y} - [F_y'(\bar{y})]^{-1}F(\bar{y}) \qquad (4.357)$$

which corresponds to one step of the Newton–Kantorovich method using an initial guess $\bar{y}$. The differential equation for $y(t)$ is in the form

$$[(1 - t)F_y'(\bar{y}) + tF_y'(\bar{y})]\frac{dy}{dt} = -F(y) + F(\bar{y}) + F_y'(\bar{y})(y - \bar{y}) \qquad (4.358)$$

The initial condition is

$$t = 0: \qquad y = y_0 \tag{4.359}$$

Since this method requires one to calculate the initial guess by the Newton–Kantorovich method (4.357), it can be expected that this method fails if the Newton–Kantorovich method is not capable of computing the solution. Hence the advantage of this modification is questionable.

Example 4.25

Consider the imbedding procedure given by (4.356) for the problem described in Example 4.24. The differential equations corresponding to (4.356) are

$$y_1' = (1 - t)y_2 + ty_2$$
$$y_2' = (1 - t)y_3 + ty_3$$
$$y_3' = (1 - t)[c\bar{y}_1\bar{y}_3 + 1 + \bar{y}_4^2 + n\bar{y}_2^2 - c\bar{y}_3 y_1 + (-2n\bar{y}_2 + s)y_2$$
$$\qquad - c\bar{y}_1 y_3 - 2\bar{y}_4 y_4] + t[-cy_1 y_3 - n y_2^2 + 1 - y_4^2 + s y_2] \tag{4.360}$$
$$y_4' = (1 - t)y_5 + ty_5$$
$$y_5' = (1 - t)[c\bar{y}_1\bar{y}_5 + (n - 1)\bar{y}_2 y_4 - s - c\bar{y}_5 y_1 - (n - 1)\bar{y}_4 y_2$$
$$\qquad + ((1 - n)\bar{y}_2 + s)y_4 - c\bar{y}_1 y_5] + t[-cy_1 y_5 - (n - 1)y_2 y_4$$
$$\qquad + s(y_4 - 1)]$$

The results of calculation are presented by Roberts and Shipman.†

Example 4.26

The problem that arises in the investigation of the confinement of a plasma column by radiation pressure is described by a nonlinear boundary value problem‡:

$$y'' = \alpha \sinh \alpha y \tag{4.361a}$$

$$y(0) = 0, \qquad y(1) = 1 \tag{4.361b}$$

The problem has been solved by making use of the imbedding formulas (4.267), (4.268), and (4.305a) as well as by the one-loop imbedding procedure

$$y'' = t\alpha \sinh \alpha y \tag{4.362}$$

The approximations of partial derivatives are the same as we have used in Examples 4.18 and 4.19. The results for $\alpha = 5$ are reported in Tables 4-45 through 4-48.

TABLE 4-45
RESULTS FOR TROESCH'S PROBLEM: $\alpha = 5$, $n = 20$, $y_0(x) = x$[a]

t	0.0	1.0	2.0	3.0	4.0	5.0
$k = 0.1$	0.5000	0.1382	0.0587	0.0575	0.0575	
$k = 0.2$	0.5000	0.1834	0.0643	0.0575	0.0575	
$k = 0.5$	0.5000	0.2537	0.0972	0.0589	0.0575	
$k = 1.0$	0.5000	0.3075	0.1536	0.0738	0.0579	0.0575

[a]The imbedding formula (4.267) is used. The values of y for $x = 0.5$ are presented.

†S. M. Roberts and J. S. Shipman, *J. Optimiz. Theory Appl. 12*, 136 (1973).
‡B. A. Troesch, *Internal Rep. NN-142*, (Redondo Beach, Calif: TRW, Inc.), 1960.

TABLE 4-46
RESULTS FOR TROESCH'S PROBLEM: $\alpha = 5$, $n = 20$, $y_0(x) = x$, $\omega = -1$[a]

t	0.0	1.0	2.0	3.0	4.0	5.0	10.0
$k = 0.1$	0.5000	0.0573	0.0574	0.0575			
$k = 0.2$	0.5000	0.0634	0.0567	0.0575			
$k = 0.5$	0.5000	0.0849	0.0499	0.0595	0.0569	0.0576	0.0575
$k = 1.0$	0.5000	0.1251	0.0181	0.0857	0.0390	0.0701	0.0559

[a]The imbedding formula (4.268) is used. The values of y for $x = 0.5$ are presented.

TABLE 4-47
RESULTS FOR THE ONE-LOOP IMBEDDING FORMULA (4.362):
$\alpha = 5$, $n = 80$, $y_0(x) = x$, $k = 1/320$[a]

	t				
x	0.2	0.4	0.6	0.8	1.0
0.2	0.070	0.039	0.024	0.015	0.010
0.4	0.155	0.094	0.063	0.044	0.032
0.5	0.209	0.135	0.095	0.071	0.054
0.6	0.276	0.190	0.143	0.112	0.091
0.7	0.362	0.269	0.215	0.179	0.152
0.8	0.480	0.386	0.329	0.288	0.257
0.9	0.658	0.577	0.525	0.486	0.455
0.95	0.793	0.733	0.692	0.661	0.636

[a]Double-precision arithmetic.

TABLE 4-48
THE DEPENDENCE OF THE ACCURACY ON THE n AND k:
IMBEDDING FORMULA (4.362), $\alpha = 5$, $y_0(x) = x$[a]

	n		
k	10	20	40
0.1	0.0423	0.0381	0.0367
0.05	0.0472	0.0428	0.0413
0.025	0.0531	0.0487	0.0474
0.0125	0.0572	0.0529	0.0518
0.00625	0.0595	0.0553	0.0545

[a]The values of $y(0.5; 1)$ are presented.

Example 4.27

The flow of a viscous fluid between two coaxial rotating discs is described by the NBVP

$$F'' = \sqrt{\text{Re}}\, HF' + \text{Re}\,(F^2 - G^2 + k) \tag{4.363a}$$

$$G'' = 2\,\text{Re}\, FG + \sqrt{\text{Re}}\, G'H \tag{4.363b}$$

$$H' = -2\sqrt{\text{Re}}\, F \tag{4.363c}$$

The boundary conditions are

$$H(0) = F(0) = H(1) = F(1) = 0 \tag{4.363d}$$

$$G(0) = 1, \qquad G(1) = s$$

Here Re is the Reynolds number. The number of boundary conditions is higher than the overall order of differential equations in (4.363). The one surplus boundary condition determines the unknown value of the parameter k. To get an agreement between the number of boundary conditions and the overall order of the set of differential equations, we can enlarge (4.363) by an extra differential equation $k' = 0$.

To solve (4.363) three special imbedding procedures may be constructed.
Imbedding procedure A:

$$F'' - \text{Re}\, k - t[\sqrt{\text{Re}}\, HF' + \text{Re}(F^2 - G^2)] = 0 \tag{4.364a}$$

$$G'' - t[2\,\text{Re}\, FG + \sqrt{\text{Re}}\, G'H] + (1 - t)\alpha = 0 \tag{4.364b}$$

$$H' + 2\sqrt{\text{Re}}\, F = 0 \tag{4.364c}$$

Here α is a parameter, $\alpha \in (-\infty, \infty)$, and the functions F, G and H satisfy the boundary conditions (4.363d). Evidently, for $t = 0$ the following initial conditions result:

$$t = 0: \quad F \equiv H \equiv 0, \quad k = 0, \quad G = 1 + \left(s - 1 - \frac{\alpha}{2}\right)x + \frac{1}{2}\alpha x^2 \tag{4.364d}$$

Imbedding procedure B:

$$F'' = \sqrt{\text{Re}}\,[tH + (1 - t)(H - \bar{H})]F' + \text{Re}\,[(tF^2 + (1 - t)(\bar{F} - F)^2) - (tG^2$$
$$+ (1 - t)(\bar{G} - G)^2) + tk + (1 - t)(k - \bar{k})] + (1 - t)\bar{F}'' \tag{4.365a}$$

$$G'' - 2\,\text{Re}\, F[tG + (1 - t)(G - \bar{G})]$$
$$- \sqrt{\text{Re}}\, G'[tH + (1 - t)(H - \bar{H})] - (1 - t)\bar{G}'' = 0 \tag{4.365b}$$

$$H' + 2\sqrt{\text{Re}}\,(tF + (1 - t)(F - \bar{F})) - (1 - t)\bar{H}' = 0 \tag{4.365c}$$

The initial conditions are

$$t = 0: \quad F = \bar{F}, \quad G = \bar{G}, \quad H = \bar{H}$$

Imbedding procedure C:

$$F'' - \alpha(1 - t) \operatorname{Re}(k - \bar{k}) = t[\sqrt{\operatorname{Re}}\, HF' + \operatorname{Re}(F^2 - G^2 + k)]$$
$$+ (1 - t)[\sqrt{\operatorname{Re}}\, \bar{H}\bar{F}' + \operatorname{Re}(\bar{F}^2 - \bar{G}^2 + \bar{k})] \qquad (4.366a)$$

$$G'' = t[2 \operatorname{Re} FG + \sqrt{\operatorname{Re}}\, G'H] + (1 - t)[2 \operatorname{Re} \bar{F}\bar{G} + \sqrt{\operatorname{Re}}\, \bar{G}'\bar{H}] \qquad (4.366b)$$

$$H' = -2\sqrt{\operatorname{Re}}\, F \qquad (4.366c)$$

The initial condition may be determined after solving the linear boundary value problem (for $t = 0$):

$$F_o'' - \alpha \operatorname{Re}(k_o - \bar{k}) = \sqrt{\operatorname{Re}}\, \bar{H}\bar{F}' + \operatorname{Re}(\bar{F}^2 - \bar{G}^2 + \bar{k})$$
$$G_o'' = 2 \operatorname{Re} \bar{F}\bar{G} + \sqrt{\operatorname{Re}}\, \bar{G}'\bar{H} \qquad (4.366d)$$
$$H_o' + 2\sqrt{\operatorname{Re}}\, F_o = 0$$

subject to boundary conditions (4.363d). Here α is an arbitrary parameter.

It can be readily shown that the imbedding procedures (4.365) and (4.366) are strongly correct multiloop procedures,[†] whereas (4.364) is a one-loop procedure.

Since continuation requires that we know the derivates with respect to t, we have to develop these relations; after differentiation of, for example (4.366) with respect to t we have

$$\frac{\partial^3 F}{\partial x^2\, \partial t} + \alpha \operatorname{Re}(k - \bar{k}) - \alpha \operatorname{Re}(1 - t)\frac{dk}{dt} - \sqrt{\operatorname{Re}}\, H\frac{\partial F}{\partial x}$$
$$- \operatorname{Re}(F^2 - G^2 + k) - t\left[\sqrt{\operatorname{Re}}\left(\frac{\partial H}{\partial t}\frac{\partial F}{\partial x} + H\frac{\partial^2 F}{\partial x\, \partial t}\right)\right.$$
$$\left. + \operatorname{Re}\left(2F\frac{\partial F}{\partial t} - 2G\frac{\partial G}{\partial t} + \frac{dk}{dt}\right)\right] \qquad (4.367a)$$
$$= -\sqrt{\operatorname{Re}}\, \bar{H}\bar{F}' - \operatorname{Re}(\bar{F}^2 - \bar{G}^2 + \bar{k})$$

$$\frac{\partial^3 G}{\partial x^2\, \partial t} - 2 \operatorname{Re} FG - \sqrt{\operatorname{Re}}\frac{\partial G}{\partial x} H - t\left[2 \operatorname{Re}\left(\frac{\partial F}{\partial t} G + F\frac{\partial G}{\partial t}\right)\right.$$
$$\left. + \sqrt{\operatorname{Re}}\left(\frac{\partial^2 G}{\partial x\, \partial t} H + \frac{\partial G}{\partial x}\frac{\partial H}{\partial t}\right)\right] = -2 \operatorname{Re} \bar{F}\bar{G} - \sqrt{\operatorname{Re}}\, \bar{G}'\bar{H} \qquad (4.367b)$$

$$\frac{\partial^2 H}{\partial x\, \partial t} + 2\sqrt{\operatorname{Re}}\frac{\partial F}{\partial t} = 0 \qquad (4.367c)$$

The boundary conditions are

$$H(0, t) = F(0, t) = H(1, t) = F(1, t) = 0$$
$$G(0, t) = 1, \qquad G(1, t) = s \qquad (4.367d)$$

[†]One parameter imbedding procedures satisfying $H(t, y^*, y^*) \equiv 0$ are called strongly correct multiloop algorithms in the paper by Kubíček and Hlaváček (1978).

The initial conditions may be written

$$H(x, 0) = H_o(x), \qquad G(x, 0) = G_o(x),$$
$$F(x, 0) = F_o(x), \qquad k(0) = k_o \qquad (4.367e)$$

Here the functions H_o, G_o, F_o and k_o may be calculated after solving the linear boundary problem (4.366d).

To solve the partial differential equations of the type (4.367) three different approaches can be adapted:

1. finite-difference approximations;
2. method of lines i.e., finite-difference approximation in the x-direction and marching integration in the t-direction;
3. method of variational variables which may be classified as the method of "perpendicular" lines.

Here the method of lines will be used to approximate (4.367).

Within the region $x \in \langle 0, 1 \rangle$ a mesh $x_i = ih$, $i = 0, 1, \ldots, N$; $h = 1/N$ will be used. Let us denote $F_i = F(x_i)$, $F_i(t) = F(x_i, t)$, etc. After replacing the derivatives in the x-direction by the finite-difference formulas we have:

$$\frac{1}{h^2}[F'_{i-1} - 2F'_i + F'_{i+1}] + \alpha \operatorname{Re}(k - \bar{k}) - \alpha \operatorname{Re}(1 - t)k' -$$

$$\frac{\sqrt{\operatorname{Re}}}{2h} H_i(F_{i+1} - F_{i-1}) - \operatorname{Re}(F_i^2 - G_i^2 + k) - t\bigg[\frac{\sqrt{\operatorname{Re}}}{2h}(H'_i(F_{i+1} - F_{i-1})$$

$$+ H_i(F'_{i+1} - F'_{i-1})) + \operatorname{Re}(2F_iF'_i - 2G_iG'_i + k')\bigg]$$

$$= -\frac{\sqrt{\operatorname{Re}}\,\bar{H}_i(\bar{F}_{i+1} - \bar{F}_{i-1})}{2h} - \operatorname{Re}(\bar{F}_i^2 - \bar{G}_i^2 + \bar{k})$$

$$i = 1, 2, \ldots, N - 1 \qquad (4.368a)$$

$$\frac{1}{h^2}(G'_{i+1} - 2G'_i + G'_{i-1}) - 2\operatorname{Re} F_iG_i - \frac{\sqrt{\operatorname{Re}}}{2h} H_i(G_{i+1} - G_{i-1})$$

$$- t\bigg[2\operatorname{Re}(F_iG'_i + F'_iG_i) + \frac{\sqrt{\operatorname{Re}}}{2h}((G'_{i+1} - G'_{i-1})H_i + (G_{i+1} - G_{i-1})H'_i)\bigg]$$

$$= -2\operatorname{Re}\bar{F}_i\bar{G}_i - \frac{\sqrt{\operatorname{Re}}}{2h}(\bar{G}_{i+1} - \bar{G}_{i-1})\bar{H}_i \qquad i = 1, 2, \ldots, N - 1 \qquad (4.368b)$$

$$\frac{H'_i - H'_{i-1}}{h} + \sqrt{\operatorname{Re}}(F'_i + F'_{i-1}) = 0 \qquad i = 1, 2, \ldots, N \qquad (4.368c)$$

The boundary conditions (4.367d) may be rewritten:

$$H'_o = F'_o = H'_N = F'_N = G'_o = G'_N = 0 \qquad (4.368d)$$

Equations (4.368) constitute a set of linear algebraic equations for the unknowns k', F'_i, G'_i, H'_i, $i = 0, 1, \ldots, N$. This set can be solved for a particular value of t, where the variables k, F_i, G_i and H_i are known, by virtue of the modified

Gaussian elimination. To integrate in the t direction the differential equations for k', F_i', G_i', H_i' any arbitrary integration method can be used. Of course, the simplest way of integration is the utilization of the Euler method, e.g., for F_i':

$$F_i(t + \Delta t) = F_i(t) + \Delta t\, F_i'(t), \qquad i = 0, 1, \ldots, N$$

The computations for all three algorithms have been performed in a double precision arithmetics (~ 15 significant digits).

The results for the one-loop algorithm, that is, the imbedding procedure A, are presented in Table 4-49. Very short step size Δt must be used to obtain satisfactory accuracy of the solution.

TABLE 4-49
RESULTS FOR THE IMBEDDING PROCEDURE A

Dependence of solution on N and Δt, Euler method,
Re $= 100$, s $= 0.5$, $\alpha = 2$

N	Δt	$H(0.5)$	$100F(0.5)$	$G(0.5)$	k
10	0.01	−0.2585	0.2017	0.7246	0.5197
	0.005	−0.2565	0.2008	0.7237	0.5199
	0.0025	−0.2554	0.2004	0.7232	0.5200
20	0.01	−0.2947	0.3275	0.7269	0.5230
	0.005	−0.2925	0.3249	0.7259	0.5231
	0.0025	−0.2914	0.3237	0.7255	0.5232
50	0.05	−0.3254	0.4102	0.7369	0.5245
	0.02	−0.3107	0.3816	0.7302	0.5247
	0.01	−0.3061	0.3744	0.7281	0.5248
	0.005	−0.3038	0.3712	0.7271	0.5249
	0.0025	−0.3026	0.3697	0.7267	0.5249
100	0.05	−0.3271	0.4182	0.7371	0.5248
	0.02	−0.3124	0.3890	0.7304	0.5249
	0.01	−0.3077	0.3816	0.7283	0.5251
Exact[a]		−0.3036	0.3776	0.7264	0.5253

[a]K. H. Well, *J. Math. Anal. Appl.* 40, 258 (1972).

For the multiloop algorithms the continuation procedure may be repeated and, in the sequel, the requirements of the integration procedures, as far as accuracy of the integration process is concerned, may be lowered. From Tables 4-50 and 4-51 the accuracy of the solution may be inferred as a function of the integration step length Δt. Based on the results of Tables 4-50 and 4-51 it is obvious that an "optimum" step-length Δt exists which is associated with the lowest computational effort. Sometimes, on the other hand, the methods may diverge if a too long step-length Δt is used. (See also Kubíček et al.[†])

[†]M. Kubíček, M. Holodniok, and V. Hlaváček, *Sci. Papers of the Prague Inst. of Chem. Technol.* K12, 47 (1977).

TABLE 4-50
RESULTS FOR THE IMBEDDING PROCEDURE C

Dependence of solution on Δt, Euler method, Re = 100, $s = 0.5$, $\alpha = 1$, N = 20.
Initial profiles: $F \equiv 0$, $H \equiv 0$, $G = 1 + (s - 1)x$, $k = 0$

		$\Delta t = 0.01$				$\Delta t = 0.05$		$\Delta t = 0.1$	
Loop	t	$H(0.5)$	$100F(0.5)$	$G(0.5)$	k	$H(0.5)$	k	$H(0.5)$	k
1	0	−3.867	−2.578	0.7500	0.5749	−3.867	0.5749	−3.867	0.5749
	1	−0.2595	0.1182	0.7226	0.5111	−0.1281	0.4576	0.0583	0.3671
2	0	−0.2039	0.8100	0.7436	0.5171	0.2090	0.4883	0.9419	0.4309
	1	−0.2914	0.3210	0.7248	0.5233	−0.3233	0.5287	−0.4410	0.5764
3	0	−0.2868	0.3045	0.7253	0.5231	−0.2046	0.5235	0.0248	0.5609
	1	−0.2903	0.3227	0.7250	0.5233	−0.2927	0.5232	0.3031	0.5175
4	0	−0.2904	0.3223	0.7250	0.5233	−0.3188	0.5230	−0.5610	0.5178
	1	−0.2903	0.3225	0.7250	0.5233	−0.2887	0.5233	−0.2617	0.5234
5	0	−0.2903	0.3225	0.7250	0.5233	−0.2903	0.5234	−0.2741	0.5255
	1	−0.2903	0.3225	0.7250	0.5233	−0.2903	0.5233	−0.2951	0.5232
6	0					−0.2892	0.5233	−0.2514	0.5231
	1					−0.2903	0.5233	−0.2937	0.5233
7	0					−0.2904	0.5233	−0.3017	0.5228
	1					−0.2902	0.5233	−0.2887	0.5233
8	0					−0.2903	0.5233	−0.2939	0.5234
	1					−0.2903	0.5233	−0.2901	0.5232
10	0							−0.2903	0.5232
	1							−0.2902	0.5233

TABLE 4-51
RESULTS FOR THE IMBEDDING PROCEDURE B

Dependence of solution on Δt, Euler method, Re = 100, $s = 0.5$, N = 20
Initial profiles: $F \equiv 0$, $H \equiv 0$, $G = 1 + (s - 1)x$, $k = 0$

	$\Delta t = 0.01$				$\Delta t = 0.05$		$\Delta t = 0.1$	
Loop	$H(0.5)$	$100F(0.5)$	$G(0.5)$	k	$H(0.5)$	k	$H(0.5)$	k
0	0.0000	0.0000	0.7500	0.0000	0.0000	0.0000	0.0000	0.0000
1	−0.2935	0.2970	0.7272	0.5232	−0.3031	0.5186	−0.0803	0.6301
2	−0.2902	0.3231	0.7250	0.5233	−0.2890	0.5235	−0.2534	0.3756
3	−0.2903	0.3225	0.7250	0.5233	−0.2903	0.5232	−0.2955	0.5128
4	−0.2903	0.3225	0.7250	0.5233	−0.2903	0.5233	−0.2896	0.5236
5					−0.2903	0.5233	−0.2903	0.5232
6							−0.2903	0.5233

4.5.7.2 Interrelations among the methods of Sections 4.1, 4.3, 4.4, and 4.5

Sections 4.1, 4.3, 4.4, and 4.5 revealed to us that there are interrelations among the methods described in these parts. A general overview is sketched in Fig. 4-27. The relations among the methods cannot be considered as a hierarchy

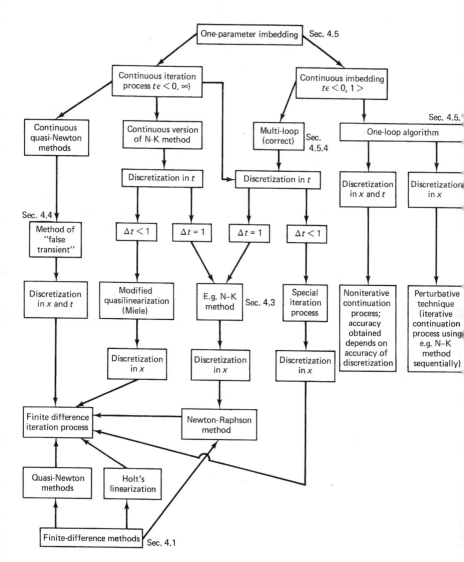

Figure 4-27 Relations among the methods.

on different levels, but the similarity among the particular procedures should be demonstrated. Of course, it is difficult to find all relations; however, the fundamental similarity is described. Hence this figure should serve as an illustration of some interrelations. For instance, let us note that (4.290) yields for $t = 0$ and $y = y_0$:

$$\frac{\partial y}{\partial t} = \omega \left[\alpha e^y - \frac{\partial^2 y}{\partial x^2} \right]$$

which for $\omega = -1$ is identical to the false transient method described in Section 4.4.

PROBLEMS

1. Some imbedding procedures may give rise to solutions that do not represent the solution of the transcendental equation

$$f(y) = 0 \qquad\qquad (4.369)$$

Consider the imbedding procedure

$$H(t, y) = (1 - t)(y - \bar{y})f(y) + tf(y) = 0$$

Differentiation of this relation yields

$$\frac{dy}{dt} = \frac{(y - \bar{y})f(y) - f(y)}{(1 - t)[f(y) + (y - \bar{y})f'(y)] + tf'(y)} \qquad (4.370)$$

The relevant initial condition is

$$t = 0: \qquad y = \bar{y}$$

After rewriting H, we get

$$H(t, y) = [(1 - t)(y - \bar{y}) + t]f(y) = 0$$

Evidently, the solution of (4.370) for $t \in \langle 0, 1 \rangle$ is

$$y(t) = \bar{y} - \frac{t}{1 - t} \qquad\qquad (4.371)$$

Of course, (4.371) does not solve (4.369).

2. Hypersonic flow around a sphere is described by a set of nonlinear differential equations

$$y_1' = -\frac{y_1}{y_5}[y_6 + 2(y_3 + y_5)] = f_1(y)$$

$$y_2' = y_1 y_3(y_3 + y_5) = f_2(y)$$

$$y_3' = y_4$$

$$y_4' = \frac{\text{Re } y_1}{y_7}[-2\sigma A y_7^2 + y_4 y_5 + y_3(y_3 + y_5)] + \frac{y_4 y_5 y_6}{2A y_7^2} + \frac{2 \text{ Re } y_2}{y_7} = f_4(y)$$

$$y_5' = y_6$$

$$y_6' = \frac{3\,\mathrm{Re}\,y_1}{4y_7}\left[\frac{\sigma A y_7^2}{y_5}(y_6 + 2(y_3 + y_5)) + y_5 y_6(1 - \sigma)\right]$$
$$\qquad - \frac{y_3 y_5 y_6}{2A y_7^2} - \frac{y_4}{2} - \frac{y_5}{2A}\left(\frac{y_6}{y_7}\right)^2 = f_6(y)$$
$$y_7' = -\frac{y_5 y_6}{2A y_7} = f_7(y)$$

Here we have denoted $y = (y_1, y_2, \ldots, y_7)$. The boundary conditions are

$$y_1(t_o) = 0.9617 \qquad y_2(t_o) = -0.1018 \qquad y_3(t_o) = 0.4078$$
$$y_5(t_o) = -0.0212 \qquad y_7(t_o) = 0.9998$$
$$y_3(t_f) = 1.0 \qquad y_5(t_f) = -1.0$$

The numerical values of the governing parameters are $\mathrm{Re} = 100$, $A = 0.515$, $\sigma = 0.400$, $t_o = 0.05$, and $t_f = 0.3303$. The imbedding procedure is:

$$y_1' = t f_1(y)$$
$$y_2' = t f_2(y)$$
$$y_3' = y_4$$
$$y_4' = t f_4(y) \qquad\qquad\qquad (4.372)$$
$$y_5' = y_6$$
$$y_6' = -\tfrac{1}{2} y_4 + t[f_6(y) + \tfrac{1}{2} y_4]$$
$$y_7' = t f_7(y)$$

Now let us try to continue the solution of (4.372) from $t = 0$ to $t = 1$. For a new value of t_k, $t_k = t_{k-1} + \Delta t$, the solution of (4.372) was calculated by the shooting method combined with the Newton correction algorithm (see Section 4.6). The resulting corrections were damped by multiplication by a factor 0.1. With $\Delta t = 0.05$ the solution can be continued up to $t = 0.55$; for higher values of t the increment Δt must be lowered. Finally, for $t = 0.91$ the maximum permissible step is as low as $\Delta t = 0.0005$. Of course, the continuation method cannot be economically used any more. However, the imbedding algorithm gives rise to a profile which is evidently close enough to the solution so that the Newton–Kantorovich method can be now successfully applied. Starting from this profile the Newton–Kantorovich method converges in 10 iterations. On the other hand, the initial profiles for the Newton–Kantorovich method calculated by continuation for $t < 0.45$ result in divergence.

Consider for this problem the multiloop imbedding procedure (4.267). Solve the resulting partial differential equations by finite-difference methods. Consider also the multiloop imbedding procedure (4.305a). Perform a study of the effect of Δt on the behavior of the solution. Try to find the solution by using the imbedding (4.325).

References: ROBERTS, S. M., SHIPMAN, J. S., AND ELLIS, W. J.: A perturbation technique for nonlinear two-point boundary value problems. *SIAM J. Numer. Anal.* 6, 347–358 (1969).

HOLT, J. F.: Numerical solution of the nonlinear two-point boundary value problems by finite-difference methods. *Commun. ACM* 7, 366–373 (1964).

3. Solve the Blasius equation

$$y''' + yy'' = 0$$

subject to

$$y(0) = y'(0) = 0, \qquad y'(\infty) = 1$$

by imbedding procedures. The asymptotic boundary conditions will be replaced at the point x_f (e.g., $x_f = 6$). Consider the one-loop imbedding procedure

$$y''' + tyy'' = 0$$

Consider also the imbedding procedures (4.305a) and (4.268). Results:
(a) The dependence $y'(x, t)$ on x is depicted in Fig. 4-28.
(b) The dependence of the missing initial condition $y''(0, t)$ on t is drawn in Fig. 4-29. The correct value $y''(0, 1)$ is $y''(0, 1) = 0.46960$.

Reference: WASSERSTROM, E.: Solving the boundary value problems by imbedding. *J. ACM 18*, 594–609 (1971).

4. For a nonlinear boundary value problem

$$y' + \frac{a}{x} y' = \phi^2 y \exp \left[\frac{\gamma\beta(1 - y)}{1 + \beta(1 - y)} \right]$$

$$y'(0) = 0, \qquad y(1) = 1$$

calculate the profile $y(x)$. Consider a one-loop imbedding formula

$$y'' + \frac{a}{x} y' = t\phi^2 y \exp \left[\frac{\gamma\beta(1 - y)}{1 + \beta(1 - y)} \right]$$

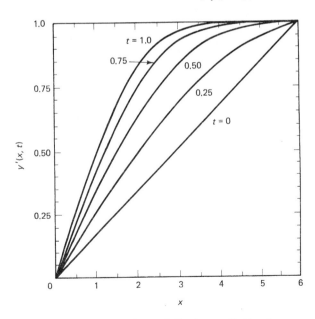

Figure 4-28 Dependence of $y'(x, t)$ on x: Problem 3.

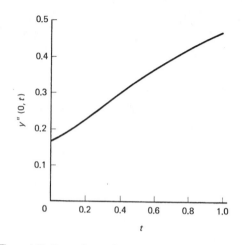

Figure 4-29 Dependence of $y''(0, t)$ on t: Problem 3.

Evidently, for $t = 0$, $y_0 \equiv 1$. The values of the parameters are (a) $\gamma = 20$, $a = 0$, $\beta = 0.4$, $\phi = 0.3637$.
Answer:

$y(0)$	1.0	0.98	0.95	0.85
t	0.0	0.2694	0.5684	1.0

(b) $\beta = 0.4$, $\gamma = 20$, $\phi = 0.4156$. Show that the function $y(x, t)$ is discontinuous with respect to t. For the sake of comparison the dependence $y(0, t)$ on t calculated by the GPM method (see Chapter 5) is presented below:

$y(0)$	0.95	0.85	0.75	0.65	0.55	0.45	0.35	0.25	0.15	0.05	0.01
t	0.4354	0.7659	0.7851	0.7086	0.6165	0.5371	0.4804	0.4525	0.4693	0.6247	1.0

Evidently, for $t \sim 0.6$, multiple solutions occur.

BIBLIOGRAPHY

One-parameter imbedding methods for systems of nonlinear algebraic equations are discussed, for example in:

ORTEGA, J. M., AND RHEINBOLDT, W. C.: *Iterative Solution of Nonlinear Equations in Several Variables*. Academic Press, New York, 1970.

Numerical and convergence aspects of multiloop imbedding techniques for systems of nonlinear algebraic equations are discussed in:

KUBÍČEK, M., HOLODNIOK M., AND MAREK, I., *J. Numer. Func. Anal. Optimiz 3*, 223–264 (1981).

The imbedding formula (4.267) of Newton–Kantorovich type for NBVPs is studied, for example, in:

BOSARGE, W. E.: *Numer. Math.* 17, 268–283 (1971).

ZHIDKOV, E. P., AND PUZYNIN, I. V.: *Zh. Vychisl. Mat. Mat. Fiz.* 7, 1086–1095 (1967) (Russian).

One-loop, one-parameter imbedding techniques are the subject of:

WASSERSTROM, E.: *J. ACM* 18, 594–602 (1971).

WASSERSTROM, E.: *SIAM Rev.* 15, 89–119 (1973).

GLASER, D.: *SIAM J. Num. Anal.* 6, 591–597 (1969).

ROBERTS, S. M., SHIPMAN, J. S., AND ELLIS, W. J.: *SIAM J. Numer. Anal.* 6, 347–358 (1969).

Some multiloop techniques are discussed in:

ROBERTS, S. M., AND SHIPMAN, J. S.: *Two-Point Boundary Value Problems: Shooting Methods.* American Elsevier, New York, 1972.

KUBÍČEK, M., HOLODNIOK, M., AND HLAVÁČEK V.: *Chem. Eng. Sci.* 34, 645 (1979).

A detailed treatment of one-loop and multiloop techniques is decribed in:

KUBÍČEK, M., AND HLAVÁČEK, V.: *Appl. Math. Comput.* 4, 317 (1978).

4.6 Shooting Methods

The numerical solution of a nonlinear two-point boundary value problem of the form

$$\frac{d^2y}{dx^2} = f\left(x, y, \frac{dy}{dx}\right) \qquad (4.373)$$

subject to the boundary conditions

$$\alpha_0 y(a) + \beta_0 y'(a) = \gamma_0 \qquad (4.374a)$$

$$\alpha_1 y(b) + \beta_1 y'(b) = \gamma_1 \qquad (4.374b)$$

may be frequently found by means of the "shooting procedure," by which is meant a family of iterative algorithms based on the solution of a corresponding initial value problem. The idea of integrating numerically an appropriate initial value problem is rather attractive because of the availability of powerful subroutines for the numerical integration of initial value problems. The method can be applied successfully to boundary value problems of any complexity as long as the initial value problem is stable and a set of good starting values is accessible.

The goal of this section is to describe various approaches to the solution of NBVP, making use of numerous variants of the shooting method. First, the problem of order 1 is discussed. Owing to the fact that only one missing condition must be guessed, the shooting approach may be clearly demonstrated; however, this type of problem is rather specific and some techniques proposed cannot be extended to problems of higher order, which are discussed later in the chapter. The methods with the first, second, and third order of convergence are developed. The Newton methods and the method of adjoints are examined in detail. The nonlinear as well as mixed boundary conditions are discussed. The particular approaches are compared and general conclusions on effectiveness (computer time expenditure, storage requirements, programming effort) are drawn. Finally, the multipoint boundary value problem is described. The multiple shooting method is suggested to integrate the highly sensitive problems.

4.6.1   Problem of Order 1

Consider the second-order differential equation (4.373) as a set of two first-order differential equations

$$\frac{dy}{dx} = z, \qquad \frac{dz}{dx} = f(x, y, z) \tag{4.375}$$

subject to the boundary conditions

$$\alpha_0 y(a) + \beta_0 z(a) = \gamma_0 \tag{4.376a}$$

$$\alpha_1 y(b) + \beta_1 z(b) = \gamma_1 \tag{4.376b}$$

After guessing

$$y(a) = \eta \tag{4.377}$$

and for $\beta_0 \neq 0$, (4.376a) yields

$$y'(a) = z(a) = \frac{1}{\beta_0}(\gamma_0 - \alpha_0 \eta) \tag{4.378}$$

Of course, the boundary conditions at the end of the integration region (i.e., for $x = b$) can be used in an analogous way. The solution of the boundary value problem requires us to solve an initial value problem; for a particular value of η the initial value problem can be integrated by reasonably adequate subroutine on the interval $x \in \langle a, b \rangle$. After integration the values $y(b, \eta)$ and $z(b, \eta)$ are obtained. On setting these values in (4.376b), a residuum $\varphi(\eta)$ results; that is,

$$\varphi(\eta) = \alpha_1 y(b, \eta) + \beta_1 z(b, \eta) - \gamma_1 \tag{4.379}$$

We now try to guess such a value η in order to satisfy

$$\varphi(\eta) = 0 \tag{4.380}$$

4.6.1.1 Inverse interpolation

To find the root of (4.380) several methods can be used (e.g., Newton or secant method, inverse interpolation, etc.) For illustration purposes we present an example where φ' need not be evaluated. We shall take advantage of the successive inverse interpolation method, which is more effective than the secant method.

Example 4.28

Axial heat and mass transfer in a tubular reactor are described by means of one second-order nonlinear differential equation:

$$\frac{1}{\text{Pe}} \frac{d^2 y}{dx^2} - \frac{dy}{dx} - p \frac{y^m}{[1 - H(1 - y)]^m} \exp\left[K - \frac{R}{1 - H(1 - y)}\right] = 0 \quad (4.381)$$

subject to the boundary conditions

$$y(0) = 1 + \frac{1}{\text{Pe}} \frac{dy(0)}{dx} \quad (4.382a)$$

$$\frac{dy(1)}{dx} = 0 \quad (4.382b)$$

The missing boundary condition will be chosen at $x = 1$:

$$y(1) = \eta \quad (4.383)$$

According to (4.375) two first-order differential equations are obtained. These equations are numerically integrated by the Runge–Kutta–Merson routine from $x = 1$ to $x = 0$. From the calculated values $y(0)$ and $y'(0)$, the residum $\varphi(\eta)$

$$\varphi(\eta) = y(0) - \frac{1}{\text{Pe}} y'(0) - 1$$

is evaluated. The results of integration are reported in Table 4-52. The first two guesses are rather good because the residuum changes the sign and if (4.381) subject to (4.382) does not exhibit multiplicity, one root of $\varphi(\eta)$ lies in the region estimated.

TABLE 4-52

CALCULATION OF THE AXIAL PROFILE OF y, INVERSE
INTERPOLATION: Pe $= 100$, $p = 0.01205$, $K = 19.3$,
$R = 10.85$, $H = 0.1437$, $m = 2$

n	$y(1) = \eta^{(n)}$	$y(0)$	$\varphi(\eta^{(n)})$	$\varphi_n^{-1}(0)$
1	0.057926	0.390133	-0.57925067	
2	0.060926	0.909082	0.15468264	0.060294
3	0.060294	0.690449	-0.18659296	0.060804
4	0.060804	0.854071	0.06375119	0.060709
5	0.060709	0.816521	0.00370431	0.060702
6	0.060702	0.814180	0.00001329	0.060702
7	0.060702	0.814172	0.00000041	0.060702

We have noticed that the inverse interpolation requires us to integrate only the original set of differential equations. The rate of convergence is quite high; roughly speaking, from the third to the seventh iteration the convergence is as high as that for the Newton method.

This technique seems to be very adequate for handling nonlinear boundary value problems for one second-order differential equation with a very complicated structure of the right-hand side.

In some problems it is an important factor to choose the appropriate direction of integration of the relevant initial value problem (i.e., to specify the point where the missing boundary conditions are to be guessed). It can happen, of course, that some physical or engineering problems are not stable in one direction (this is not to be confused with instability of the numerical method). The integration in this direction is practically impossible due to the inherent instability. For instance, the equation considered in Example 4.28 can be successfully integrated only from $x = 1$ to $x = 0$; an attempt to integrate this equation from $x = 0$ to $x = 1$ gives rise to an unstable initial value problem even if a very precise initial guess has been used. Some insight can be obtained from the fact that after neglecting the nonlinear term in (4.381), the linear equation

$$\frac{d^2 y}{dx^2} - \text{Pe} \frac{dy}{dx} = 0$$

has the solution $y = C_1 + C_2 e^{Pex}$. This solution has in the positive integration direction a strongly growing component; for example, for $\text{Pe} = 100$ any arbitrary small deviation from the correct initial value exponentially grows up during the integration process. Thus due to rounding errors, errors of approximation, and so on, integration in this direction is impossible.

4.6.1.2 Newton and Richmond methods

So far we have dealt only with the method which does not require us to evaluate the derivatives of φ. Of course, (4.380) can be also solved by means of the Newton or Richmond method. For problems where the differentiation of the differential equations does not lead to very cumbersome relations, the Newton and Richmond procedures are convenient methods.

The Newton method for the solution of a single equation (4.380) can be written

$$\eta^{k+1} = \eta^k - \frac{\varphi(\eta^k)}{\varphi'(\eta^k)} \tag{4.384}$$

where φ' denotes the derivative with respect to η. This method exhibits the quadratic rate of convergence. The cubic rate can be reached by making use of the Tschebyshev or Richmond iteration formula (see Chapter 2).

For one iteration step in the Newton procedure, the values $\varphi(\eta^k)$ and $\varphi'(\eta^k)$ have to be known. These values can be obtained on the basis of parallel integration of (4.373), (4.377), and (4.378) and the auxiliary equation†

$$\Omega_1'' = \frac{\partial f}{\partial y} \Omega_1 + \frac{\partial f}{\partial y'} \Omega_1' \tag{4.385}$$

---

†Such equations are often called "variational equations," as the variable Ω_1 characterizes the effect of the small variation in η.

subject to the initial conditions

$$\Omega_1(0) = 1 \tag{4.386}$$

$$\Omega_1'(0) = -\frac{\alpha_0}{\beta_0} \tag{4.387}$$

We have denoted, for the sake of brevity, $\Omega_1 = \partial y/\partial \eta$ and $\eta = \eta^k$. The first derivative of the residuum φ' can be written

$$\varphi'(\eta^k) = \alpha_1 \Omega_1(1) + \beta_1 \Omega_1'(1) \tag{4.388}$$

Equation (4.385) is obtained by differentiation of (4.373) with respect to η and obviously

$$\frac{\partial}{\partial \eta}\left(\frac{dy}{dx}\right) = \frac{d}{dx}\left(\frac{\partial y}{\partial \eta}\right)$$

(see Chapter 2.) However, for the application of methods with the cubic rate of convergence we have to evaluate the second derivative $\varphi''(\eta^k)$. This value can be computed when the auxiliary equation of higher order

$$\Omega_2'' = \frac{\partial f}{\partial y}\Omega_2 + \frac{\partial^2 f}{\partial y^2}\Omega_1^2 + \frac{\partial f}{\partial y'}\Omega_2' + \frac{\partial^2 f}{\partial y'^2}(\Omega_1')^2 + 2\frac{\partial^2 f}{\partial y\,\partial y'}\Omega_1\Omega_1'. \tag{4.389}$$

subject to the initial conditions

$$\Omega_2(0) = \Omega_2'(0) = 0 \tag{4.390}$$

is integrated parallel with (4.373), (4.377), (4.378), and (4.385)–(4.387).

The new dependent variable Ω_2 is defined by (4.391),

$$\Omega_2 = \frac{\partial^2 y}{\partial \eta^2} \tag{4.391}$$

The second derivative of the residuum is given by

$$\varphi''(\eta^k) = \alpha_1 \Omega_2(1) + \beta_1 \Omega_2'(1) \tag{4.392}$$

The Newton algorithm requires an integration of two differential equations of second order, (4.373) and (4.385); on the other hand, the method of third-order convergence demands the simultaneous solution of three differential equations of second order, (4.373), (4.385), and (4.389). On the basis of an example it will be shown that the method may be advantageous when after differention of f the extent of computational work does not increase rapidly.

The development of this algorithm can be extended to problems subject to nonlinear boundary conditions. In the same way methods of higher order of convergence can be developed making use of the Tschebyshev iteration formulas of fourth and higher orders.

Example 4.29

Heat and mass transfer in a porous catalyst of the plate shape for steady-state conditions is described by (4.393):

$$\frac{d^2y}{dx^2} = \phi^2 y \exp\left[\frac{\gamma\beta(1-y)}{1+\beta(1-y)}\right] \tag{4.393}$$

subject to the linear boundary conditions

$$x = 0: \quad \frac{dy}{dx} = 0$$
$$x = 1: \quad y = 1 \tag{4.394}$$

After choosing

$$y(0) = \eta \tag{4.395}$$

we have to integrate (4.393) and to test the residuum

$$\varphi(\eta) = y(1) - 1 \tag{4.396}$$

as long as the equation $\varphi(\eta) = 0$ is not satisfied. The auxiliary equation (4.385) takes the form

$$\Omega_1'' = \phi^2\Omega_1 \exp\left[\frac{\gamma\beta(1-y)}{1+\beta(1-y)}\right]\left[1 - \frac{\gamma\beta y}{[1+\beta(1-y)]^2}\right] \tag{4.397}$$

$$\Omega_1(0) = 1, \quad \Omega_1'(0) = 0 \tag{4.398}$$

The auxiliary equation of the second order (4.389) may be written

$$\Omega_2'' = \phi^2 \exp\left[\frac{\gamma\beta(1-y)}{1+\beta(1-y)}\right]\left\{\left[\frac{\gamma\beta}{(1+\beta(1-y))^2}\left(\frac{\gamma\beta y}{(1+\beta(1-y))^2} - 1\right)\right.\right.$$
$$\left.\left. - \frac{\gamma\beta(1+\beta(1-y)) + 2\gamma\beta^2 y}{(1+\beta(1-y))^3}\right]\Omega_1^2 + \left(1 - \frac{\gamma\beta y}{(1+\beta(1-y))^2}\right)\Omega_2\right\} \tag{4.399}$$

$$\Omega_2(0) = \Omega_2'(0) = 0 \tag{4.400}$$

A comparison of both third-order iteration methods versus the Newton formula has been made. The transport equation (4.393) has been solved for different values of the parameters γ, β, and ϕ, and furthermore, for various initial guesses η^0. In Table 4-53 the number of iterations k necessary to meet the prescribed accuracy in the boundary condition

$$|\varphi(\eta^k)| < \epsilon$$

is presented. As an illustration of the rate of convergence of the method of third order, Table 4-54 is presented. In this table the difference $\eta^k - \eta^*$ is tabulated, where η^* denotes the exact value of $y(0)$. On the basis of the computed results, it is obvious that the acceleration of the rate of convergence is substantial; on the other hand, the computational time increases only slightly. It is necessary to note that the Newton method for some initial values of η^0 can diverge; however, the method of third order often results in a convergent procedure.

TABLE 4-53

COMPARISON OF METHODS: $\gamma = 20$, $\beta = 0.1$, $\phi = 1$[a]

η^0	Method	ϵ 10^{-2}	10^{-3}	10^{-4}	10^{-5}
1	N	3	4	4	5
	R	2	3	3	3
	T	3	3	3	3
0.5	N	2	2	3	3
	R	1	1	2	2
	T	1	2	2	2
0.1	N	3	3	4	4
	R	1	2	2	2
	T	2	2	3	3

[a]N, Newton; R, Richmond; T, Tschebyshev.

TABLE 4-54

ILLUSTRATION OF THE RATE OF THE CONVERGENCE:

$\eta^* = 0.374533$, $\gamma = 20$, $\beta = 0.1$, $\phi = 1$[a]

	k 0	1	2	3	4	5
N	0.625467	−0.227060	−0.073084	−0.007387	−0.000075	0
R	0.625467	−0.193342	0.001135	0		
N	0.125467	−0.020375	−0.000568	−0.000001	0	
R	0.125467	−0.000620	0			
N	−0.274533	−0.107950	−0.016212	−0.000359	0	
R	−0.274533	0.006331	0			

[a]N, Newton; R, Richmond.

4.6.2 Problems of Higher Order

4.6.2.1 Newton method

Consider a set of ordinary differential equations

$$\frac{dy_i}{dx} = f_i(x, y_1, y_2, \ldots, y_n) \qquad i = 1, 2, \ldots, n \tag{4.401}$$

subject to the linear two-point boundary conditions

$$\sum_{j=1}^{n} a_{ij} y_j(a) = c_i \qquad i = 1, 2, \ldots, r \tag{4.402}$$

$$\sum_{j=1}^{n} b_{ij} y_j(b) = d_i \qquad i = 1, 2, \ldots, n - r \tag{4.403}$$

For the sake of simplicity, separated boundary conditions have been chosen. We try to reduce the boundary value problem to the initial value problem at the point $x = a$ by guessing the $n - r$ missing initial conditions [problem of the $(n - r)$th order at the point $x = a$]:

$$y_1(a) = \eta_1, \quad y_2(a) = \eta_2, \quad \ldots, \quad y_{n-r}(a) = \eta_{n-r} \tag{4.404}$$

Let us denote the matrices

$$A_1 = \begin{pmatrix} a_{1,1} & a_{1,2} & \cdots & a_{1,n-r} \\ a_{2,1} & \cdots\cdots\cdots & & a_{2,n-r} \\ \cdot & & & \\ \cdot & & & \\ \cdot & & & \\ a_{r,1} & a_{r,2} & \cdots & a_{r,n-r} \end{pmatrix} \tag{4.405}$$

$$A_2 = \begin{pmatrix} a_{1,n-r+1} & a_{1,n-r+2} & \cdots & a_{1,n} \\ a_{2\ n-r+1} & \cdots\cdots\cdots\cdots\cdots & a_{2,n} \\ \cdot & & \\ \cdot & & \\ \cdot & & \\ a_{r,n-r+1} & \cdots\cdots\cdots\cdots\cdots & a_{r,n} \end{pmatrix} \tag{4.406}$$

If the matrix A_2 is regular, the set of missing initial conditions (4.404) together with (4.402) yields a complete set of initial conditions because

$$\begin{pmatrix} y_{n-r+1}(a) \\ y_{n-r+2}(a) \\ \cdot \\ \cdot \\ \cdot \\ y_n(a) \end{pmatrix} = A_2^{-1}c - A_2^{-1}A_1\eta \tag{4.407}$$

Here we have denoted $\eta = (\eta_1, \eta_2, \ldots, \eta_{n-r})^T$ and $c = (c_1, c_2, \ldots, c_r)^T$. For a singular matrix A_2, supposing that the problem is correctly formulated, a reordering of variables yields a regular matrix A_2. Hence (4.404) and (4.407) determine a complete set of initial conditions at the point $x = a$.

Let us assume that the right-hand sides of differential equations (4.401) are continuously differentiable with respect to all y_j and are continuous with respect to x in a sufficiently large region. Based on this assumption the auxiliary equations can be developed as in Section 2.2:

$$\frac{dp_{ij}(x)}{dx} = \sum_{k=1}^{n} \frac{\partial f_i(x, y_1, \ldots, y_n)}{\partial y_k} p_{kj} \tag{4.408}$$

for $i = 1, 2, \ldots, n$ and $j = 1, 2, \ldots, n - r$. Here we have denoted

$$p_{kj}(x) = \frac{\partial y_k(x)}{\partial \eta_j} \tag{4.409}$$

The initial conditions for (4.408) are

$$p_{kj}(a) = \delta_{kj} \qquad k, j = 1, 2, \ldots, n - r \qquad (4.410a)$$

and

$$P = -A_2^{-1}A_1 \qquad (4.410b)$$

where

$$P = \begin{pmatrix} p_{n-r+1,1}(a) & p_{n-r+1,2}(a) & \cdots & p_{n-r+1,n-r}(a) \\ p_{n-r+2,1}(a) & \cdots\cdots\cdots\cdots & & p_{n-r+2,n-r}(a) \\ \cdots\cdots\cdots\cdots\cdots\cdots\cdots\cdots\cdots\cdots\cdots \\ p_{n,1}(a) & \cdots\cdots\cdots\cdots\cdots\cdots\cdots & p_{n,n-r}(a) \end{pmatrix} \qquad (4.410c)$$

The relation (4.410b) results after differentiation of (4.407) with respect to η.

The solution $y(x)$ of the initial value problem (4.401) with initial conditions (4.404) and (4.407) depends on the guessed initial values η:

$$y = y(x, \eta_1, \eta_2, \ldots, \eta_{n-r}) = y(x, \eta) \qquad (4.411)$$

This solution is also a solution of the BVP (4.401)–(4.403) if BC (4.403) is satisfied [the conditions (4.402) are automatically fulfilled for (4.407)]; that is, the calculated values y at the point $x = b$ obey

$$\sum_{j=1}^{n} b_{ij}y_j(b, \eta) = d_i \qquad i = 1, 2, \ldots, n - r \qquad (4.412)$$

This relation represents a set of $n - r$ nonlinear algebraic equations for $n - r$ variables $\eta_1, \ldots, \eta_{n-r}$ and can be written in a concise form:

$$F_i(\eta) = 0 \qquad i = 1, 2, \ldots, n - r \qquad (4.413)$$

To establish the roots of (4.413) by means of Newton's method, the Jacobian matrix is to be evaluated

$$\Gamma(\eta) = \begin{pmatrix} \dfrac{\partial F_1}{\partial \eta_1} & \dfrac{\partial F_1}{\partial \eta_2} & \cdots & \dfrac{\partial F_1}{\partial \eta_{n-r}} \\ \cdot \\ \cdot \\ \cdot \\ \dfrac{\partial F_{n-r}}{\partial \eta_1} & \cdots\cdots & & \dfrac{\partial F_{n-r}}{\partial \eta_{n-r}} \end{pmatrix} \qquad (4.414)$$

where

$$\frac{\partial F_i}{\partial \eta_j} = \sum_{k=1}^{n} b_{ik}p_{kj}(b) \qquad (4.415)$$

A corrected value of η can then be readily calculated:

$$\eta^{k+1} = \eta^k - \Gamma^{-1}(\eta^k)F(\eta^k) \qquad (4.416)$$

Fox seems to be the first (in 1960) to use the Newton method toward correction of the residuum in BC, and thus we shall call it the Newton–Fox procedure. The general program for getting a solution of a NBVP makes use of this procedure. An application of the Newton method for two second-order differential equations is presented in Example 4.30.

4.6.2.2 Method of adjoints

The method of adjoints is an important practical tool for solving linear boundary value problems. For linear problems we have shown that this technique is capable of finding the missing initial conditions. Goodman and Lance also proposed to use this method to solve nonlinear boundary value problems.

The set of differential equations (4.401) subject to boundary conditions

$$y_i(a) = c_i \qquad i = 1, 2, \ldots, r \tag{4.417}$$

$$y_{i_m}(b) = d_m \qquad m = 1, 2, \ldots, n - r \tag{4.418}$$

will be considered. For a guess,

$$y_{i+r}(a) = \eta_i \qquad i = 1, 2, \ldots, n - r \tag{4.419}$$

Equation (4.401) with initial conditions (4.417) and (4.419) can be integrated to determine $y(x)$. Now let us consider a nearby solution $y(x) + \delta y(x)$, where $\delta y(x)$ is often referred to as the variation. The differential equation for the nearby solution is

$$y'(x) + \delta y'(x) = f(x, y(x) + \delta y(x)) \tag{4.420}$$

After expanding the nonlinear function on the right-hand side in a Taylor series and on truncation of all higher-order terms, we have

$$\delta y_i'(x) = \sum_{j=1}^{n} \frac{\partial f_i(x, y(x))}{\partial y_j} \delta y_j(x) \qquad i = 1, 2, \ldots, n \tag{4.421}$$

These equations, often called the variational equations, form a set of ordinary linear differential equations with nonconstant coefficients. The relevant set of adjoint differential equations may be written (see Chapter 3)

$$z_i'(x) = -\sum_{j=1}^{n} \frac{\partial f_j(x, y(x))}{\partial y_i} z_j(x) \qquad i = 1, 2, \ldots, n \tag{4.422}$$

The matrix coefficient of the adjoint equations is the negative transpose of that given by (4.421). Referring to (3.72) the Green's identity may be written in the form

$$\sum_{i=1}^{n} z_i(b) \, \delta y_i(b) - \sum_{i=1}^{n} z_i(a) \, \delta y_i(a) = 0 \tag{4.423}$$

The aim is now to evaluate $\delta y(x)$ in such a way so that $y(x) + \delta y(x)$ is the solution to the particular NBVP. Clearly,

$$\delta y(x) = y^*(x) - y(x) \tag{4.424}$$

Here $y^*(x)$ is the exact solution to the given NBVP and $\delta y(x)$ is an approximation obtained after integration of (4.401) with initial conditions (4.417) and (4.419). From (4.417) it is obvious that

$$\delta y_i(a) = 0 \qquad i = 1, 2, \ldots, r \tag{4.425}$$

Equation (4.423) can be used to calculate the corrections $\delta y_i(a), i = r + 1, \ldots, n$, to the set of missing initial conditions $y_i(a)$. The Kronecker delta terminal conditions will be chosen for the adjoint variables $z_i(b)$ (see Chapter 3):

$$z_i^{(m)}(b) = \delta_{ii_m} \qquad m = 1, 2, \ldots, n - r \qquad (4.426)$$

This choice is not necessary but if used, the extent of calculations is lower in comparison with other possibilities.

For each m the adjoint system (4.422) must be integrated backward from $x = b$ to $x = a$ with terminal conditions (4.426). After integration the values of adjoint variables $z^{(m)}(a)$ will be obtained and inserting $z^m(a)$'s in (4.423), a set of linear algebraic equations results:

$$
\begin{bmatrix}
z_{r+1}^{(1)}(a) & z_{r+2}^{(1)}(a) & \ldots & z_n^{(1)}(a) \\
z_{r+1}^{(2)}(a) & \ldots\ldots\ldots\ldots & & z_n^{(2)}(a) \\
& \cdot & & \\
& \cdot & & \\
& \cdot & & \\
z_{r+1}^{(n-r)}(a) & \ldots\ldots\ldots\ldots & & z_n^{(n-r)}(a)
\end{bmatrix}
\begin{bmatrix}
\delta y_{r+1}(a) \\
\delta y_{r+2}(a) \\
\cdot \\
\cdot \\
\cdot \\
\delta y_n(a)
\end{bmatrix}
$$

$$
=
\begin{bmatrix}
\delta y_{i_1}(b) \\
\delta y_{i_2}(b) \\
\cdot \\
\cdot \\
\cdot \\
\delta y_{i_{n-r}}(b)
\end{bmatrix}
=
\begin{bmatrix}
d_1 - y_{i_1}(b) \\
d_2 - y_{i_2}(b) \\
\cdot \\
\cdot \\
\cdot \\
d_{n-r} - y_{i_{n-r}}(b)
\end{bmatrix}
\qquad (4.427)
$$

Solution of (4.427) yields a new trial initial value

$$\eta_i^{\text{new}} = \eta_i + \delta y_{i+r}(a) \qquad i = 1, 2, \ldots, n - r \qquad (4.428)$$

The procedure is as follows:

1. Guess the missing initial conditions η_i, $i = 1, 2, \ldots, n - r$ and integrate (4.401) with η_i and (4.417). The profiles of $y(x)$ and $\partial f_i(x, y(x))/\partial y_j$ are stored.
2. Integration of $n - r$ sets of adjoint equations from $x = b$ to $x = a$ with initial conditions (4.426). This integration makes use of profiles stored.
3. Solution of linear algebraic equations (4.427).
4. Form a new trial value according to (4.428). Terminate the calculation process if $\| \delta y(a) \|$ is less than the predetermined tolerance. If tolerance test is not passed, go to item 1.

If the profiles $\partial f_i(x, y(x))/\partial y_j$ are not stored, it is necessary to calculate them for each backward integration from stored profiles $y(x)$. It seems reasonable under such circumstances to integrate all sets of adjoint equations simultaneously. It should be pointed out here, however, that there is also another possibility of integration. We shall suppose that (4.401) is not unstable in the backward direction inherently. Integration of (4.401) from $x = a$ to $x = b$ yields $y(b)$. Now (4.401) is integrated simultaneously with adjoint equations from

$x = b$ to $x = a$, making use of terminal conditions $y(b)$. Of course, we can integrate simultaneously (4.401) with all sets of adjoint equations. As a consequence this modification enables one to make use of integration subroutines with automatic step-size control. The step-size control algorithm may be used only for (4.401).

Roberts and Shipman have shown that the Newton method and method of adjoints are entirely equivalent, supposing that all integrations can be performed accurately. Of course, the approximation and round-off errors make this statement invalid and differences may be expected. It should be noted that the main advantage of the Newton method in comparison with the method of adjoints consists in the fact that the original and auxiliary equations are integrated in one direction and it is not necessary to store the profiles.

4.6.2.3 Nonlinear and mixed boundary conditions

Consider a general type of two-point boundary conditions

$$g_i[y_1(a), \ldots, y_n(a), y_1(b), \ldots, y_n(b)] = 0 \qquad i = 1, 2, \ldots, n \qquad (4.429)$$

For this type of boundary conditions the methods presented must be slightly modified. We now guess all n missing initial conditions

$$y_i(a) = \eta_i \qquad i = 1, 2, \ldots, n \qquad (4.430)$$

Using derivative-free formulas (e.g., Warner's backward linear interpolation) the procedure remains unchanged; after integration of (4.401) with (4.430) the calculated values $y_i(b)$, $i = 1, 2, \ldots, n$, are inserted into (4.429) and a system of nonlinear equations results:

$$F_i(\eta) = 0 \qquad i = 1, 2, \ldots, n \qquad (4.431)$$

An application of the Newton–Fox method requires us to integrate (4.401) together with the auxiliary equations for $p_{ij}(x) = \partial y_i(x)/\partial \eta_j$. The initial conditions for these variables are clearly $p_{ij}(a) = \delta_{ij}$. Now (4.431) can be solved by the Newton method; the elements in the Jacobian matrix are

$$\frac{\partial F_i}{\partial \eta_j} = \frac{\partial g_i}{\partial y_j(a)} + \sum_{k=1}^{n} \frac{\partial g_i}{\partial y_k(b)} p_{kj}(b) \qquad i, j = 1, 2, \ldots, n \qquad (4.432)$$

If some equations (4.429) do not contain $y(b)$, the dimension of η_i (i.e., the number of missing conditions) can be reduced after presolving these boundary conditions. This approach also diminishes the dimension of the set of auxiliary differential equations for $p_{ij}(x)$.

To solve this problem by the method of adjoints, the following procedure must be used. For each i_m, $m = 1, 2, \ldots, n - r$, where $y_{i_m}(b)$ occurs in (4.429), the adjoint equations are solved with the Kronecker delta terminal conditions ($n - r$ backward-integrated systems). We have

$$\delta y_{i_m}(b) = \sum_{i=1}^{n} z_i^{(m)}(a)\, \delta y_i(a) \qquad m = 1, 2, \ldots, n - r \qquad (4.433)$$

Equation (4.429) requires that

$$g_i(\boldsymbol{\eta} + \delta y(a), y(b) + \delta y(b)) = 0 \qquad i = 1, 2, \ldots, n \qquad (4.434)$$

We can write with respect to (4.433)

$$\frac{\partial g_i}{\partial \delta y_k(a)} = \frac{\partial g_i}{\partial y_k(a)} + \sum_{m=1}^{n-r} \frac{\partial g_i}{\partial y_{i_m}(b)} z_k^{(m)}(a) \qquad i, k = 1, 2, \ldots, n \qquad (4.435)$$

If (4.434) is considered as a relation to evaluate $\delta y(a)$, one Newton iteration can be used because for $\delta y(a) = 0$ we have calculated the residua $g_i(\boldsymbol{\eta}, y(b))$ as well as the partial derivative (4.432). To perform additional iteration steps the modified Newton method can be adopted supposing that a "stationary" value of partial derivatives, calculated in the first step, will be used. For a case where the nonlinear equation contains only weak nonlinearity and the boundary conditions are strongly nonlinear, this modification may be successful.

A great deal of other types of boundary conditions are discussed in detail in the book by Roberts and Shipman (1972).

4.6.2.4 Method of third order

To adopt the root-finding methods of higher order (e.g., the Tschebyshev formulas or tangent hyperbolas) to find the roots of (4.413), the second derivatives $\partial^2 F_i / \partial \eta_k \, \partial \eta_m$ must be evaluated. For the variables

$$q_{ijm} = \frac{\partial^2 y_i}{\partial \eta_j \, \partial \eta_m} \qquad (4.436)$$

a set of differential equations can be developed:

$$\frac{dq_{ijm}}{dx} = \sum_{k=1}^{n} \left(\sum_{s=1}^{n} \frac{\partial^2 f_i(x, y_1, \ldots, y_n)}{\partial y_k \, \partial y_s} p_{sm} p_{kj} + \frac{\partial f_i(x, y_1, \ldots, y_n)}{\partial y_k} q_{kjm} \right)$$

$$i = 1, 2, \ldots, n; \quad j, m = 1, 2, \ldots, n-r \qquad (4.437)$$

This technique is analogous to that described in Section 2.2. The initial conditions are

$$q_{ijm}(a) = 0 \qquad (4.438)$$

After inserting this in (4.412), we get

$$\frac{\partial^2 F_i}{\partial \eta_k \, \partial \eta_m} = \sum_{j=1}^{n} b_{ij} q_{jkm}(b) \qquad (4.439)$$

A comparison of the third-order technique with the Newton method is presented in the following example.

Example 4.30

An application of the shooting method combined with the second- and third-order correcting procedures will be presented for two nonlinear differential equations describing longitudinal heat and mass transfer in a tubular reactor. The equations describing this phenomenon may be written[†]

$$y'' - \text{Pe}_M y' + \text{Pe}_M \text{Da}(1 - y) \exp\left(\frac{\theta}{1 + \epsilon\theta}\right) = 0$$

$$\theta'' - \text{Pe}_H \theta' - \text{Pe}_H \beta(\theta - \theta_c) + \text{Pe}_H B \text{Da}(1 - y) \exp\left(\frac{\theta}{1 + \epsilon\theta}\right) = 0 \tag{4.440}$$

subject to the boundary conditions

$$x = 0: \qquad \theta' = \text{Pe}_H\theta, \qquad y' = \text{Pe}_M y \tag{4.441}$$

$$x = 1: \qquad \theta' = 0, \qquad y' = 0 \tag{4.442}$$

For the chosen values

$$\eta_1 = y(1), \qquad \eta_2 = \theta(1) \tag{4.443}$$

an integration from $x = 1$ to $x = 0$ has to be performed. The numerical integration of (4.440), (4.442), and (4.443) requires a simultaneous solution of four differential equations of first order. For the second-order Newton–Fox process, eight additional ($n = 4$, $r = 2$) equations (4.408) of first order have to be integrated simultaneously with (4.440); that is, a system of 12 equations results. However, after utilizing the third-order procedure, 12 additional equations of first order [i.e., (4.437)] must be solved. As a result, in this case 24 equations have to be integrated. The factor 2 in the number of the equations does not mean that the computing time increases two times, because the computing time is affected mainly by the nonlinearity of the exponential function in (4.440). Clearly, in the auxiliary system the same nonlinear functions appear (i.e., the increase in the overall computing time may be expected to be small). A detailed comparison of both second- and third-order methods for this system has been performed. Roughly speaking, the number of iterations required decreases to $\frac{2}{3}$ to $\frac{1}{2}$ of those needed using the Newton method.

A comparison of the rate of convergence of both methods is presented in Table 4-55, where E is an Euclidean deviation between the kth approximation η^k and the exact solution η^*:

$$E = \sqrt{(\eta_1^k - \eta_1^*)^2 + (\eta_2^k - \eta_2^*)^2} \tag{4.444}$$

The integration of the initial value problem has been performed by the Runge–Kutta–Merson method and the built-in error control procedure has been used only for (4.440).

TABLE 4-55
ILLUSTRATION OF THE RATE OF THE CONVERGENCE:
$\text{Pe}_M = \text{Pe}_H = 2$, $\text{Da} = 0.12$, $\epsilon = \theta_c = 0$, $\beta = 2$, $B = 12$

η^0	$k =$	0	1	2	3	4	η^*
0;1	Newton	0.2536	0.1127	0.0114	0.0001	0	
	third order	0.2536	0.0023	0			
0;0	Newton	1.122	0.1992	0.0678	0.0032	0	0.2346
	third order	1.122	0.2107	0.0056	0		1.0963
0;0.9999	Newton	1.338	0.6157	0.1606	0.0166	0	
	third order	1.338	0.1389	0.0343	0		

[†]V. Hlaváček and H. Hofmann, *Chem. Eng. Sci.* 25, 173 (1970).

4.6.2.5 Comparison of the methods

We have shown above how to take advantage of auxiliary equations to establish the solution of a particular NBVP. In principle, two different procedures can be applied to solve the NBVP by shooting methods: Newton or Newton-like methods and inverse linear interpolation as well. Now we are faced with the serious problem of which procedure should be preferred. Sometimes the differentiation of the right-hand sides is rather tedious and the development of the auxiliary equations results after arduous differentiation. It is obvious, therefore, that the inverse interpolation ,e.g., the Warner backward interpolation (a generalization of the secant method), is convenient. This algorithm has been also adopted in the general program. For orientation purposes, as to the dimension of differential equations to be solved by different approaches, Table 4-56 is presented. Some general conclusions can be drawn based on the results of this table. The third-order methods seem not to be convenient for problems where $n - r > 3$ because of the high number of auxiliary equations for calculation of first and second derivatives. For $n - r > 3$ the second-order procedures (Newton) or interpolation should be used. A compromise between the Warner and Newton–Fox method is a modification of the Newton procedure where the derivatives of the function F are calculated numerically. This procedure is called the discretized Newton method.

The finite-difference approximation of derivatives has some shortcomings. If the value of y_i is perturbed by only a small increment the error of approximation of the integration formula strongly affects the accuracy of the derivative. On the other hand, for a large value of the increment, the finite-difference approximation is inaccurate. It is obvious that an optimum value of the increment exists; unfortunately, it is difficult to estimate it. Although the Newton–Fox and discretized Newton methods require us to integrate $n(n - r + 1)$ equations, the overall computer time can be different; namely, in the Newton–Fox method one set of $n(n - r + 1)$ equations is simultaneously integrated; on the other hand, in the discretized version n differential equations are integrated $(n - r + 1)$ times. If the right-hand sides of differential equations contain many times such functions as exp or log, the former procedure should be preferred because only one evaluation of such functions is required in each step; in the discretized method this nonlinear function must be calculated repeatedly in each integration. In turn, the complicated algebraic terms on the right-hand sides of auxiliary equations may result in higher computer time expenditure, and in such circumstances the discretized or similar method should be preferred.

A comparison of the Newton–Fox method with the method of adjoints indicates the superiority of the former technique. Let us consider the separated (nonmixed) BC in the form (4.402) and (4.403). Further, let us denote by s the number of variables $y_j(b)$ which are in (4.403) (i.e., where $\sum_{i=1}^{n-r} b_{ij}^2 \neq 0$). For the correctly formulated problem, it is obvious that $s \geq n - r$.

TABLE 4-56

Shooting method + Warner interpolation	n[b]
Newton–Fox shooting procedure	$n(n - r + 1)$ simultaneously
	or
	$(n - r)$ times ($2n$ simultaneously)
	or
	n — profiles $y(x)$ are to be stored $+ (n - r)$ times (n simultaneously)
Method of third order	$n(n - r + 1) + \dfrac{n(n - r)(n - r + 1)}{2}$ simultaneously
	or
	$\dfrac{(n - r)(n - r + 1)}{2}$ times ($4n$ simultaneously)
	or
	n — profiles $y(x)$ are to be stored $+ \dfrac{(n - r)(n - r + 1)}{2}$ times ($3n$ simultaneously)
Method of adjoints (s terminal values of y appear in BC; s often equals to $n - r$)	n — profiles $y(x)$ are to be stored $+ s$ times (n simultaneously)
	or
	$n + n(s + 1)$ simultaneously
	or $\quad n$ $+ s$ times ($2n$ simultaneously)
Discretized Newton in BC	$(n - r + 1)$ times (n simultaneously)
Modified Newton–Fox shooting procedure	n[c]

[a]The original NBVP consist of n first-order differential equations; $n - r$ missing initial conditions are to be guessed.

[b]It is necessary to integrate $(n - r + 1)$ times (n simultaneously) to start the iteration.

[c]In the first iteration it is necessary to integrate $n(n - r + 1)$ equations simultaneously to construct the Jacobian matrix for the initial guess.

For the case $s = n - r$, both methods are comparable; however, the method of adjoints requires us to store the profiles $y(x)$. For $s > n - r$, the superiority of the Newton methods is obvious. For instance, the computer time expenditure to solve Example 4.30 by the method of adjoints is approximately higher by the factor 2.

The modified Newton method can sometimes be used; that is, the elements in the Jacobian matrix are evaluated only in the first iteration. The Jacobian matrix is then used in each iteration; the algorithm requires only integration of the original set of differential equations. For a good initial guess this modification can be quite successful. Very often, however, the Jacobian matrix must be reevaluated after performing two to four iterations.

4.6.2.6 Multipoint boundary value problems

The shooting method can also be successfully adopted to solution of multipoint BVP. Of course, we have more possibilities of the point at which the missing initial conditions should be guessed. For illustration purposes a simple example of a tubular reactor with two recycles will be considered. The mass balance can be written as an ordinary differential equation:

$$\frac{dy}{dx} = R(y) \tag{4.445}$$

and the balance of the inlet stream yields a three-point boundary condition

$$C_o y(x_o) + C_1 y(x_1) + C_2 y(x_2) = D \tag{4.446}$$

where C_i are nonzero coefficients (or matrices in a more general case).

An unknown initial condition $y(x_0) = \eta$ can be guessed and (4.445) integrated from $x = x_0$ to $x = x_1$; with the value $y(x_1, \eta)$ the integration is continued until $x = x_2$ is reached. After inserting η, $y(x_1, \eta)$ and $y(x_2, \eta)$ into the mixed boundary condition (4.446), a residuum is obtained which depends only on the guess η. The same routines for numerical solution of the nonlinear equation (4.446) may be used to determine the correct guess η. Clearly, we can also integrate (4.445) starting at $x = x_1$. On guessing $y(x_1) = \eta$, integration of (4.445) from $x = x_1$ to $x = x_2$ and from $x = x_1$ to $x = x_0$ yields $y(x_2, \eta)$ and $y(x_0, \eta)$, respectively.

On inserting these results in (4.446) a function of a single variable η again results. It is obvious that as long as the particular equations are stable in both directions, the direction of integration is quite arbitrary.

Let us note that BVPs where more than three points are used in BC are encountered very seldom in engineering problems; however, in physical problems they may be frequent (three- and four-body systems). A close relation to multipoint boundary value problems is the method of multiple shooting, which is sometimes used to integrate sensitive BVPs.

4.6.2.7 Multiple shooting method

So far we have chosen the initial conditions at the point $x = a$ when $n - r \leq n/2$. Of course, this condition is not restrictive and we may guess the missing initial conditions at the point $x = b$; the dimension of the auxiliary equations in such circumstances grows. Moreover, the initial conditions can be guessed in any arbitrary point $x_0 \in (a, b)$ and equations are then integrated in both directions (i.e., in the direction $x_0 \rightarrow a$ and $x_0 \rightarrow b$ as well). For instance, for the BVP (4.401)–(4.403) the initial conditions are guessed to be

$$y_1(x_0) = \eta_1, \qquad y_2(x_0) = \eta_2, \ldots, y_n(x_0) = \eta_n \qquad (4.447)$$

Integration of (4.401) from $x = x_0$ to $x = a$ yields $y_i(a, \eta)$ and after inserting in (4.402), we have

$$F_i(\eta) = \sum_{j=1}^{n} a_{ij} y_j(a, \eta) - c_i = 0 \qquad i = 1, 2, \ldots, r \qquad (4.448)$$

Integration of (4.401) in $\langle x_0, b \rangle$ yields $y_i(b, \eta)$, which on inserting in (4.403) gives

$$F_{i+r}(\eta) = \sum_{j=1}^{n} b_{ij} y_j(b, \eta) - d_i = 0 \qquad i = 1, 2, \ldots, n - r \qquad (4.449)$$

Equations (4.448) and (4.449) enable us to calculate n unknowns, $\eta_1, \eta_2, \ldots, \eta_n$.

There are boundary value problems which are not suitable for the shooting method because of the high sensitivity of the corresponding initial value problem. If the initial value problem is unstable in both directions of integration, Morrison, Riley, and Zancanaro (1962) have devised a new procedure—the multiple shooting method—which may overcome this difficulty. Recently, the method was successfully applied by Roberts and Shipman (1972). The basic idea of the method will be elucidated on a simple two-point BVP for one first-order differential equation. A generalization to a set of differential equations is obvious and the reader is referred to the book of Roberts and Shipman.

Consider a BVP

$$\frac{dy}{dx} = f(x, y) \qquad (4.450)$$

subject to a BC

$$g(y(a), y(b)) = 0 \qquad (4.451)$$

Let us choose N interior points x_i, $a < x_1 < x_2 \ldots < x_N < b$. The subintervals x_i, x_{i+1} need not be equidistant. After denoting $x_0 = a$ and $x_{N+1} = b$, the functions $y_i(x)$, $i = 0, 1, \ldots, N$, are introduced, which obey both the differential equation

$$\frac{dy_i}{dx} = f(x, y_i) \qquad i = 0, 1, \ldots, N \qquad (4.452)$$

and the boundary conditions

$$y_i(x_{i+1}) - y_{i+1}(x_{i+1}) = 0 \qquad i = 0, 1, \ldots, N - 1 \qquad (4.453a)$$

$$g(y_0(x_0), y_N(x_{N+1})) = 0 \qquad (4.453b)$$

Equation (4.453a) merely expresses the requirement of continuity of the solution. On guessing

$$y_i(x_i) = \eta_i \qquad i = 0, 1, \ldots, N \qquad (4.454)$$

each equation (4.452) can be integrated from x_i to x_{i+1}, which is shorter than the original interval $\langle a, b \rangle$ so that the shooting method can be successful (i.e., no overflow occurs). For $N + 1$ guesses of initial values (4.454), $N + 1$ boundary conditions (4.453a) and (4.453b) must be satisfied. The multiple shooting method can be regarded as a hybrid between classical shooting and finite-difference methods. It is obvious that equations (4.452) are mutually independent and that a perturbation of one value of η_i, 0, 1, . . . , N, changes two residuals, (4.453), only. Hence the set of N auxiliary equations consist of N independent auxiliary equations (i.e., the problem can be solved serially). The number of internal mesh points depends on the sensitivity of the original problem. According to the character of the expected profile, the particular subintervals can be selected nonequidistant. It should be pointed out that the initial guesses η_i must be chosen adequately; that is, after integration in the subinterval $\langle x_i, x_{i+1} \rangle$ the final value y_{i+1} ought to be modified in such a manner so that the "blow-up" effect does not occur. This strategy depends, of course, on preliminary knowledge of the behavior of the particular problem and on the skill and ingenuity of the programmer.

4.6.3   Discussion

For the shooting method the investigator has to answer some questions, such as:

1. How can an adequate integration routine be selected?
2. How can reasonable initial values be guessed?
3. How can a particular shooting procedure be applied to solve large sets of differential equations?
4. How can problems having multiple solutions be handled?

A brief discussion of each of these questions follows.

4.6.3.1 Integration method

The selection of an appropriate integration technique should meet various criteria, among them a high order of approximation, and low computer storage requirements and computer-time expenditure. There exist a great number of efficient step-by-step numerical integration routines for solution of initial value problems. Frequently, the standard one-step methods such as Runge–Kutta and multistep methods such as methods of the Adams type are recommended in textbooks. The great accuracy and low computer-time expenditure can be obtained simultaneously only by procedures with step-size control. Thus such techniques should be preferred. A great many practical problems exhibit marked changes in the shape of the profile (combustion and explosion problems, hydrodynamic problems, etc.) which may be economically handled only by step-size control routines. As we have already pointed out (Section 2.1.3), the multistep methods have their limitations. There are problems with the starting of such routines; moreover, the step-size control mechanism is heavily adapted. On the other hand, the Runge–Kutta-like method does not suffer from these disadvantages. It should be mentioned here again that the step-size control may be used for the original differential equations only, the auxiliary equations are integrated simultaneously uncontrolled. This modification is very convenient because the approximation to a Jacobian matrix is usually sufficient, but the integration step can be longer. In practical cases, however, one extra iteration may result compared with the case where both the original and auxiliary equations are controlled.

With the step-size control algorithm, the following strategy can be used. For some first iterations, low accuracy is assigned; in the vicinity of the solution, the maximum admissible error, which is controlled by the step-size adjustment subroutine, is progressively lowered. This strategy can also be applied for constant-step-size techniques. The following example illustrates this procedure.

Example 4.31

Mass transfer in a tubular reactor where an isothermal second-order reaction occurs is described by a second-order equation with split boundary conditions,

$$\frac{1}{Pe}\frac{d^2y}{dx^2} - \frac{dy}{dx} - Da\, y^2 = 0$$

$$y(0) = 1 + \frac{1}{Pe}y'(0), \qquad y'(1) = 0$$

$$(4.455)$$

On guessing $y(1) = \eta$, the NBVP is reduced to an IVP at the point $x = 1$; after integration from $x = 1$ to $x = 0$, a residuum

$$\varphi(\eta) = y(0) - \frac{1}{Pe}y'(0) - 1$$

results. The fourth-order Runge–Kutta method was used for the integration of the relevant IVP. During the iteration process the length of the integration step was successively changed. The Newton–Fox procedure was used to correct the missing boundary condition η. The results of calculation are reported in Table 4-57.

TABLE 4-57
RESULTS OF EXAMPLE 4.31 (Pe = 2, Da = 2)

h	Iteration k	η^k	$\varphi(\eta^k)$
0.1	0	1.000	5.654
	1	0.722	1.773
	2	0.528	0.408
	3	0.451	0.042
	4	0.441	0.000
	5	⌐0.441	
0.04	0	⌊0.441	0.020
	1	0.436	0.000
	2	⌐0.436	
0.02	0	⌊0.436	0.008
	1	0.434	0.000
	2	0.434	

Source: Values taken from the paper by P. H. McGinnis, Jr., *Chem. Eng. Progr. Symp. Ser. No. 55, 61,* 1 (1965).

4.6.3.2 Estimation of initial guesses

A correct estimation of the initial conditions may guarantee convergence of of the Newton shooting procedure and substantially reduces the computer time. In this section we wish to draw attention to some systematic approaches as to how to choose the missing boundary conditions. Generally, it appears to be convenient to reduce the BVP to an IVP at a point where the number of missing boundary conditions to be guessed for (4.401)–(4.403) is less, i.e., at the point $x = a$ when $n - r \leq n/2$, and vice versa.

Very often, reasonable initial guesses may be obtained from a simplified physical model. For instance, for the problem of axial mixing (see Example 4.28) with high values of Peclet number a good initial guess can be obtained from a piston flow description (Pe $\rightarrow \infty$); this means mathematically that the term $(1/\mathrm{Pe})y''$ in the differential equation (4.381) is omitted, the resulting initial value problem

$$\frac{dy}{dx} = -p \frac{y^m}{[1 - H(1 - y)]^m} \exp\left[K - \frac{R}{1 + H(1 - y)}\right] \qquad (4.456)$$

with the initial condition

$$y(0) = 1 \qquad (4.457)$$

is readily solved. The calculated value $y(1)$ is used along with (4.382b) for backward integration. On the other hand, for low values of parameter Pe, Pe $\rightarrow 0$, a flat shape can be expected for the profile. This gives rise to an approximation:

$$p \frac{y^m}{[1 - H(1 - y)]^m} \exp\left[K - \frac{R}{1 + H(1 - y)}\right] = 1 - y \qquad (4.458)$$

The quality of both these approximations can be inferred from Fig. 4-30, where profiles of y for different values of Pe are drawn. We know in advance, because of the physical meaning of the dimensionless concentration y, that $y \in (0, 1\rangle$.

For problems where such simplifying assumptions are impossible, a simple mathematical approximation is always possible; for instance, the differential equation can be replaced on a sparse grid by its finite-difference analogy. The resulting set of nonlinear algebraic equations can be readily solved and a first rough guess may be obtained. Of course, such approximations may fail; for example, in the vicinity of a branching point two solutions can occur, and the rough finite-difference approximation cannot distinguish them. We present for the illustration a simple example describing the nonisothermal diffusion:

$$y'' = \phi^2 y \exp\left[\frac{\gamma\beta(1 - y)}{1 + \beta(1 - y)}\right]$$
$$y(1) = 1, \qquad y'(0) = 0 \qquad [\sim y(-1) = 1] \qquad (4.459)$$

The second derivative can be approximated from three points: $x_0 = -1$, $x_1 = 0$,

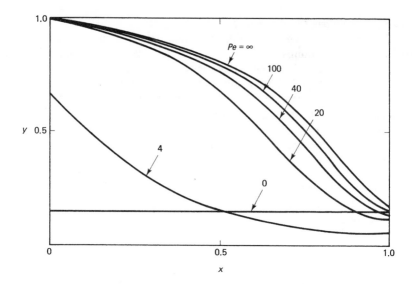

Figure 4-30 Dependence of the solution of NBVP described by (4.381) and (4.382) on the parameter Pe ($m = 1.66$, $H = 0.656$, $R = 14.57$, $p \exp(K - R) = 0.218$).

$x_2 = 1$. The finite-difference approximation yields

$$\frac{y_0 - 2y_1 + y_2}{h^2} = 2 - 2y_1 = \phi^2 y_1 \exp\left[\frac{\gamma\beta(1 - y_1)}{1 + \beta(1 - y_1)}\right] \qquad (4.460)$$

Here $y_1 \sim y(0)$.

Similarly, for three mesh points located at $x_0 = 0$, $x_1 = 0.5$, and $x_2 = 1$ and after making use of the asymmetrical three-point approximation for the first derivative at the point x_0,

$$\frac{-3y_0 + 4y_1 - y_2}{2h} = 0$$

we obtain

$$1 - y_1 = \frac{3}{8}\phi^2 y_1 \exp\left[\frac{\gamma\beta(1 - y_1)}{1 + \beta(1 - y_1)}\right] \qquad (4.461)$$

Here we have denoted $y_1 \sim y(0.5)$. The results from both approximations together with the exact values obtained by the shooting method are reported in Table 4-58. For the variety of physical and engineering problems, a universal approach for getting a reasonable set of initial guesses does not exist, but depends largely on the skill and experience of the investigator.

It is obvious that the missing initial conditions can be better estimated for low-dimensional problems, where the number of mutual interactions is small.

TABLE 4-58
INITIAL GUESSES $y(0)$ OBTAINED FROM ROUGH
FINITE-DIFFERENCE APPROXIMATION: $\gamma = 20$, $\beta = 0.2$

ϕ	Approximation (4.460)	Approximation (4.461)	Exact value
0.1	0.9949	0.9949	0.9949
0.3	0.9475	0.9497	0.9490
0.6	0.3781	0.6241	0.5520
1.5	0.0337	−0.2722	0.0023

Sometimes we are able to find relations between the missing initial conditions in an analytical form. Consider again, for instance, the diffusion problem

$$\frac{1}{Pe_\theta} \theta'' - \theta' + B\,Da(1-y)^n \exp\left(\frac{\theta}{1+\epsilon\theta}\right) = 0 \qquad (4.462a)$$

$$\frac{1}{Pe_y} y'' - y' + Da(1-y)^n \exp\left(\frac{\theta}{1+\epsilon\theta}\right) = 0 \qquad (4.462b)$$

$$x = 0: \qquad \theta' = Pe_\theta\,\theta, \qquad y' = Pe_y y$$
$$x = 1: \qquad y' = \theta' = 0 \qquad\qquad\qquad (4.463)$$

Linear combination of (4.462) yields

$$B\left(\frac{1}{Pe_y} y'' - y'\right) = \frac{1}{Pe_\theta} \theta'' - \theta' \qquad (4.464)$$

On integration of this equation and after inserting boundary conditions (4.463) a simple relation results:

$$\theta(1) = By(1) \qquad (4.465)$$

Thus it is sufficient to guess only the value $y(1) = \eta$ because $\theta(1) = B\eta$. After integration of (4.462) from $x = 1$ to $x = 0$, a residuum at $x = 0$ results:

$$R_1(\eta) = y'(0) - Pe_y y(0)$$

For a correct guess η is $R_1(\eta) = 0$ and automatically $R_2(\eta) = \theta'(0) - Pe_\theta\theta(0) = 0$. A systematic search in the one-dimensional problem is more convenient than to guess independently two values $\eta_1 = y(1)$ and $\eta_2 = \theta(1)$. Unfortunately, there is no general procedure on how to reduce the dimensionality of the problem considered.

Frequently, a whole family of solutions is required for engineering calculations, depending on the value of the parameter in question. Of course, these solutions are rather similar and after calculation of a solution for a given set of parameters this can be used as a first approximation for a new set of parameters where, for example, only one parameter is slightly changed.

It is obvious that the dependence between values of missing initial conditions and the value of the parameter considered can be used for interpolation.

A detailed discussion is presented later in the section dealing with the dependence on parameters.

To calculate the solution of a sensitive problem the continuation procedure (one-parameter operator imbedding) or a perturbation technique can be used. Both methods have been thoroughly studied by Roberts and Shipman (1972). The perturbation technique was devised for numerical solution of a sensitive NBVP of the type

$$Ly = R(y) \tag{4.466}$$

where L is a linear differential operator and R a nonlinear operator. A modified perturbed problem will be considered:

$$Ly = \epsilon R(y) \tag{4.467}$$

For $\epsilon_0 = 0$ the solution of (4.467) is known. The slightly perturbed equation is solved and the solution is used for another perturbed equation and so on; that is, a sequence of problems

$$Ly_k = \epsilon_k R(y_k) \qquad k = 1, 2, \ldots, N \tag{4.468}$$

is successively solved where $1 = \epsilon_N > \epsilon_{N-1} > \ldots > \epsilon_1 > \epsilon_0 = 0$. Hence the solution y_{k-1} obtained for ϵ_{k-1} will be used as the first approximation to y_k. This procedure is dealt with in detail in Section 4.5.

The continuation method makes use of a solution calculated on a short interval. After calculation of the solution on this short interval the interval is enlarged, the solution recalculated, and so on. The calculation is finished when we have reached the prescribed interval length. A more detailed treatment of this algorithm is presented in Section 3.5 for linear cases, and in Lee[†], and Scott[‡] for nonlinear cases.

A simple method that yields a sufficiently good initial guess is the method of solution of the relevant transient equations by the finite-difference method on a very sparse grid (see Section 4.4). This method can be used if only one stable steady-state solution exists. Frequently, the profile is sufficiently good after a small number of time steps.

4.6.3.3 Problems of large systems

For problems of higher order where a large number of missing boundary conditions must be guessed, a number of questions arise:

1. What are the computer storage requirements?
2. What is the computer time expenditure?
3. What type of algorithmization which might be used for a large family of BVP should be adopted?

[†] E. S. Lee, *Quasilinearization and Invariant Imbedding* (New York: Academic Press, 1968).
[‡] M. R. Scott, *Invariant Imbedding and Its Applications to Ordinary Differential Equations: An Introduction* (Reading, Mass.: Addison-Wesley, 1973).

In Table 4-59 a comparison of computer storage requirements for the finite-difference and Newton–Kantorovich methods (where a finite-difference approximation is used) with the shooting methods is reported.

TABLE 4-59

COMPARISON OF COMPUTER STORAGE REQUIREMENTS FOR A pTH-ORDER
PROBLEM FOR A SET OF n SECOND-ORDER DIFFERENTIAL EQUATIONS

Finite-difference approach ($N-1$ internal mesh points, three-point approximations)	$n(N+1)(3n+2)$[a]
Shooting method + Warner interpolation	$14n + 4p^2 + 5p$
Discretized Newton method	$14n + p^2 + 3p$
Newton–Fox shooting procedure	$14n(p+1) + p^2 + 2p$

[a]This value can be substantially lowered by using some appropriate linearization technique other than the Newton–Kantorovich method.

For integration of the relevant initial value problem the Runge–Kutta method with five evaluations of the right-hand side in one step is considered. For Merson's modification of the Runge–Kutta method, the coefficient 14 in Table 4-59 can be replaced by 10 because the results of all five evaluations need not be stored. The results show that the finite-difference methods are not convenient from the point of view of computer storage requirements if a large number of mesh points is to be used. The method of introduced parameters as well as the solution of parabolic partial differential equations for evaluating the steady-state profile have comparable computer storage requirements. The shooting method with Warner's interpolation and the discretized version of the Newton method demand the lowest computer core capacity. The Newton–Fox procedure for large systems also has high computer storage requirements, which can be reduced by making use of other integration routines with a lower order of approximation (e.g., the Euler method) or by means of the COMMON statement.

It is apparent that the shooting method with Warner's interpolation can be readily algorithmized. As far as simplicity is concerned, the discretized version of the Newton procedure can compete with the Warner technique. For the classical Newton–Fox method, the auxiliary equations must be developed; however, the procedure is still quite general. On the other hand, the finite-difference methods require a specific approach for each problem.

To reduce the computer time the quasi-Newton methods (or modified Newton method) can be used; here the Jacobian matrix can be constructed without integration of auxiliary equations in each iteration. The Jacobian matrix is calculated by means of auxiliary equations only once (in the first iteration); later during the iteration process the Jacobian matrix is corrected by the residuum vector (see Chapter 2). This modification reduces the calculation time because the auxiliary equations are considered only in the first iteration.

4.6.3.4 Problems of multiple solutions

If a nonlinear BVP exihibits multiple solutions, it is not simple to establish all of them because we usually do not known in advance the type of multiplicity. For a problem of order 1 we can make use of systematic exploration of the admissible interval for selection of the missing boundary condition. A more complicated situation occurs for problems of higher order, where systematic exploration of the region of values of missing boundary conditions may be rather complicated. Sometimes the physical considerations may be helpful. Because the convergence (attracting) region of initial guesses corresponding to a particular profile may be rather small, there is no guarantee that some of the solutions to the NBVP have not been omitted. Sometimes a random selection of initial guesses in the admissible region must be made. A more detailed discussion of multiple solutions is presented in Chapter 5. For illustration purposes an NBVP having five solutions will be investigated.

Example 4.32

Consider the NBVP described in Example 4.30 for values of governing parameters reported in Table 4-55. The NBVP may be converted to an IVP at the point $x = 1$ by choosing

$$\theta(1) = \eta_1, \qquad y(1) = \eta_2$$

The Newton–Fox method has been used. In Table 4-60 the course of the iteration process is reported starting from different initial guesses. Five different solutions of the NBVP result, and the profiles are drawn in Fig. 4-31. Of course, it is a tedious task to find these profiles; a systematic search in the admissible region is necessary. There are indeed other methods which may cope with this situation more rationally, and a detailed analysis of such techniques is presented in Chapter 5.

TABLE 4-60
FIVE DIFFERENT SOLUTIONS OF THE NBVP GIVEN BY (4.440)–(4.442):
COURSE OF ITERATIONS

Itera-tion	a η_1	a η_2	b η_1	b η_2	c η_1	c η_2	d η_1	d η_2	e η_1	e η_2
0	2.0000	0.0000	4.0000	0.7500	2.9000	0.9800	3.6000	0.9500	0.0000	0.0000
1	2.1644	0.4378	3.1441	0.6149	3.2155	0.9815	3.6781	0.9396	0.7395	0.1570
2	4.4148	0.8706	3.1447	0.6189	3.2114	0.9853	3.6919	0.9374	1.0299	0.2206
3	4.1768	0.8817	3.1448	0.6189	3.2132	0.9848	3.6926	0.9373	1.0932	0.2340
4	4.1098	0.8949			3.2133	0.9848	3.6926	0.9373	1.0963	0.2346
5	4.0792	0.8971							1.0963	0.2346
6	4.0775	0.8973								
7	4.0774	0.8973								

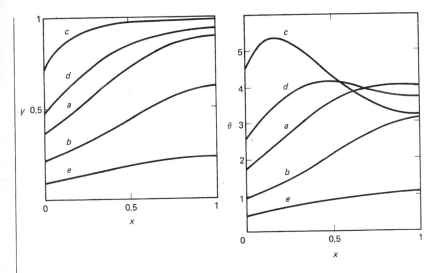

Figure 4-31 Five different solutions of the NBVP given by (4.440)–(4.442).

However, in the overwhelming majority of cases, for reasonable guesses of missing initial conditions, the shooting method is superior to any other direct procedures. To illustrate the advantages of the shooting procedure for a case with multiplicity, an example will be presented.

Example 4.33

The steady-state heat and mass balance and the first-order exothermic chemical reaction occurring within a porous spherical catalyst without film resistance is described by the highly nonlinear differential equation

$$\frac{d^2y}{dx^2} + \frac{2}{x}\frac{dy}{dx} = \phi^2 y \exp\left[\frac{\gamma\beta(1-y)}{1+\beta(1-y)}\right]$$

subject to the boundary conditions

$$x = 0: \quad \frac{dy}{dx} = 0$$

$$x = 1: \quad y = 1$$

Here γ, β, and ϕ are parameters. Copelowitz and Aris[†] have shown that for high values of γ and β, multiple solutions may occur. For $\gamma = 60$ and $\beta = 2.5$, 15 steady states have been calculated. The dependence of the effectiveness factor $\eta = (3/\phi^2)$ $(dy/dx)_{x=1}$ as a function of parameter λ [$\lambda^2 = \phi^2 \exp(-\gamma)$] is drawn in Fig. 4-32. To demonstrate the extreme sensitivity of the profile for $(\gamma, \beta) = (80, 2.5)$ it suffices to sketch the qualitative behavior when $y(0) = 3 \times 10^{-5}$. The first derivative y' is zero for $x = 0$ but at $x = 1.14 \times 10^{-10}$, y has increased to 0.179. The dependent

†I. Copelowitz, and R. Aris, *Chem. Eng. Sci. 25*, 906 (1970).

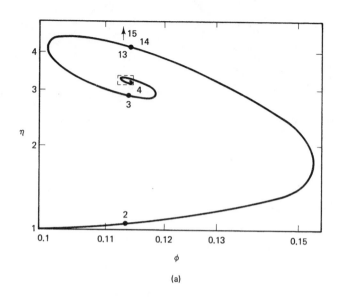

(a)

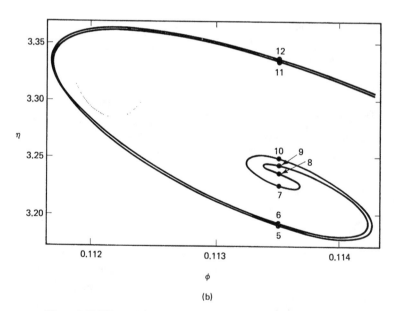

(b)

Figure 4-32 Fifteen different solutions of NBVP in Example 4.33.

variables y reached a value of 0.9996 at $x = 6.47 \times 10^{-6}$ and from this point until $x = 1$, the profile is indistinguishable from the profile obtained with $y(0) = 0.9996$.

It seems that the shooting method is the only procedure that is capable of calculating the relevant profiles. For the integration of the balance equation, an implicit collocation procedure has been used.† Explicit integration methods result in an extremely high computer time.

PROBLEMS

1. The optimum temperature profile in a tubular reactor for a simple first-order consecutive reaction $A \longrightarrow B \longrightarrow C$ is described by a set of equations

$$\frac{dx_1}{dt} = -k_1 x_1$$

$$\frac{dx_2}{dt} = k_1 x_1 - k_2 x_2$$

$$\frac{dz_1}{dt} = k_1(z_1 - z_2)$$

$$\frac{dz_2}{dt} = k_2 z_2$$

subject to the boundary conditions

$$x_1(0) = x_1^0 \qquad z_1(t_f) = 0$$
$$x_2(0) = x_2^0 \qquad z_2(t_f) = 1$$

where the reaction-rate constant is of the form

$$k_i = G_i \exp\left(-\frac{E_i}{RT}\right)$$

The temperature T giving rise to the maximum yield of B is given by

$$T = \frac{(E_1 - E_2)/R}{\ell n \left[\dfrac{E_1 G_1 x_1 (z_2 - z_1)}{E_2 G_2 x_2 z_2}\right]}$$

The parameters of the governing equation are:

$$E_1 = 18{,}000 \text{ cal/mol} \qquad G_1 = 5.35 \times 10^{10} \text{ min}^{-1}$$
$$E_2 = 30{,}000 \text{ cal/mol} \qquad G_2 = 4.61 \times 10^{17} \text{ min}^{-1}$$
$$R = 1.987 \text{ cal/mol K} \qquad x_1^0 = 0.53$$
$$t_f = 8 \text{ min} \qquad x_2^0 = 0.43$$

   Calculate the maximum yield of B, $x_2(t_f)$. Evaluate $x_1(t_f)$, $z_1(0)$, and $z_2(0)$. Show that the usual Newton–Fox method using auxiliary equations is inconvenient and that the direct shooting with Warner interpolation is superior to the Newton–Fox method.

   $[x_2(t_f) = 0.6794;\ x_1(t_f) = 0.1704;\ z_1(0) = -0.61000;\ z_2(0) = -0.82824]$

   Reference: ROTHENBERGER, B. F., AND LAPIDUS, L.: *AICHE J. 13*, 114 (1967).

†M. L. Michelsen, and J. Villadsen, *Chem. Eng. Sci. 27*, 751 (1972).

2. The equation describing the behavior of a boundary layer is given by

$$y''' = -yy''$$

subject to the boundary conditions

$$y(0) = y'(0) = 0$$
$$y(1) = 0.492$$

To calculate the unknown value $y''(0)$, use the Newton–Fox method together with the Richmond correcting procedure. [$y''(0) = 1.000$.]

3. The small oscillation of a mass which is attached by two springs is described by

$$y'' = -y^3$$

subject to the boundary conditions

$$y(0) = 0.2000$$
$$y(2) = 0.1846$$

Making use of a simple three-point formula, estimate the missing boundary condition. With this value apply the Newton–Fox shooting method. Calculate the amplitude $y(1)$. [$y(1) = 0.1960$.]

4. The steady-state simulation of continuous contact countercurrent processes involving simultaneous heat and mass transfer may by described as a nonlinear boundary value problem. For instance, for a continuous adiabatic gas absorption contactor unit, the model can be written in the following form:

$$\frac{dY_A}{dt} = N\left[\frac{x_A J_A}{P}\exp\left(-\frac{A_A}{T_L}\right) - \frac{Y_A}{1 + Y_A + Y_B}\right]$$

$$\frac{dY_B}{dt} = GN\left[\frac{(1 - x_A)J_B}{P}\exp\left(-\frac{A_B}{T}\right) - \frac{Y_B}{1 + Y_A + Y_B}\right]$$

$$\frac{dT_G}{dt} = HN(T_L - T_G)$$

$$\frac{dT_L}{dt} = \frac{\phi}{RC_L}\frac{dY_A}{dt} + \frac{r_{B_0} + \mu}{RC_L}\frac{dY_B}{dt} + \frac{C_G}{C_L}\frac{dT_G}{dt}$$

$$\frac{dR}{dt} = \frac{dY_A}{dt} + \frac{dY_B}{dt}$$

$$\frac{dx_A}{dt} = \frac{1}{R}\frac{dY_A}{dt} - \frac{x_A}{R}\frac{dR}{dt}$$

Thermodynamic and physical property data for the system ammonia–air–water are:

$J_A = 1.36 \times 10^{11}$ N/m²	$\phi = 1.08 \times 10^5$ J/kmol
$A_A = 4.212 \times 10^3$ K	$r_{B_0} = 1.36 \times 10^5$ J/kmol
$J_B = 6.23 \times 10^{10}$ N/m²	$\mu = 0.0$ J/kmol
$A_B = 5.003 \times 10^3$ K	$C_L = 232$ J/kmol
$G = 1.41$	$C_G = 93$ J/kmol
$H = 1.11$	$P = 10^5$ N/m²
$N = 10$	

The inlet stream conditions are:

$$Y_A(0) = 0.05 \qquad T_L(1) = 293$$
$$Y_B(0) = 0.0 \qquad R(1) = 1.0$$
$$T_G(0) = 298 \qquad x_A(1) = 0.0$$

By the Newton–Fox procedure, calculate the profiles of all dependent variables.

5. The temperature profile in a tube that is cooled inside by a cooling medium and on the external surface of which a surface exothermic reaction occurs can be described by a linear differential equation with nonlinear boundary conditions:

$$\frac{d^2\theta}{d\xi^2} + \frac{1}{\xi}\frac{d\theta}{d\xi} = 0$$

$$\xi = \xi_1: \qquad \theta = \theta_c$$

$$\xi = 1: \qquad Nu\,\theta + \frac{d\theta}{d\xi} = \frac{1}{e^{-\theta} + \mu}$$

Solve by the Newton–Fox shooting procedure for the following parameter values: $\theta_c = 0$, $\xi_1 = 0.5$, $\delta = 8$, $\mu = 0.01$, and $Nu = 10$.

6. Holt presented a very sensitive nonlinear boundary value problem from the boundary layer theory:

$$y''' = -\left(\frac{3-n}{2}\right)yy'' - ny'^2 + 1 - z^2 + sy'$$

$$z'' = -\left(\frac{3-n}{2}\right)yz' - (n-1)y'z + s(z-1)$$

$$y(0) = 0 \qquad y'(t_f) = 0$$
$$y'(0) = 0 \qquad z(t_f) = 1$$
$$z(0) = 0$$

The parameters of the problem under consideration are: $n = -0.1$, $s = 0.2$, and $t_f = 6$. To solve this problem, use the multiple shooting method with $N = 2$. For integration of the relevant initial value problems, use the Runge–Kutta–Merson routine.

References: HOLT, J.: *Commun. ACM* 7, 6, (1964).

ROBERTS, S. M., AND SHIPMAN, J. S.: *Two-Point Boundary Value Problems: Shooting Methods.* American Elsevier, 1972, p. 158.

7. Concentration and temperature profiles in a tubular nonisothermal nonadiabatic reactor are described by two nonlinear differential equations:

$$\frac{dy}{dx} = Da(1-y)\exp\left(\frac{\theta}{1+\theta/\gamma}\right)$$

$$\frac{d\theta}{dx} = Da\,B(1-y)\exp\left(\frac{\theta}{1+\theta/\gamma}\right) - \beta(\theta - \theta_c)$$

For the prescribed amount of heat transferred, the integral boundary condition results:

$$x = 0: \qquad y = 0, \qquad \beta \int_0^1 (\theta - \theta_c)\, dx = C$$

Calculate the profiles of y and θ for the parameters:
(a) $Da = 0.1$, $B = 1$, $\beta = 0.05$, $\gamma = 40$, $\theta_c = 0$, $C = 0.005$.
(b) $Da = 0.2$, $B = 5$, $\beta = 0.1$, $\gamma = 40$, $\theta_c = 0$, $C = 0.1$.
Solve this problem by the Newton–Fox method. [*Hint:* Find the inlet temperature $\theta(0)$ which satisfies the integral boundary condition by means of the Newton method.]

8. Diffusion in a biological floc can be described by

$$\frac{d^2y}{dx^2} - \frac{M^2 y}{1 + By} = 0$$

$$x = 0 \quad : \quad y = 1$$

$$x = 1 \quad : \quad y' = 0$$

Find a solution of this equation for
(a) $M = 1$, $B = 10$.
(b) $M = 4$, $B = 1$.
(c) $M = 10$, $B = 100$.

9. The two-body equations of motion are

$$x'' = -\frac{kx}{r^3}$$

$$y'' = -\frac{ky}{r^3}$$

$$z'' = -\frac{kz}{r^3}$$

where $r = \sqrt{x^2 + y^2 + z^2}$ and $k = 1$. Calculate the trajectory of x, y, and z for a three-point boundary conditions:

$$x(0) = 1 \qquad\qquad x(1) = 1.35649$$
$$y(0) = 0 \qquad\qquad y(2) = 0.17807$$
$$z(1) = 1.26011 \qquad z(2) = 1.31388$$

[Results:

t	x	y	z
0.0	1.0000	0.0000	1.0000
0.5	1.2108	0.0494	1.1613
1.0	1.3565	0.0964	1.2601
1.5	1.4485	0.1396	1.3090
2.0	1.4920	0.1781	1.3139

$x'(0) = 0.5000$, $y'(0) = 0.1000$, $z'(0) = 0.4000$.]

10. Heat and mass transfer in a porous spherical catalytic particle is described by the equation given in Example 4.33. Calculate the dependence of effectiveness factor η given by

$$\eta = \frac{3}{\phi^2}\left(\frac{dy}{dx}\right)_{x=1}$$

on the parameter ϕ for the parameter values $\gamma = 20$ and $\beta = 0.4$. Show that in the region $\phi \in (0.58, 0.72)$, three solutions occur.

11. Differential equations describing heat and mass transfer in a porous catalyst particle and incorporating the dependence of concentration and diffusion coefficient on temperature are

$$\frac{1}{\xi^2}\frac{d}{d\xi}\left[\frac{\xi^2\zeta^a}{1+\sigma Y}\frac{dY}{d\xi}\right] = \phi^2\frac{Y}{\zeta}\exp\left[\gamma\left(1 - \frac{1}{\zeta}\right)\right]$$

$$\frac{1}{\xi^2}\frac{d}{d\xi}\left[\xi^2\frac{d\zeta}{d\xi}\right] = -\beta\phi^2\frac{Y}{\zeta}\exp\left[\gamma\left(1 - \frac{1}{\zeta}\right)\right]$$

subject to the boundary conditions

$$\xi = 0: \qquad \frac{dY}{d\xi} = \frac{d\zeta}{d\xi} = 0$$

$$\xi = 1: \qquad Y = 1, \qquad \zeta = 1$$

Show that an algebraical equation between both differential equations exists:

$$\zeta = \left[1 + (1-a)\frac{\beta}{\sigma}\ell n\left(\frac{1+\sigma}{1+\sigma Y}\right)\right]^{\frac{1}{1-a}}, \qquad a \neq 1$$

$$\zeta = \left(\frac{1+\sigma}{1+\sigma Y}\right)^{\beta/\sigma} \qquad\qquad a = 1$$

Find the profiles of Y and ζ for parameter values $a = 1$, $\sigma = 0.1$, $\phi^2 = 0.5$. $\beta = 0.2$, and $\gamma = 20$. (Hint: Combine both differential equations in such a manner as to eliminate the nonlinear right-hand side. The resulting equation is then analytically integrated.)

12. The concentration (f) and temperature (ϑ) profiles in a tubular reactor are described by two nonlinear differential equations:

$$\frac{d\vartheta}{dx} = \frac{1}{a}f\exp\left(\frac{\vartheta}{1+\epsilon\vartheta}\right) - \vartheta$$

$$\frac{df}{dx} = -\frac{1}{aB}f\exp\left(\frac{\vartheta}{1+\epsilon\vartheta}\right)$$

The initial conditions are

$$x = 0: \qquad f = f_0, \qquad \vartheta = \vartheta_0$$

Calculate the inlet dimensionless concentration f_0 if the following specifications are given: $B = 10$, $a = 1.40$, $\epsilon = 0.0$, $\vartheta_0 = 0.0$, and $\max_{x \in (0,5)} \vartheta(x) = 2.0$.

BIBLIOGRAPHY

A systematic treatment of shooting methods for NBVP making use of the method of adjoints and the multiple shooting technique has been presented in:

ROBERTS, S. M., AND SHIPMAN, J. S.: *Two-Point Boundary Value Problems: Shooting Methods*. American Elsevier, New York, 1972.

A comprehensive paper dealing with various aspects of shooting methods is:

OSBORNE, M. R.: On shooting methods for boundary value problems. *J. Math. Anal. Appl.* 27, 417–433 (1969).

An application of a derivative-free shooting method is:

WARNER, F. J.: On the solution of "jury problems" with many degrees of freedom. *Math. Tables and Other Aids Comput.* 11, 268–271 (1957).

A method of adjoints has been published in:

GOODMAN, T. R., AND LANCE, G. N.: The numerical integration of two-point boundary value problems. *Math. Tables Other Aids Comput.* 10, 82–86 (1956).

The shooting technique using Newton's method with auxiliary equations has been presented in:

FOX, L., AND LANGER, R. E., EDS.: *Boundary Value Problems in Differential Equations*. University of Wisconsin Press, Madison, Wis., 1960.

FOX, L., ED.: *Numerical Solution of Ordinary and Partial Differential Equations*. Pergamon Press, Elmsford, N.Y., 1962.

MCGINNIS, P. H., JR.: Numerical solution of boundary value nonlinear ordinary differential equations. *Chem. Eng. Prog. Symp. Ser. No. 55, 61*, 1–7 (1965).

For the shooting technique using third-order correction algorithms, see:

KUBÍČEK, M., AND HLAVÁČEK, V.: Method of third order convergence for solution of nonlinear two-point boundary value problems. Part I: *Chem. Eng. Sci. 25*, 1833–1836 (1970); Part II, *ibid. 26*, 321–324 (1971).

Solution of a highly sensitive NBVP by the multiple shooting technique has been suggested in:

MORRISON, D. D., RILEY, J. D., AND ZANCANARO, J. F.: Multiple shooting method for two-point boundary value problems. *Commun. ACM 5*, 613–614 (1962).

A good survey of methods (mostly of the shooting type) for nonlinear boundary value problems is:

KELLER, H. B.: Numerical solution of two-point boundary value problems. *SIAM Reg. Conf. Ser. Appl. Math. No.* 24, Philadelphia, 1976.

A survey paper on nonlinear boundary value problems is:

DANIEL, J. W.: A road map of methods for approximating solutions of two-point boundary value problems, in *Codes for Boundary Value Problems in Ordinary Differential Equations* (Ed. B. Childs et al.), Lecture Notes in Computer Science. Springer-Verlag, Berlin, 1978.

Shooting codes for nonlinear boundary value problems:

GLADWELL, I.: Shooting codes in the NAG library; also SCOTT, M. L., AND WATTS, H. A.: Superposition, orthonormalization, quasilinearization and two-point boundary value problems; in Eds. B. Childs et al, Lecture Notes in Computer Science 76. Springer-Verlag, New York, 1979.

DIEKOFF, H. J. ET AL: Comparing routines for the numerical solution of initial value problems of ordinary differential equations in multiple shooting, *Numer. Math 27*, 449 (1977).

Numerical Realization **5**
of Parametric Studies
in Nonlinear Boundary
Value Problems

In physical and engineering practice the need for repeated calculations for a sequence of values of governing parameters is great. Since the values of these parameters are evaluated from experiments or calculated from correlations, we cannot determine the exact values of the parameters. As a result, we perform our calculations with the expected value; however, we should also investigate the solution for parameter values occurring in the confidence region. Sometimes the physical problem requires us to construct the dependence of solution on certain parameters.

Sections 5.1 to 5.5 are devoted to the methodology of calculation of the dependence of a solution of a NBVP on a parameter. In Section 5.1 the specific features of certain problems are utilized to perform appropriate transformations which may lead to explicit relations. Section 5.2 is devoted to a sequential use of standard methods for a sequence of values of the parameter in question. Sections 5.3 and 5.4 are developed on the base of differentiation of original equations with respect to parameter or boundary condition. The shortcomings of these methods are also discussed. In Section 5.5 a general approach to evaluation of the dependence of a solution of a NBVP on a parameter is developed which makes use of marching procedures. So far this method makes it possible to calculate a number of difficult problems and may be recommended for problems that can be integrated by marching. Nonlinear boundary value problems may give rise to multiple solutions and as a result evaluation of branching points is necessary. In Section 5.6 certain methods are described which allow us to calculate the branching points in a straightforward way. These methods are based on the GPM concept.

5.1 Conversion of Boundary Value Problems to Initial Value Problems and Parameter Mapping Techniques

The general concept of this technique is to transform the original nonlinear boundary value problem into an initial value problem which can be easily solved by any numerical marching integration routine. Evidently, this transformation makes it possible to solve the nonlinear boundary value problem in a straightforward way without any trial-and-error work. It is, of course, of interest to be able to convert a boundary value problem to an initial value problem; unfortunately, there is no general standard transformation that is capable of doing it. For a certain group of NBVPs the appropriate transformations can be found which are able to transform the boundary value problems into initial value problems; however, it is very difficult to find an algorithm. We can note that this technique is advantageous for performing a parametric study.

The conversion of boundary value to initial value problems is a problem with an old history. At the beginning of this century Blasius converted the nonlinear two-point boundary value problem

$$y''' + yy'' = 0, \qquad y(0) = y'(0) = 0, \qquad y'(\infty) = 2$$

into the relevant initial value problem. Recently, this problem has been extensively studied by Na (1968), and Klamkin (1962, 1970). To illustrate the application of this method, an example will be solved.

Example 5.1

Examination of the steady-state two-dimensional boundary layer flow of a power-law fluid past a flat plate and the following similarity transformation yield a nonlinear boundary value problem

$$y''' + y(y'')^{2-\alpha} = 0 \tag{5.1}$$

subject to the boundary conditions

$$y(0) = 0, \qquad y'(0) = 0 \tag{5.2a}$$

$$y'(\infty) = 1 \tag{5.2b}$$

Consider the new variables $\bar{y}$ and $\bar{x}$:

$$y = A^{\alpha_1}\bar{y} \tag{5.3}$$

$$x = A^{\alpha_2}\bar{x} \tag{5.4}$$

where A, α_1, and α_2 are constants to be determined. After substituting (5.3) and (5.4) into (5.1), we obtain

$$\bar{y}''' + \bar{y}(\bar{y}'')^{2-\alpha}A^{\alpha_1(2-\alpha)+\alpha_2(2\alpha-1)} = 0 \tag{5.5}$$

If we choose

$$\alpha_1(2 - \alpha) + \alpha_2(2\alpha - 1) = 0 \tag{5.6}$$

(5.5) becomes independent on A:

$$\bar{y}''' + \bar{y}(\bar{y}'')^{2-\alpha} = 0 \tag{5.7}$$

The boundary conditions (5.2a) are

$$\bar{y}(0) = 0, \qquad \bar{y}'(0) = 0 \tag{5.8}$$

To determine the unknown initial condition $y''(0)$, we set

$$y''(0) = A \tag{5.9}$$

On rewriting (5.9) in the new variables, we get

$$A^{\alpha_1-2\alpha_2}\bar{y}''(0) = A \tag{5.10}$$

By letting

$$\alpha_1 - 2\alpha_2 = 1 \tag{5.11}$$

(5.10) yields

$$\bar{y}''(0) = 1 \tag{5.12}$$

The parameters α_1 and α_2 can be evaluated after solving (5.6) and (5.11):

$$\alpha_1 = \frac{2\alpha - 1}{3}, \qquad \alpha_2 = \frac{\alpha - 2}{3} \tag{5.13}$$

Evidently, we have transformed the original BVP, (5.1)–(5.3), into a relevant initial value problem given by (5.7) and initial conditions (5.8) and (5.12). To establish the unknown constant A, we make use of the terminal asymptotic condition (5.2b), which after introducing the new variables gives rise to

$$A^{\alpha_1-\alpha_2}\bar{y}'(\infty) = 1 \tag{5.14}$$

On substituting for α_1 and α_2 from (5.13), we finally get

$$A = [\bar{y}'(\infty)]^{-(\alpha+1)/3} \tag{5.15}$$

To integrate the transformed problem, the following steps must be performed:

1. The calculation is done for a given value of α.
2. Integrate the transformed problem, (5.7), with the initial conditions (5.8) and (5.12) as long as $\bar{y}'$ reaches its asymptotic value $\bar{y}'(\infty)$. The calculation can be performed by any marching technique (e.g., by the Runge–Kutta method).
3. Determine A from (5.15). Now the calculated profile $\bar{y}(\bar{x})$ can be easily transformed into $y(x)$ making use of (5.3) and (5.4) because the values of A, α_1, and α_2 are already known.
4. Repeat steps 2 and 3 for the next value of α.

The calculated results are tabulated for various values of α in Table 5-1.

TABLE 5-1
PARAMETRIC STUDY OF THE BLASIUS EQUATION

α	$y''(0)$	α	$y''(0)$
0.1	0.11711	1.1	0.50511
0.2	0.15483	1.2	0.53533
0.3	0.19487	1.3	0.56556
0.4	0.23612	1.4	0.59629
0.5	0.27790	1.5	0.61885
0.6	0.31901	1.6	0.64319
0.7	0.35913	1.7	0.66606
0.8	0.39779	1.8	0.68575
0.9	0.43502	1.9	0.70781
1.0	0.47032	2.0	0.72757

Source: Results taken from the paper by S. H. Lin and L. T. Fan, *AICHE J. 18*, 654 (1972).

We have noted that we can calculate the problem for a predetermined value of the governing parameter α. Sometimes, however, we need to establish a dependence between parameters, a dependence of the dependent variables on the values of governing parameters, and so on. There are physical problems where we can perform a transformation of the boundary value problem to an initial value problem; however, the value of the parameter under question is the result of integration. Hence we cannot determine the solution of the NBVP for the preassigned values of the parameter, but we are able to calculate the interrelations between parameters. The following example illustrates this situation.

Example 5.2

Axial mass dispersion in an isothermal tubular reactor for an nth-order reaction is described by

$$\frac{1}{\text{Pe}}y'' - y' - ky^n = 0 \tag{5.16}$$

$$x = 0: \quad \frac{1}{\text{Pe}}y' = y - 1 \tag{5.17a}$$

$$x = 1: \quad y' = 0 \tag{5.17b}$$

We define the new dependent and independent variables by (5.3) and (5.4):

$$y = A^{\alpha_1}\bar{y} \tag{5.3}$$

$$x = A^{\alpha_2}\bar{x} \tag{5.4}$$

The differential equation (5.16) is transformed to

$$\frac{1}{\text{Pe}}\frac{d^2\bar{y}}{d\bar{x}^2} - A^{\alpha_2}\frac{d\bar{y}}{d\bar{x}} - kA^{\alpha_1(n-1)+2\alpha_2}\bar{y}^n = 0 \tag{5.18}$$

To guarantee the independence of (5.18) of A, we let

$$\alpha_2 = 0 \qquad \text{(i.e., } \bar{x} = x\text{)} \tag{5.19}$$

To eliminate A from (5.18), we let

$$\bar{k} = kA^{\alpha_1(n-1)} \tag{5.20}$$

Obviously, after this transformation (5.18) yields

$$\frac{1}{\text{Pe}} \frac{d^2\bar{y}}{dx^2} - \frac{d\bar{y}}{dx} - \bar{k}\bar{y}^n = 0 \tag{5.21}$$

The boundary condition (5.17b) becomes

$$\frac{d\bar{y}(1)}{dx} = 0 \tag{5.22}$$

For the missing terminal condition, we let

$$y(1) = A \tag{5.23}$$

On comparing with (5.3), we have

$$\bar{y}(1) = A^{-\alpha_1}y(1) = A^{1-\alpha_1}$$

To have the value $\bar{y}(1)$ independent of A, we let $\alpha_1 = 1$, that is,

$$\bar{y}(1) = 1 \tag{5.24}$$

The initial conditions (5.22) and (5.24) are sufficient to integrate (5.21) from $x = 1$ to $x = 0$ by a marching procedure (e.g., the Runge–Kutta method). The boundary condition (5.17a) after transformation yields

$$\frac{1}{\text{Pe}} \frac{dy(0)}{dx} - y(0) + 1 = \frac{A}{\text{Pe}} \frac{d\bar{y}(0)}{dx} - A\bar{y}(0) + 1 = 0$$

For the unknown value of A, we get an explicit relationship:

$$A = \left[\bar{y}(0) - \frac{1}{\text{Pe}} \frac{d\bar{y}(0)}{dx} \right]^{-1} \tag{5.25}$$

The procedure to solve the transformed problem is as follows:

1. Choose a value of $\bar{k}$.
2. Integrate (5.21) with the initial conditions (5.22) and (5.24) from $x = 1$ to $x = 0$. The profile $\bar{y}(x)$ is calculated.
3. Making use of calculated values $\bar{y}(0)$ and $\bar{y}'(0)$, the parameter A may be easily determined from (5.25).
4. Expression (5.20) yields

$$k = \bar{k}A^{1-n} \tag{5.26}$$

5. Using (5.3), the profile $y(x)$ can be readily calculated:

$$y(x) = A\bar{y}(x) \tag{5.27}$$

6. Repeat steps 2 through 5 for a different value of $\bar{k}$. The results for $n = \frac{1}{2}$ and $n = 2$ are reported in Tables 5-2 and 5-3.

TABLE 5-2

TABLE 5-2
EXAMPLE 5.2: RESULTS FOR $n = 0.5$, Pe $= 5$

$\bar{k}$	1.0000	2.0000	3.0000	6.0000	10.0000
$\bar{y}(0)$	1.9062	3.0103	4.2999	9.2096	18.0295
$\bar{y}'(0)$	−1.2839	−3.1102	−5.4424	−15.2825	−34.6610
$A = y(1)$	0.4623	0.2753	0.1856	0.0815	0.0401
k	0.6799	1.0494	1.2924	1.7132	2.0015
$y(0.8)$	0.4966	0.3164	0.2275	0.1191	0.0717
$y(0.6)$	0.5702	0.4073	0.3224	0.2103	0.1544
$y(0.4)$	0.6627	0.5264	0.4516	0.3456	0.2874
$y(0.2)$	0.7671	0.6674	0.6104	0.5252	0.4753
$y(0.0)$	0.8813	0.8287	0.7980	0.7508	0.7223

k	$y(1)$	$y(0)$
0.0953	0.9077	0.9814
0.1821	0.8287	0.9651
0.4027	0.6487	0.9259
0.8868	0.3495	0.8510
1.1817	0.2234	0.8117
1.3868	0.1570	0.7868
1.4686	0.1348	0.7774
1.8791	0.0552	0.7341

TABLE 5-3
EXAMPLE 5.2: RESULTS FOR
$n = 2$, Pe $= 5$

k	$y(1)$	$y(0)$
0.1107	0.9032	0.9797
0.2461	0.8125	0.9584
0.4122	0.7279	0.9362
0.6163	0.6490	0.9130
0.8685	0.5757	0.8887
1.1815	0.5078	0.8634
1.5723	0.4452	0.8370
2.0637	0.3876	0.8095
2.6868	0.3350	0.7809
3.4840	0.2870	0.7510
5.8653	0.2046	0.6877
10.0789	0.1389	0.6193
24.8583	0.0684	0.5068

Evidently, this method makes it possible to calculate in the straightforward way the dependence of $y(1)$ [and $y(x)$] on k for a given value of Peclet number. The approach usually adopted would require us to solve iteratively the NBVP for a given k. It is apparent from the examples calculated above that the transformation to an initial value problem is a very powerful technique, especially if a parametric study should be performed.

To formulate the aforementioned principles in a more general way, consider the ordinary differential equation

$$F\left(\frac{d^2y}{dx^2}, \frac{dy}{dx}, y, x\right) = 0 \tag{5.28}$$

subject to the boundary conditions

$$y(0) = 0, \qquad y(x_f) = a \tag{5.29}$$

To transform (5.28), the following transformations of dependent and independent variables will be considered:

$$x = f(\bar{x}, A, \alpha_1, \alpha_2) \tag{5.30}$$

$$y = g(\bar{y}, A, \alpha_1, \alpha_2) \tag{5.31}$$

Here α_1 and α_2 are parameters that must be established before the transformed differential equation is integrated by a marching technique, while A is a constant that is determined a posteriori (i.e., after integration). To calculate the values of α_1 and α_2, two conditions are imposed:

1. The differential equation in the new variables must be independent of A.
2. The transformed value of the missing initial condition $d\bar{y}(0)/d\bar{x}$ must be independent of A for a reasonable choice of the function $dy(0)/dx = f(A)$.

Evidently, if the parameters α_1 and α_2 meet these two conditions, the transformed differential equation

$$G\left(\frac{d^2\bar{y}}{d\bar{x}^2}, \frac{d\bar{y}}{d\bar{x}}, \bar{y}, \bar{x}\right) = 0 \tag{5.32}$$

can be integrated by marching routines as an IVP with the initial conditions

$$\bar{y}(0) = 0, \quad \frac{d\bar{y}(0)}{d\bar{x}} = b \tag{5.33}$$

Here b results from the condition 2. The solution of the problem is $\bar{y} = h(\bar{x})$. The unknown value of the parameter A is sought as a solution of the set of equations

$$\bar{y} = h(\bar{x}), \qquad x_f = f(\bar{x}, A, \alpha_1, \alpha_2), \qquad a = g(\bar{y}, A, \alpha_1, \alpha_2) \tag{5.34}$$

Obviously the last two equations result from the terminal boundary condition at $x = x_f$. The method fails if we are not capable of determining any value of A.

Recently, Klamkin (1970) has devised a method for converting a boundary value problem of the type

$$A_{mnrs}(y'')^m(y')^n y^r x^s = 0 \tag{5.35}$$

to the corresponding initial value problem. The NBVP (5.35) is subjected to linear boundary conditions

$$y'(0) = ay(0) + b, \qquad y^{(e)}(L) = k \tag{5-36}$$

Here m, n, r, and s are arbitrary indices, e is an arbitrary integer, and L is the length of the region (can also be ∞).

Consider a transformation

$$y = \lambda F(\mu x) \tag{5.37}$$

where $F(x)$ satisfies (5.35) for the initial conditions

$$F(0) = F'(0) = 1 \tag{5.38}$$

In order that both y and $F(x)$ meet (5.35), the differential equation must be invariant under the two-parameter group of homogeneous transformations

$$x_1 = \mu x, \qquad y_1 = y/\lambda$$

After inserting (5.37) into (5.35), we get

$$c = m + n + r, \qquad d = 2m + n - s \tag{5.39}$$

Of course, the constants c and d must be constant for all set of indices m, n, r, s. On inserting (5.37) into (5.36), we have [$y(0) = \lambda$]

$$y'(0) = a\lambda + b = \lambda\mu \tag{5.40}$$

$$\lambda\mu^e F^{(e)}(\mu L) = k \tag{5.41}$$

The algorithm consists of integrating (5.35) for $y = F(x)$ subject to (5.38). The goal of this integration is to determine $F^{(e)}(\mu L)$. On inserting this value into (5.41), a set of two equations for λ and μ results which can be easily solved for $L = \infty$. For finite values of L, these equations must be solved by trial and error. The reader who is interested in a detailed treatment of these methods is referred to Klamkin's and Na's papers.

A very powerful tool for performing a parametric study is a method that has been referred to as a parametric mapping method. The spirit of this method will be illustrated for a single second-order differential equation (two-point boundary value problem)

$$\frac{d^2 y}{dx^2} = f\left(x, y, \frac{dy}{dx}, \alpha\right) \tag{5.42}$$

subject to the boundary conditions

$$G_0[y(x_0), y'(x_0), \alpha] = 0 \tag{5.43}$$

$$G_1[y(x_1), y'(x_1), \alpha] = 0 \tag{5.44}$$

Here α is a parameter.

Suppose that we are able to find substitutions

$$\xi = \xi(x, \alpha) \tag{5.45}$$

and

$$Y = Y(x, y, \alpha) \tag{5.46}$$

which transform (5.42)–(5.44) into

$$\frac{d^2Y}{d\xi^2} = F\left(\xi, Y, \frac{dY}{d\xi}\right) \tag{5.47}$$

subject to the boundary conditions

$$\bar{G}_0[Y(\xi_0), \ Y'(\xi_0)] = 0 \tag{5.48}$$

$$\bar{G}_1[Y(\xi_1), \ Y'(\xi_1), \alpha] = 0 \tag{5.49}$$

Here the coordinate ξ_0 is not dependent on α, while ξ_1 can be a function of α. Supposing that such transformation can be found, the following procedure (parameter mapping) may be used:

1. For a reasonable value of $Y(\xi_0)$ or $Y'(\xi_0)$, (5.48) makes it possible to calculate $Y'(\xi_0)$ or $Y(\xi_0)$, respectively.
2. The initial value problem (5.47) can be integrated with the initial conditions calculated in step 1 as long as the second boundary condition (5.49) is satisfied. This gives rise to a value of α.
3. Making use of substitutions (5.45) and (5.46), $y(x_0)$ and $y'(x_0)$ can be determined from (5.43). The profile $Y(\xi)$ can be recalculated in $y(x)$ or the initial value problem for (5.42) can be integrated with the already calculated initial conditions $y(x_0)$ and $y'(x_0)$ for a given value of α.

Let us note that the value of α cannot be chosen a priori; this value is the result of computation. To apply this method it is necessary that (5.47) and one boundary condition do not contain the parameter α.

Obviously, the parameter mapping technique is a very versatile concept which can be successfully used for a number of physical problems (e.g., for the diffusion-like differential equations). To illustrate the parameter mapping method, three examples are presented.

Example 5.3

The heat and mass transfer in a porous catalyst is described by

$$\frac{d^2y}{dx^2} + \frac{a}{x}\frac{dy}{dx} = \phi^2 y \exp\left[\frac{\gamma\beta(1-y)}{1+\beta(1-y)}\right] \tag{5.50}$$

subject to the boundary conditions

$$x = 0: \quad \frac{dy}{dx} = 0$$

$$x = 1: \quad y = 1 \tag{5.51}$$

On introducing a new independent variable

$$\xi = \phi x \tag{5.52}$$

we obtain the following equation (clearly, the new dependent variable Y is identical with y, i.e., $Y = y$)

$$\frac{d^2 Y}{d\xi^2} + \frac{a}{\xi} \frac{dY}{d\xi} = Y \exp\left[\frac{\gamma\beta(1 - Y)}{1 + \beta(1 - Y)}\right] \tag{5.53}$$

subject to the boundary conditions

$$\begin{aligned} \xi = 0: \quad & \frac{dY}{d\xi} = 0 \\ \xi = \phi: \quad & Y = 1 \end{aligned} \tag{5.54}$$

Guessing the missing boundary condition

$$Y(0) = Y_0 \tag{5.55}$$

we can integrate the relevant initial value problem given by (5.53) with initial conditions

$$\xi = 0: \quad Y = Y_0, \quad \frac{dY}{d\xi} = 0 \tag{5.56}$$

The point ξ_1 where

$$Y(\xi_1) = 1 \tag{5.57}$$

determines the value of the parameter ϕ because

$$\xi_1 = \phi$$

To satisfy the terminal condition (5.57), a trial-and-error approach for determination of ξ_1 must be adopted. However, a straightforward algorithm can be constructed which is based on the interchange of dependent and independent variables.

For purposes of numerical calculation, (5.53) can be rewritten

$$\frac{dY}{d\xi} = g, \quad \frac{dg}{d\xi} = -\frac{a}{\xi} g + Y \exp\left[\frac{\gamma\beta(1 - Y)}{1 + \beta(1 - Y)}\right] \tag{5.58}$$

Starting from a certain point $\bar{\xi} > 0$, where $Y = \bar{Y} < 1$ and $dY/d\xi = \bar{Y}'$, it is possible to integrate instead of differential equation (5.58) a new set of differential equations

$$\begin{aligned} \frac{dg}{dY} &= -\frac{a}{\xi} + \frac{Y}{g} \exp\left[\frac{\gamma\beta(1 - Y)}{1 + \beta(1 - Y)}\right] \\ \frac{d\xi}{dY} &= \frac{1}{g} \end{aligned} \tag{5.59}$$

from $Y = \bar{Y}: g = \bar{Y}', \xi = \bar{\xi}$ to $Y = 1$, where $\xi(Y = 1) = \phi$. The results calculated for $\beta = 0.2, 0.3$ and 0.4 are reported in Table 5-4. To integrate the relevant initial value problem, the Runge–Kutta–Merson method has been used. After inspection of the results it is clear that (5.50) exhibits multiple solutions. This procedure can also be used for problems with boundary conditions of the third kind, that is, for

$$x = 1: \quad y = 1 - \text{Sh} \frac{dy}{dx} \tag{5.60}$$

	ϕ		
$y(0)$	$\beta = 0.2$	$\beta = 0.3$	$\beta = 0.4$
0.98	0.27660	0.27252	0.26852
0.95	0.42222	0.40803	0.39354
0.85	0.66088	0.59534	0.53795
0.75	0.77614	0.65723	0.56082
0.65	0.84326	0.67465	0.54708
0.55	0.88693	0.67324	0.52144
0.45	0.92012	0.66488	0.49382
0.35	0.95209	0.65656	0.46901
0.25	0.99279	0.65418	0.45029
0.15	1.0607	0.66712	0.44244
0.05	1.2334	0.73358	0.46563
0.01	1.5213	0.87007	0.53558
0.001	1.9579	1.0908	0.65775
0.0001	2.4028	1.3206	0.78780

[a]Mapping of the parameter in dependence on
$Y_0 = y(0)$

Example 5.4

Troesch's equation, which describes confinement of a plasma column by radiation pressure, has the form

$$\frac{d^2 y}{dx^2} = n \sinh ny \tag{5.61}$$

subject to the boundary conditions

$$y(0) = 0, \qquad y(1) = 1 \tag{5.62}$$

After substitution of

$$Y = ny, \qquad \xi = nx \tag{5.63}$$

the parameter n can be eliminated from (5.61),

$$\frac{d^2 Y}{d\xi^2} = \sinh Y \tag{5.64}$$

However, it appears in the boundary conditions

$$Y(0), \qquad Y(n) = n \tag{5.65}$$

The parameter mapping technique is very simple; on choosing the missing initial condition

$$\frac{dY(0)}{d\xi} = \eta \qquad \left[= \frac{dy(0)}{dx} \right] \tag{5.66}$$

(5.64) is integrated using the initial conditions (5.66) and $Y(0) = 0$ until

$$Y(n) = n \tag{5.67}$$

One has to iterate to satisfy (5.67) within a preassigned tolerance ϵ; for example, if $Y(n) > n$, the integration procedure returns to the preceding value and the integration step is halved. This simple bisection procedure is repeated as long as the difference $|Y(n) - n| > \epsilon$. However, a straightforward method of satisfying (5.67) exists. Starting from a certain value $\bar{\xi}$, where $dY(\bar{\xi})/d\xi > 1$, one can switch from the integration of

$$\frac{dY}{d\xi} = z, \qquad \frac{dz}{d\xi} = \sinh Y$$

$$Y(0) = 0, \qquad z(0) = \eta \tag{5.68}$$

which yields $z(\bar{\xi}) = \bar{z}$ and $Y(\bar{\xi}) = \bar{Y}$, to a new system

$$\frac{d\xi}{d\varphi} = \frac{1}{z-1}, \qquad \frac{dY}{d\varphi} = \frac{z}{z-1}, \qquad \frac{dz}{d\varphi} = \frac{\sinh Y}{z-1} \tag{5.69}$$

subject to the initial conditions

$$\varphi = \bar{Y} - \bar{\xi}: \qquad \xi = \bar{\xi}, \qquad Y = \bar{Y}, \qquad z = \bar{z} \tag{5.70}$$

Clearly, the value of Y for $\varphi = 0$ is the unknown value of the parameter n. For illustration purposes, a course of one integration is presented in Table 5-5. It is

TABLE 5-5
COURSE OF ONE INTEGRATION IN EXAMPLE 5.4:
NONITERATIVE INTEGRATION

ξ	Y	$z = \dfrac{dY}{d\xi}$
0.0000	0.0000	0.0010
3.5933	0.0182	0.0182
6.0402	0.2102	0.2106
7.0135	0.5594	0.5667
7.7081	1.1433	1.2065
8.1071	1.7653	2.0036
8.5071	2.8918	4.0101
8.7886	4.6259	10.0053

φ	ξ	Y	z
−4.1627	8.7886	4.6259	10.005
−3.6010	8.8414	5.2404	13.666
−2.8809	8.8879	6.0070	20.107
−2.0032	8.9244	6.9212	31.804
−1.0061	8.9495	7.9435	53.057
0.0000	8.9646	8.9646	88.426

obvious that this algorithm for a given value of η calculates a posteriori the value of n. The dependence of n calculated for a sequence of η is presented in Table 5-6. The integration was performed in double-precision arithmetic (~ 15 significant digits) using the Runge–Kutta–Merson marching integration technique. It is very simple to interpolate in Table 5-6 if one needs to know the value $y'(0)$ for a given value of n.

$\eta = y'(0)$	n	$\eta = y'(0)$	n
0.9	0.792	0.000356	10.01
0.8	1.151	0.0001	11.28
0.6	1.753	$1 \cdot 10^{-5}$	13.59
0.4	2.394	$1 \cdot 10^{-6}$	15.89
0.2	3.308	$1 \cdot 10^{-7}$	18.20
0.1	4.129	$1 \cdot 10^{-8}$	20.50
0.0457	5.000	$1 \cdot 10^{-9}$	22.80
0.01	6.611	$1 \cdot 10^{-10}$	25.10
0.003	7.849	$1 \cdot 10^{-11}$	27.41
0.001	8.965	$1 \cdot 10^{-12}$	29.71

Example 5.5

Temperature and conversion profiles in a nonisothermal-nonadiabatic tubular reactor with recycle are described by two first-order differential equations subject to mixed boundary conditions:

$$\frac{dy}{ds} = \text{Da} \, (1 - y) \exp\left(\frac{\theta}{1 + \epsilon\theta}\right) \tag{5.71}$$

$$\frac{d\theta}{ds} = \text{Da} \, B(1 - y) \exp\left(\frac{\theta}{1 + \epsilon\theta}\right) - \beta(\theta - \theta_c) \tag{5.72}$$

$$(1 - \mu)y(1) = y(0), \qquad (1 - \lambda)\theta(1) = \theta(0) \tag{5.73}$$

For the adiabatic case ($\beta = 0$) and for $\mu = \lambda$, a combination of (5.71) and (5.72) yields

$$\theta = By \tag{5.74}$$

Hence (5.72) can be rewritten

$$\frac{d\theta}{ds} = \text{Da} \, B\left(1 - \frac{\theta}{B}\right) \exp\left(\frac{\theta}{1 + \epsilon\theta}\right) \tag{5.75}$$

Application of the parameter mapping method will be illustrated on the three different situations:

1. Parametric mapping of the dependence $\theta(0)$ versus λ. For fixed values of B, Da, and ϵ, the procedure is as follows:
   (a) Guess $\theta(0)$, $0 < \theta(0) < B$.
   (b) Integrate the initial value problem (5.75) from $s = 0$ to $s = 1$.
   (c) Calculate λ, from (5.73) $\lambda = 1 - \theta(0)/\theta(1)$.
   (d) Repeat steps 1 through 3 for a different guess of $\theta(0)$. The results are shown in Fig. 5-1. We can note that for some values of the parameter λ, three solutions exist. The multiple profiles calculated for Da $= 0.0625$ and $\lambda = 0.718$ are drawn in Fig. 5-2.
2. Parameter mapping of the dependence $\theta(0)$ versus Da. For fixed values of B, ϵ, and λ, the procedure is as follows:
   (a) For the transformation

$$t = s \, \text{Da} \tag{5.76}$$

   (5.75) yields

$$\frac{d\theta}{dt} = B\left(1 - \frac{\theta}{B}\right) \exp\left(\frac{\theta}{1 + \epsilon\theta}\right) \tag{5.77}$$

   (b) Guess $\theta(0)$, $0 < \theta(0) < B$.

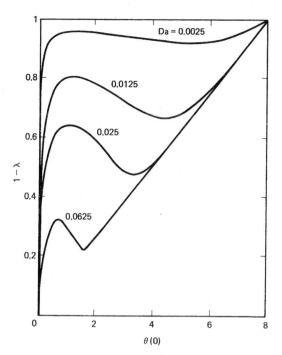

Figure 5-1 Dependence of $(1 - \lambda)$ on $\theta(0)$. $B = 8$, $\epsilon = 0.05$: Example 5.5.

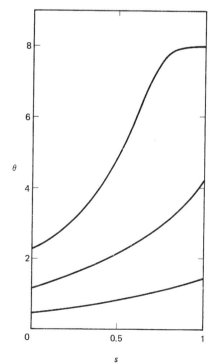

Figure 5-2 Three resulting profiles for $B = 8$, $\epsilon = 0.05$, Da = 0.0625, $\lambda = 0.718$: Example 5.5.

(c) Integrate (5.77) from $t = 0$ to the point $\bar{t}$ where

$$\theta(\bar{t}) = \frac{\theta(0)}{1 - \lambda} \tag{5.78}$$

To avoid iterations, (5.79) can be integrated

$$\frac{dt}{d\theta} = \left[B\left(1 - \frac{\theta}{B}\right) \exp\left(\frac{\theta}{1 + \epsilon\theta}\right) \right]^{-1} \tag{5.79}$$

instead of (5.77). The initial condition is $\theta = \theta(0)$: $t = 0$. Equation (5.79) can also be solved easily on the interval $\theta \in \langle\theta(0), \theta(0)/(1 - \lambda)\rangle$ by quadrature formulas because the right-hand side of (5.79) does not depend on t.

(d) The solution of (5.79) gives

$$\text{Da} = \bar{t} = t\left[\frac{\theta(0)}{1 - \lambda}\right]$$

(e) Repeat steps 1 through 4 for another guess of $\theta(0)$. In Fig. 5-3 a dependence $\theta(0) = f(\text{Da})$, which was calculated in this way, is presented.

3. Parameter mapping of the dependence $\theta(1)$ versus $\theta(0)$. For the sake of brevity let us denote $\theta_0 = \theta(0)$, $\theta_1 = \theta(1)$. Equation (5.75) gives rise after simple rearrangement to

$$\frac{d\theta_1}{d\theta_0} = \frac{B - \theta_1}{B - \theta_0} \exp\left(\frac{\theta_1}{1 + \epsilon\theta_1} - \frac{\theta_0}{1 + \epsilon\theta_0}\right) \tag{5.80}$$

because $d\theta(0)/ds \neq 0$. For given values of B, Da, and ϵ, the dependence $\theta_1(\theta_0)$ can be calculated by integration of (5.80) with the initial condition

$$\theta_0 = 0: \qquad \theta_1 = \theta_1^0$$

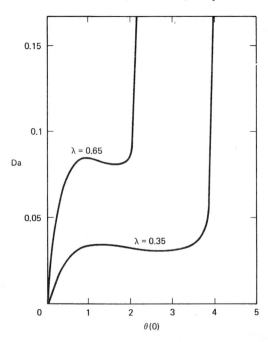

Figure 5-3 Dependence of $\theta(0)$ on Da; $B = 6$, $\epsilon = 0.05$: Example 5.5.

This initial condition can be calculated from (5.75). On integration of (5.75) with the initial condition $\theta(0) = 0$, a terminal condition $\theta_1^0 = \theta(1)$ results.

A number of these dependences calculated for different values of Da are drawn in Fig. 5-4. Evidently, the boundary condition (5.73) is the straight line

$$\theta_1 = \frac{\theta_0}{1 - \lambda} \tag{5.81}$$

An intersection of this line with the curve $\theta_1(\theta_0)$ yields the values θ_1 and θ_0 for given values of the parameters λ and Da.

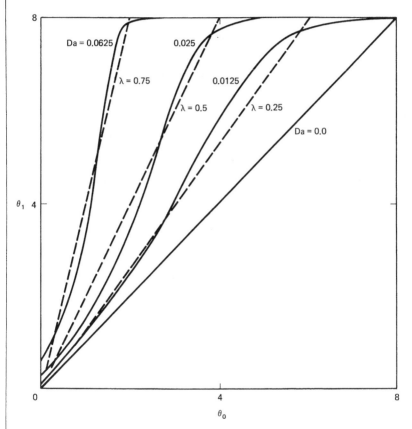

Figure 5-4 Dependence of θ_1 on θ_0; $B = 8$, $\epsilon = 0.05$: Example 5.5.

Number of solutions:

Da	$\lambda = 0.75$	$\lambda = 0.5$	$\lambda = 0.25$
0.0125	1	1	3
0.025	1	3	1
0.0625	3	1	1

Example 5.6

Axial heat and mass transfer in a tubular reactor is described by

$$\frac{1}{\text{Pe}_\theta} \frac{d^2\theta}{dx^2} - \frac{d\theta}{dx} + B \, \text{Da} \, (1 - y) \exp\left(\frac{\theta}{1 + \epsilon\theta}\right) = 0 \tag{5.82}$$

$$\frac{1}{\text{Pe}_y} \frac{d^2y}{dx^2} - \frac{dy}{dx} + \text{Da} \, (1 - y) \exp\left(\frac{\theta}{1 + \epsilon\theta}\right) = 0 \tag{5.83}$$

subject to the boundary conditions

$$x = 0: \quad \text{Pe}_\theta \theta = \frac{d\theta}{dx}, \quad \text{Pe}_y \, y = \frac{dy}{dx} \tag{5.84}$$

$$x = 1: \quad \frac{d\theta}{dx} = \frac{dy}{dx} = 0 \tag{5.85}$$

1. First, the case $\text{Pe}_\theta = \text{Pe}_y = \text{Pe}$ will be considered. After combination of (5.82) and (5.83) a simple invariant can be developed:

$$\theta = By \tag{5.86}$$

Hence it is sufficient to integrate one single equation, for example,

$$\frac{1}{\text{Pe}} \frac{d^2\theta}{dx^2} - \frac{d\theta}{dx} + \text{Da}(B - \theta) \exp\left(\frac{\theta}{1 + \epsilon\theta}\right) = 0 \tag{5.87}$$

subject to the boundary conditions

$$\begin{aligned} x = 0: \quad &\text{Pe}\theta = \frac{d\theta}{dx} \\ x = 1: \quad &\frac{d\theta}{dx} = 0 \end{aligned} \tag{5.88}$$

After transformation of the independent variable

$$t = \text{Pe}(x - 1) \tag{5.89}$$

and on reparametrization

$$r = \frac{\text{Da}}{\text{Pe}} \tag{5.90}$$

a transformed differential equation results:

$$\frac{d^2\theta}{dt^2} - \frac{d\theta}{dt} + r(B - \theta) \exp\left(\frac{\theta}{1 + \epsilon\theta}\right) = 0 \tag{5.91}$$

The boundary conditions are

$$t = 0: \quad \frac{d\theta}{dt} = 0 \tag{5.92a}$$

$$t = -\text{Pe}: \quad \theta = \frac{d\theta}{dt} \tag{5.92b}$$

Equation (5.91) can be solved easily as an initial value problem.

(a) At $t = 0$ a value of θ is chosen:

$$t = 0: \qquad \theta = \theta_1 \tag{5.93}$$

Equation (5.91) is intergrated with the initial conditions (5.92a) and (5.93) until the relation (5.92b) is satisfied.

(b) The negative value of the independent variable determines the corresponding value of Pe.

(c) From the definition of r [see (5.90)] a relation between the parameters Pe and Da can be established. Hence for a given value of B a relation $\theta_1 = \theta(x = 1) \sim \text{Pe} \sim \text{Da}$ is evaluated. The results calculated for $B = 10$ and for a sequence of values of r are drawn in Fig. 5-5.

2. The case $\text{Pe}_\theta \neq \text{Pe}_y$. After some simple rearrangements of (5.82) and (5.82), the following differential equation can be written:

$$B\left(\frac{1}{\text{Pe}_y}\frac{d^2y}{dx^2} - \frac{dy}{dx}\right) = \frac{1}{\text{Pe}_\theta}\frac{d^2\theta}{dx^2} - \frac{d\theta}{dx}$$

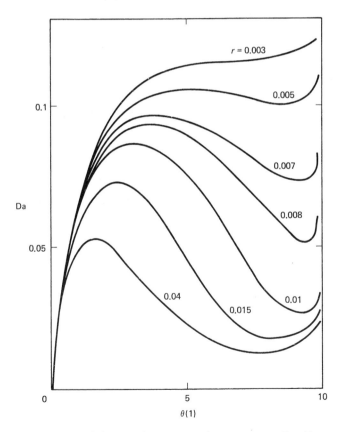

Figure 5-5 Damköhler number versus outlet temperature; $B = 10$, $\epsilon = 0.05$, $\text{Pe}_\theta = \text{Pe}_y$. Example 5.6.

On integration of this equation and after inserting boundary conditions (5.84) and (5.85), a simple relation results:

$$\theta(1) = By(1) \tag{5.94}$$

In contrast to the case of equal Peclet numbers the relation (5.86) for the adiabatic temperature rise holds exclusively at the reactor outlet (i.e., at $x = 1$).

To transform the set of (5.82)–(5.85), the following substitutions will be adopted:

$$t = \text{Pe}_\theta(x - 1), \qquad r = \frac{\text{Da}}{\text{Pe}_\theta}, \qquad q = \frac{\text{Pe}_y}{\text{Pe}_\theta} \tag{5.95}$$

This reparametrization yields the equations

$$\frac{d^2\theta}{dt^2} - \frac{d\theta}{dt} + rB(1 - y)\exp\left(\frac{\theta}{1 + \epsilon\theta}\right) = 0 \tag{5.96}$$

$$\frac{d^2y}{dt^2} - q\frac{dy}{dt} + qr(1 - y)\exp\left(\frac{\theta}{1 + \epsilon\theta}\right) = 0 \tag{5.97}$$

The relevant boundary conditions are

$$t = 0: \qquad \frac{d\theta}{dt} = \frac{dy}{dt} = 0 \tag{5.98}$$

$$t = -\text{Pe}_\theta: \qquad \theta = \frac{d\theta}{dt} \tag{5.99a}$$

$$qy = \frac{dy}{dt} \tag{5.99b}$$

The expression (5.94) can be rewritten

$$t = 0: \qquad \theta = By \tag{5.100}$$

The algorithm is as follows:

1. For given values of parameters ϵ, B, r, and q, the initial condition

$$\theta_1 = \theta(t = 0) \tag{5.101a}$$

is chosen. The missing initial condition for y can be easily calculated from (5.100):

$$y(t = 0) = \frac{\theta_1}{B} \tag{5.101b}$$

2. The set of equations (5.96) and (5.97) can be integrated across as long as the boundary condition (5.99a) is not matched. The second boundary condition (5.99b) is automatically fulfilled.

3. The value of the independent variable t determines the corresponding value of Pe_θ.

4. Making use of definition equations (5.95), the corresponding values of Pe_y and Da can be calculated.

5. Repeat steps 2 through 4 for a different guess θ_1. The calculated dependence $\text{Da} = f(\theta_1)$ is drawn in Fig. 5-6.

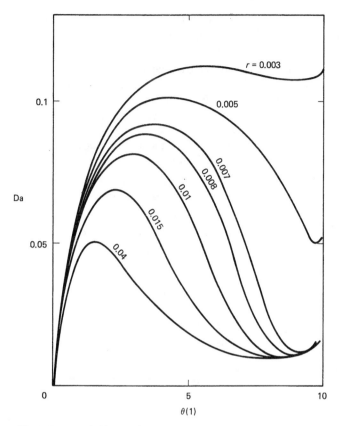

Figure 5-6 Damköhler number versus outlet temperature; $B = 10, q = 8$, $\epsilon = 0.05$. Example 5.6.

5.2 Sequential Use of Standard Methods

With the development of powerful computers the engineer is able to investigate a large number of feasible solutions to problems, to manipulate the values of parameters, relegating to the computer all but the most creative aspects of mathematical development of a method tailored for a particular problem—those which require the engineer's judgment and decision-making capabilities. In other words, to perform a parametric study it is frequently far more convenient to use standard methods of solution to nonlinear boundary value problems instead of developing a new, sophisticated technique tailored to the problem in question. The chief merit of the direct use of standard methods is that the solution could be obtained in a relatively short period of time. The major criticism of this approach is the computer-time expenditure. The sequential use of a standard method consists of the following human–machine interaction.

1. Choose an initial value of the parameter α.
2. For an appropriate initial approximation, calculate the solution by a standard method for solving nonlinear boundary value problems.
3. After inspection of the solution, guess a new value of α which is not far from the former value.
4. As an initial approximation, use the profile calculated in the last step and solve the problem. Go to step 3.

Of course, if multiple solutions do not occur, the procedure described above can be completely automated. For a general operator equation

$$F(y, \alpha) = 0 \qquad (5.102)$$

the calculation procedure is shown in a flowsheet (Fig. 5-7). To illustrate the strategy, an example exhibiting multiple solutions is presented next.

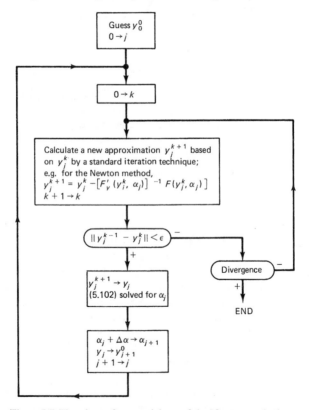

Figure 5-7 Flowchart of sequential use of the Newton method.

Example 5.7

Heat and mass transfer in a porous catalyst is described by

$$y'' + \frac{a}{x}y' = \alpha^2 y \exp\left[\frac{\gamma\beta(1-y)}{1+\beta(1-y)}\right] \qquad (5.103)$$

subject to the boundary conditions

$$x = 0: \quad y' = 0 \qquad (5.104)$$
$$x = 1: \quad y = 1$$

Evidently, for $\alpha = 0$ we get a trivial solution $y \equiv 1$. Hence we shall attempt to change the values of α starting from $\alpha = 0$. The Newton–Kantorcvich method can be adequate to do this. This procedure can be used both for the Newton–Kantorovich method realized by the difference analogy (see Section 4.3) and for the shooting method with a Newton correction algorithm (see Section 4.6). For the former case the solution of (5.103) and (5.104) is given in Table 5-7. The interval $\langle 0, 1 \rangle$ was divided into n parts; the second and first derivatives in (5.103) were replaced by three-point difference formulas of the order $O(h^2)$, where $h = 1/n$. In the last column of Table 5-7 the number of iterations in the Newton–Kantorovich method (the number of calculations of the tridiagonal matrix) necessary to pass from $\alpha - \Delta\alpha$ to α is given. When multiple solutions exist, difficulties may arise in the vicinity of branching points. However, by virtue of the algorithm given above, we can find an approximative value of α at the branching point. The case exhibiting multiple solutions is presented in column (b) in Table 5-7.

TABLE 5-7
SEQUENTIAL CALCULATION BY THE NEWTON–KANTOROVICH
METHOD (EXAMPLE 5.7):[a]

(a) $a = 2, \gamma = 20, \beta = 0.05, h = 0.1$
(b) $a = 0, \gamma = 20, \beta = 0.4, \quad h = 0.025$

(a)			(b)		
α	$y(0)$	N	α	$y(0)$	N
0.2	0.9933	2	0.05	0.9987	2
0.4	0.9733	3	0.10	0.9948	2
0.6	0.9401	3	0.15	0.9879	3
0.8	0.8937	3	0.20	0.9772	3
1.0	0.8348	3	0.25	0.9609	3
1.2	0.7645	3	0.30	0.9349	3
1.4	0.6849	3	0.35	0.8827	3
1.6	0.5990	3	0.40	0.0132	16[b]
1.8	0.5110	3	0.45	0.0054	4
2.0	0.4257	4	0.50	0.0023	4

[a] N, number of iterations; initial approximation $y \equiv 1$.
[b] For $h = 0.05$, the Newton–Kantorovich method does not converge.

Example 5.8†

The behavior of a diffusion flame is described by a set of differential equations.

$$g'' + fg' + \frac{\tau_1}{g} \exp\left(-\frac{\gamma_5}{g}\right)\Big\{[x_3(f' - \Lambda) + x_4 - g][x_1(f' - \Lambda) + x_2 - g]$$

$$- \tau_2 g \exp\left(-\frac{\alpha_6}{g}\right)\Big[1 - \frac{1}{F}(x_3(f' - \Lambda) + x_4 - g)$$

$$- \frac{1}{\alpha_0}(x_1(f' - \Lambda) + x_2 - g)\Big]\Big\} = 0 \qquad (5.105)$$

where

$$x_1 = \frac{\bar{g}_2 - \bar{g}_1 + \alpha_0}{1 - \Lambda}, \quad x_2 = \bar{g}_1, \quad x_3 = \frac{\bar{g}_2 - \bar{g}_1 - \alpha_F}{1 - \Lambda},$$

$$x_4 = \bar{g}_1 + \alpha_F$$

The boundary conditions are

$$g(-\infty) = \bar{g}_1 \qquad (5.106a)$$

$$g(\infty) = \bar{g}_2 \qquad (5.106b)$$

where g is the unknown dependent variable and f and f' are known functions obtained by solving the Blasius problem

$$f''' + ff'' = 0 \qquad (5.107)$$

with given boundary conditions

$$f'(-\infty) = \Lambda \qquad (5.108a)$$

$$f'(\infty) = 1 \qquad (5.108b)$$

$$f(0) = 0 \qquad (5.108c)$$

The nonlinear three-point boundary value problem defined by (5.107) and (5.108) was presolved by using a multiple shooting method (see Section 4.6). The third-order differential equation (5.107) was replaced by a system of first-order equations and solved by the shooting method. Since the boundary conditions are given at $\pm\infty$ and 0, it is rather convenient to start the integration at $\zeta = 0$ and integrate across in both directions. The asymptotic boundary conditions can be approximated at $\zeta = \pm 4$. These values of ζ were evaluated from the analysis of the Blasius problem [i.e., from (5.107)]. Since this relation does not contain the g variable, the resulting profiles of f and f' can be used to solve (5.105). It was found that for some parameter values, (5.105) and (5.106) exhibit multiple solutions and

†This example is taken from J. F. Holt, Numerical solution of nonlinear two point boundary value problems by finite differences using Newton's method, *Report No. TR-0158 (3307-02)-10*, Aerospace Corp., El Segundo, Los Angeles, Calif., 1968.

only some of them can be calculated by shooting because of the high sensitivity of the particular initial value problems with respect to the initial condition guessed at the point $\xi = -4$. Hence the finite-difference method was used and the resulting finite-difference equations were solved by the Newton method. A step length of $h = 0.03125$ was used.

Below the results are presented for the following values of parameters: $\bar{g}_1 = \bar{g}_2 = 0.07$, $\alpha_F = 3$, $\alpha_0 = 1.5$, $\gamma_5 = 2.5$, $\gamma_6 = 3.75$, $\Lambda = 0.9$, $\tau_2 = 0$. The profiles calculated for $\tau_1 = 173$ are shown in Fig. 5-8. In the following text each profile

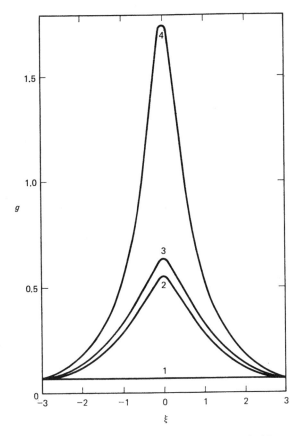

Figure 5-8 Four solutions at $\tau_1 = 173$. Example 5.8.

will be characterized by the value g_{max} corresponding to the maximum on the curve $g(\xi)$. To determine the parametric dependence of $g(\xi)$ or g_{max} on the value τ_1, a sequential use of the Newton method for the finite difference analogy was adopted. The sequential use of the Newton method makes it possible to calculate the entire branch of the solution (see Fig. 5-9). Some numerical results for $\tau_1 = 173$ are presented in Table 5-8.

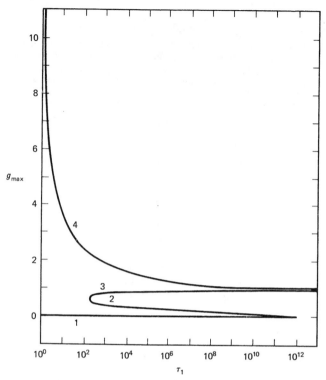

Figure 5-9 Dependence of g_{max} on τ_1. Example 5.8.

TABLE 5-8
MULTIPLE SOLUTIONS CORRESPONDING TO FIG. 5-8:
$\tau_1 = 173$, $h = 0.03125$

ζ	Second solution		Third solution		Fourth solution		Solution of Eqs. (5.107) and (5.108)	
	g	g'	g	g'	g	g'	f	f'
−4.0	0.0700	0.0000	0.0700	0.0000	0.0700	0.0000	−3.642	0.9000
−3.0	0.0722	0.0069	0.0725	0.0077	0.0733	0.0103	−2.742	0.9002
−2.0	0.1009	0.0682	0.1045	0.0761	0.1163	0.1024	−1.841	0.9028
−1.0	0.2599	0.2725	0.2819	0.3041	0.3554	0.4137	−0.9324	0.9169
−0.5	0.4208	0.3459	0.4615	0.3878	0.6255	0.7090	−0.4705	0.9318
−0.25	0.5010	0.2767	0.5527	0.3226	0.8414	1.0627	−0.2364	0.9409
0.00	0.5495	0.0926	0.6128	0.1374	1.1814	1.6831	0.0000	0.9505
0.125	0.5535	−0.0299	0.6220	0.0075	1.4064	1.8562	0.1191	0.9553
0.25	0.5419	−0.1538	0.6143	−0.1306	1.6245	1.4975	0.2389	0.9601
0.375	0.5157	−0.2613	0.5898	−0.2586	1.7508	0.4005	0.3591	0.9647
0.5	0.4778	−0.3381	0.5508	−0.3590	1.7124	−0.9924	0.4800	0.9692
1.0	0.2940	−0.3325	0.3403	−0.3979	0.8258	−1.4034	0.9685	0.9839
2.0	0.1032	−0.0777	0.1101	−0.0939	0.1740	−0.2437	1.9609	0.9976
3.0	0.0720	−0.0066	0.0724	−0.0080	0.0762	−0.0208	2.9600	0.9999
4.0	0.0700	0.0000	0.0700	0.0000	0.0070	0.0000	3.9600	1.0000

Roberts and Shipman called the aforementioned technique the perturbation technique.[†] They have used it to solve the sequence of problems occurring in the continuation methods (see, e.g., Problem 1 in Section 4.5).

Generally speaking, the sequential approach can be used if:

1. The right-hand sides of the differential equations are too complicated and it is difficult to calculate the derivatives.
2. The extent of the parametric study is small.

Sequential use of standard methods may fail if on a particular branch bifurcation points occur (problems in flame theory, combustion, hydrodynamics, reaction engineering, enzymatic, and biological problems, snapping and buckling, aircraft maneuvers, etc.) To circumvent these difficulties arclength or pseudoarclength continuation method must be used.[‡]

5.3   Method of Parametric Differentiation

5.3.1   Description of the Method

The basic idea of the method of parametric differentiation (also method of differentiation with respect to an actual parameter) will be shown on a simple operator equation

$$F(y, \alpha) = 0 \tag{5.109}$$

depending on the parameter α. According to the implicit function theorem[§] a differential equation can be derived describing the dependence of the solution y on the parameter

$$\frac{dy}{d\alpha} = -[F'_y(y, \alpha)]^{-1} F'_\alpha(y, \alpha) \tag{5.110}$$

Here we have supposed that the Fréchet derivatives exist. If one solution of the operator equation (5.109) is known (i.e., for $\alpha = \alpha_0$), we have already determined the solution $y = y_0$ for which

$$F(y_0, \alpha_0) = 0 \tag{5.111}$$

Then (5.110) can be considered as a differential equation with the initial condition

$$y(\alpha_0) = y_0 \tag{5.112}$$

describing the dependence $y(\alpha)$.

[†]S. M. Roberts, J. S. Shipman, and W. I. Ellis, *SIAM J. Numer. Anal.* 6, 347 (1969).

[‡]H. B. Keller, "Numerical Solution of Bifurcation and Nonlinear Eigenvalue Problems", *Applications of Bifurcation Theory*, ed., P. Rabinowitz (New York: Academic Press, 1977), pp. 359–384; also W. C. Rheinboldt, "Numerical Analysis of Continuation Methods for Nonlinear Structural Problems, *Computers and Structures 13*, 103–114 (1981).

[§]See, for example, in J. M. Ortega and W. C. Rheinboldt, *Iterative Solution of Nonlinear Equations in Several Variables* (New York: Academic Press, 1970).

Consider now a single second-order differential equation

$$y'' - f(x, y, y', \alpha) = 0 \qquad (5.113)$$

subject to boundary conditions that are not dependent on α, for example,

$$\begin{aligned}
\bar{\alpha}_0 y(0) + \bar{\beta}_0 y'(0) &= \bar{\gamma}_0 \\
\bar{\alpha}_1 y(1) + \bar{\beta}_1 y'(1) &= \bar{\gamma}_1
\end{aligned} \qquad (5.114)$$

Differentiation of Eq. (5.113) with respect to α leads to a special form of Eq. (5.110):

$$\frac{\partial^3 y(x, \alpha)}{\partial x^2\, \partial \alpha} - \frac{\partial f\left[x, y(x, \alpha), \dfrac{\partial y(x, \alpha)}{\partial x}, \alpha\right]}{\partial y} \frac{\partial y}{\partial \alpha}$$

$$- \frac{\partial f\left[x, y(x, \alpha), \dfrac{\partial y(x, \alpha)}{\partial x}, \alpha\right]}{\partial y'} \frac{\partial^2 y}{\partial x\, \partial \alpha}$$

$$- \frac{\partial f\left[x, y(x, \alpha), \dfrac{\partial y(x, \alpha)}{\partial x}, \alpha\right]}{\partial \alpha} = 0 \qquad (5.115)$$

We may note that a certain third order partial differential equation with two independent variables x and α results. Solution of this partial differential equation satisfying boundary conditions (5.114) and the initial condition

$$\alpha = \alpha_0: \qquad y(x, \alpha_0) = \varphi(x), \qquad x \in \langle 0, 1 \rangle \qquad (5.116)$$

is for the corresponding α solution of the original problem (5.113) and (5.114) if $\varphi(x)$ is the solution of (5.113) and (5.114) for $\alpha = \alpha_0$, that is:

$$\begin{aligned}
\varphi'' - f(x, \varphi, \varphi', \alpha_0) &= 0 \\
\bar{\alpha}_0 \varphi(0) + \bar{\beta}_0 \varphi'(0) &= \bar{\gamma}_0 \\
\bar{\alpha}_1 \varphi(1) + \bar{\beta}_1 \varphi'(1) &= \bar{\gamma}_1
\end{aligned}$$

and when the dependence $y(x, \alpha)$ is continuously differentiable with respect to α.

5.3.2 Numerical Realization of the Method

Numerical realization of the parametric differentiation method will be divided into three different groups of methods. The methods of the first group approximate the partial differential equations by the finite-difference analogy; the methods of the second group are based on the partial discretization [i.e., the space derivatives alone are approximated by the finite-difference formulas (method of lines)]. Obviously, a set of ordinary differential equations (initial value problem) results. Finally, (5.115) may be approximated by the solution of a sequence of linear problems for "variational" variables. Here a linear boundary value problem for the "variational" variables must be solved. A detailed discussion of all these techniques is presented below. At first we address ourselves to methods that use the finite-difference approximation.

For the sake of simplicity, consider a single second-order differential equation. To approximate (5.115) by a finite-difference analogy we make use of a rectangular grid (x_i, α_j); $x_i = ih$, $i = 0, \ldots, n$; $\alpha_j = jk + \alpha_0$, $j = 0, 1, 2, \ldots$. For brevity we use the notation $y_i^j \sim y(x_i, \alpha_j)$. The finite-difference analogy of (5.115) is

$$\frac{1}{kh^2}(y_{i-1}^{j+1} - 2y_i^{j+1} + y_{i+1}^{j+1} - y_{i-1}^j + 2y_i^j - y_{i+1}^j) - \frac{\bar{f}_y}{k}(y_i^{j+1} - y_i^j)$$

$$- \frac{\bar{f}_{y'}}{2kh}(y_{i+1}^{j+1} - y_{i-1}^{j+1} - y_{i+1}^j + y_{i-1}^j) - \bar{f}_\alpha = 0 \qquad (5.117)$$

$$i = 1, 2, \ldots, n - 1$$

Here we have denoted by $\bar{f}_y, \bar{f}_{y'}, \bar{f}_\alpha$ the corresponding coefficients (partial derivative of f) of (5.115) which are determined for the values of y_i at the jth profile already calculated, for example,

$$\bar{f}_y = \frac{\partial f\left(x_i, y_i^j, \dfrac{y_{i+1}^j - y_{i-1}^j}{2h}, \alpha_j\right)}{\partial y}$$

The boundary conditions (5.114) are approximated by formulas which have the same order of approximation as (5.117).

The six-point scheme represented by (5.117) gives rise to a set of linear algebraic equations for y_i^{j+1}, $i = 0, 1, \ldots, n$, with a tridiagonal matrix. To determine a new profile y^{j+1} the Thomas algorithm is to be used only once. Next, an example is presented which illustrates some typical features of the method described above.

Example 5.9

Heat and mass transfer in a porous catalyst is described by a single nonlinear second-order differential equation

$$y'' + \frac{a}{x}y' = \phi^2 y \exp\left[\frac{\gamma\beta(1-y)}{1 + \beta(1-y)}\right] \qquad (5.118)$$

subject to the boundary conditions

$$y'(0) = 0, \qquad y(1) = 1 \qquad (5.119)$$

After differentiation of (5.118) with respect to a parameter α, we get

$$\alpha = \phi:$$

$$\frac{\partial^3 y}{\partial x^2 \, \partial \phi} + \frac{a}{x}\frac{\partial^2 y}{\partial x \, \partial \phi} = \phi^2\left[1 - \frac{\gamma\beta y}{(1 + \beta(1-y))^2}\right]R(y)\frac{\partial y}{\partial \phi} + 2\phi y R(y) \qquad (5.120)$$

$\alpha = \gamma\beta$:

$$\frac{\partial^3 y}{\partial x^2 \, \partial(\gamma\beta)} + \frac{a}{x} \frac{\partial^2 y}{\partial x \, \partial(\gamma\beta)} = \phi^2 \left[1 - \frac{\gamma\beta y}{[1 + \gamma\beta(1-y)/\gamma]^2} \right] R(y)$$

$$\times \frac{\partial y}{\partial(\gamma\beta)} + \phi^2 (1-y) y R(y)[1 + \gamma\beta(1-y)/\gamma]^{-2} \qquad (5.121)$$

where we have denoted

$$R(y) = \exp\left[\frac{\gamma\beta(1-y)}{1 + \gamma\beta(1-y)/\gamma} \right]$$

For (5.120) and (5.121) the boundary conditions

$$x = 0: \qquad \frac{\partial y}{\partial x} = 0$$

$$x = 1: \qquad y = 1 \qquad\qquad (5.122)$$

have to be satisfied. The initial condition for (5.120) can be found easily:

$$\phi = 0: \qquad y(x) \equiv 1 \qquad\qquad (5.123)$$

On the other hand, the initial condition for (5.121) [i.e., for $(\gamma\beta)_0 = 0$] can be calculated from the linear differential equation resulting from (5.118) for $\gamma\beta = 0$:

$$y'' + \frac{a}{x} y' - \phi^2 y = 0 \qquad\qquad (5.124)$$

with boundary conditions given by (5.119). The replacement of (5.120) by its finite-difference analogy yields

$$\sum_{r=i-1}^{i+1} \sum_{s=j}^{j+1} C_r^s y_r^s = 2h^2 k \phi y_i^j R(y_i^j) \qquad i = 0, 1, \ldots, n-1$$

where

$$C_{i-1}^{j+1} = -C_{i-1}^j = 1 - \frac{a}{2x_i}$$

$$C_{i+1}^{j+1} = -C_{i+1}^j = 1 + \frac{a}{2x_i}$$

$$C_i^j = -C_i^{j+1} = 2 + h^2\phi^2 \left\{ 1 - \frac{\gamma\beta y_i^j}{[1 + \gamma\beta(1-y_i^j)/\gamma]^2} \right\} R(y_i^j)$$

and

$$y_{-1}^{j+1} = y_1^{j+1}, \qquad y_{-1}^j = y_1^j, \qquad y_n^{j+1} = y_n^j = 1$$

The accuracy of the numerical solution of (5.120) and (5.121) depends on the chosen step sizes h and k. The effect of different values of h used for integration of (5.120) is presented in Table 5-9. In this table the values of $y(x = 0)$ for various values of h and two different values of k are reported. The effect of different values of the step size k (for two values of h) can be followed from Table 5-10. The parameter ϕ changes again from $\phi = 0$ to $\phi = 1$ for a constant value of $\gamma\beta$.

A conclusion may be drawn that the error of the solution $y(x, \phi)$ increases with increasing of ϕ. This is caused by the error of approximation, depending on the type of the grid used. An illustration of the dependence of the error on the grid size is given in Table 5-11. [The values $y(x = 0)$ are presented.] Here (5.120) and (5.121) are integrated alternatively according to Fig. 5-10 (parameter plane $\phi - \gamma\beta$); for

TABLE 5-9

EFFECT OF THE STEP SIZE h ON THE ACCURACY OF THE
SOLUTION, EXAMPLE 5.9: $a = 0$, $\gamma = 20$, $\gamma\beta = 1$

	$k_\phi = 0.01$		$k_\phi = 0.002$	
h	$\phi = 0.1$	$\phi = 1.0$	$\phi = 0.1$	$\phi = 1.0$
0.2	0.9955	0.5543	0.9951	0.5520
0.1	0.9955	0.5549	0.9951	0.5525
0.05	0.9955	0.5548	0.9951	0.5521
0.025	0.9950	0.5545	0.9951	0.5503
	0.9950[a]	0.5521[a]	0.9950[a]	0.5521[a]

[a]Accurate value of $y(0)$.

TABLE 5-10

EFFECT OF THE STEP SIZE k ON THE ACCURACY OF THE
SOLUTION, EXAMPLE 5.9: $a = 0$, $\gamma = 20$, $\gamma\beta = 1$

	$h = 0.1$		$h = 0.025$	
k_ϕ	$\phi = 0.1$	$\phi = 1.0$	$\phi = 0.1$	$\phi = 1.0$
0.05	0.9975	0.5668	0.9975	0.5668
0.02	0.9960	0.5578	0.9960	0.5576
0.01	0.9955	0.5549	0.9955	0.5545
0.002	0.9951	0.5525	0.9951	0.5503
	0.9950[a]	0.5521[a]	0.9950[a]	0.5521[a]

[a]Accurate value of $y(0)$.

TABLE 5-11

GROWTH OF THE ERROR OF THE SOLUTION,
EXAMPLE 5.9: $a = 0$, $\gamma = 20$, $h = 0.05$,
$k_{\gamma\beta} = 0.01$, $k_\phi = \pm 0.01$

$\gamma\beta$	$\phi = 0$	$\phi = 1$	$\phi = 1$[a]
1.0	1.0000	0.5548	0.5521
1.1	1.0070	0.5418	0.5390
1.2	1.0070	0.5336	0.5249
1.3	1.0145	0.5188	0.5100
1.4	1.0145	0.5093	0.4940
1.5	1.0225	0.4924	0.4770
1.6	1.0225	0.4812	0.4588
1.7	1.0310	0.4610	0.4394
1.8	1.0310	0.4487	0.4189

[a]Accurate value of $y(0)$.

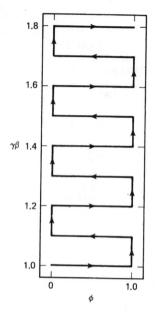

Figure 5-10 Parameter plane ϕ–$\gamma\beta$: the course of integration (see Table 5-11).

$\phi = 0$ the solution ought to be $y \equiv 1$. A similar case is shown in Table 5-12, where the integration proceeds according to Fig. 5-11. It would probably be useful to correct repeatedly the profiles $y(x, \phi)$ after calculation of a certain number of profiles $y(x, \phi)$ by the finite-difference method by making use of the Newton's method applied to the original boundary value problem. Certain difficulties may arise if for a given value of ϕ, $\phi = \phi_1$, the solution branches. It is clear that for this value of ϕ the number of solutions of the original boundary value problem changes suddenly. It is evident that the parametric differentiation method is capable of calculating only the solutions occurring on a particular branch. Sometimes the solutions pertaining to other branches may be calculated by an appropriate strategy, which is, for instance, based on combined alternate integrations along the ϕ or $\gamma\beta$ paths† (see Fig.

TABLE 5-12
GROWTH OF THE ERROR OF THE SOLUTION, EXAMPLE 5.9:
$a = 0$, $\gamma = 20$, $k_{\gamma\beta} = 0.01$, $k_\phi = \pm 0.01$

	$h = 0.05$		$h = 0.025$		
$\gamma\beta$	$\phi = 0$	$\phi = 1$	$\phi = 0$	$\phi = 1$	$\phi = 1^a$
2.0	1.0000	0.3778	1.0000	0.3778	0.3745
2.5	—	0.2546	—	0.2542	0.2521
3.0	1.0012	0.1453	0.9996	0.1450	0.1441

[a]Accurate value of $y(0)$.

†M. Kubíček and V. Hlaváček, *Chem. Eng. Sci.* 26, 705 (1971).

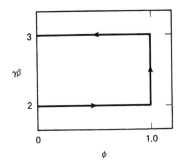

Figure 5-11 Parameter plane ϕ–$\gamma\beta$: the course of integration (see Table 5-12).

5-12 and Table 5-13). In Fig. 5-13 the dependence of $y(0)$ on ϕ is plotted. In this way we may arrive at a solution that would be otherwise inaccessible by a sole integration of (5.120) starting at $\phi = 0$. For a set of differential equations of second order, the development of the corresponding partial differential equations is more complicated.

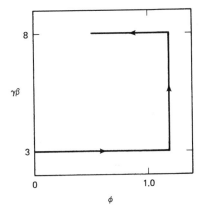

Figure 5-12 Parameter plane ϕ–$\gamma\beta$: the course of integration (see Table 5-13).

TABLE 5-13
PROCESS OF APPROACHING THE LOWER SOLUTION, EXAMPLE 5.9:
$a = 2,\ \gamma = 40$

$\gamma\beta$	Note	$\phi = 0$	$\phi = 0.2$	$\phi = 0.4$	$\phi = 0.8$	$\phi = 1.2$
3	a	1.0000	0.9936	0.9730	0.8757	0.6271
	b	1.0000	0.9933	0.9723	0.8738	0.6216
	c	1.0000	0.9933	0.9723	0.8737	0.6211

$\gamma\beta$	Note	$\phi = 0.5$	$\phi = 0.6$	$\phi = 0.8$	$\phi = 1.0$	$\phi = 1.2$
8	a	-0.0012	-0.0013	-0.0008	-0.0005	-0.0003
	b	0.0010	-0.0005	-0.0003	-0.0002	-0.0001
	c	0.0032	0.0001	0.0000	0.0000	0.0000

[a] $k_\phi = \pm 0.01,\ k_{\gamma\beta} = 0.025,\ h = 0.05$.
[b] $k_\phi = \pm 0.002,\ k_{\gamma\beta} = 0.01,\ h = 0.05$.
[c] Accurate value of $y(0)$.

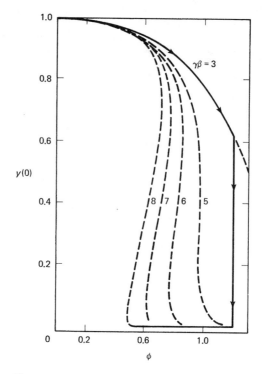

Figure 5-13 Dependence of $y(0)$ on ϕ ($a = 2$, $\gamma = 40$).

Example 5.10

Axial heat and mass transfer accompanied by a chemical reaction occurring in a tubular reactor is described by the dimensionless differential equations

$$\frac{1}{\text{Pe}}\theta'' - \theta' - \beta(\theta - \theta_c) + B\,\text{Da}\,(1 - y)\exp\left(\frac{\theta}{1 + \epsilon\theta}\right) = 0 \quad (5.125a)$$

$$\frac{1}{\text{Pe}}y'' - y' + \text{Da}\,(1 - y)\exp\left(\frac{\theta}{1 + \epsilon\theta}\right) = 0 \quad (5.125b)$$

subject to the boundary conditions

$$x = 1: \quad \theta' = y' = 0 \quad (5.126a)$$

$$x = 0: \quad \frac{1}{\text{Pe}}\theta' = \theta, \quad \frac{1}{\text{Pe}}y' = y \quad (5.126b)$$

On differentiation of (5.125) with respect to the parameter Da, we obtain the following third-order partial equations:

$$\frac{1}{\text{Pe}}\frac{\partial^3\theta}{\partial x^2\,\partial\text{Da}} - \frac{\partial^2\theta}{\partial x\,\partial\text{Da}} + \left(B\frac{\partial R}{\partial\theta} - \beta\right)\frac{\partial\theta}{\partial\text{Da}} + B\frac{\partial R}{\partial y}\frac{\partial y}{\partial\text{Da}} = -\frac{BR}{\text{Da}}$$

$$\frac{1}{\text{Pe}}\frac{\partial^3 y}{\partial x^2\,\partial\text{Da}} - \frac{\partial^2 y}{\partial x\,\partial\text{Da}} + \frac{\partial R}{\partial y}\frac{\partial y}{\partial\text{Da}} + \frac{\partial R}{\partial\theta}\frac{\partial\theta}{\partial\text{Da}} = -\frac{\partial R}{\partial\text{Da}} \quad (5.127)$$

where we have denoted

$$R = \text{Da}\,(1 - y)\exp\left(\frac{\theta}{1 + \epsilon\theta}\right)$$

Equations (5.127) must satisfy the boundary conditions

266

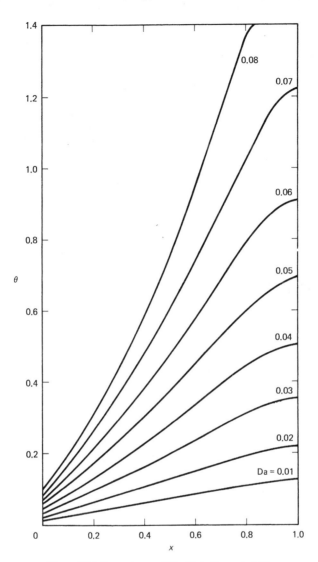

Figure 5-14 Dependence of $\theta(x, \text{Da})$: Example 5.10.

$$x = 1: \qquad \frac{\partial \theta}{\partial x} = \frac{\partial y}{\partial x} = 0$$

$$x = 0: \qquad \frac{1}{\text{Pe}}\frac{\partial \theta}{\partial x} = \theta, \qquad \frac{1}{\text{Pe}}\frac{\partial y}{\partial x} = y \tag{5.128}$$

The initial conditions for (5.127) may be obtained by solving (5.125) and (5.126) for a particular value of Da; for Da = 0, an analytical solution is possible. We choose a finite-difference analogy in the same way as we have described in Example 5.9 (the six-point scheme). However, the resulting set of linear algebraic equations is no longer tridiagonal, but after an appropriate ordering of variables a pentadiagonal band matrix results. The results for two different step lengths are reported in Table 5-14. The dependence $\theta(x, \text{Da})$ is depicted in Fig. 5-14.

TABLE 5-14
SOLUTION OF (5.127), EXAMPLE 5.10:
$B = 10$, Pe $= 10$, $\epsilon = 0.05$, $\beta = 0$

Step length		$\theta(x, \mathrm{Da})$	
h	k	$\theta(1, 0.01)$	$\theta(1, 0.05)$
0.05	0.005	0.1005	0.6588
0.02	0.001	0.1040	0.6952
		0.1057^a	0.7108^a

[a]Accurate solution.

The third-order partial differential equation (5.115) can be solved, of course, by the method of lines. After denoting $y_i(\alpha) \sim y(x_i, \alpha)$, a three-point finite-difference approximation yields

$$\frac{1}{h^2}[y'_{i-1} - 2y'_i + y'_{i+1}] - \frac{\partial f}{\partial y}y'_i - \frac{1}{2h}\frac{\partial f}{\partial y'}(y'_{i+1} - y'_{i-1}) - \frac{\partial f}{\partial \alpha} = 0 \qquad (5.129)$$

Here the prime represents differentiation with respect to α. The approximation of boundary conditions must be of the same order as the approximation of differential equation. Evidently, approximation (5.129) yields a set of linear algebraic equations for y'_i, $i = 0, 1, \ldots, n$. The Euler method of integration gives rise to

$$y_i(\alpha + k) = y_i(\alpha) + ky'_i(\alpha) \qquad (5.130)$$

If the Runge–Kutta method with four enumerations of the right-hand side is used, then in one integration step the set of linear algebraic equations must be solved four times.

The third possibility for solving the differential equations resulting from parametric differentiation takes advantage of the differential equations for the "variational" variable $\Omega(x)$:

$$\Omega(x) = \frac{\partial y(x)}{\partial \alpha} \qquad (5.131)$$

Evidently, (5.115) may be easily rewritten in terms of the variational variable $\Omega(x)$:

$$\Omega'' - \frac{\partial f}{\partial y}\Omega - \frac{\partial f}{\partial y'}\Omega' - \frac{\partial f}{\partial \alpha} = 0 \qquad (5.132)$$

Here the arguments of the functions $\partial f / \partial y$, $\partial f / \partial y'$, and $\partial f / \partial \alpha$ are x, $y(x)$, $y'(x)$, and α. The boundary conditions pertaining to (5.132) can be developed by differentiation of boundary conditions (5.114) with respect to α:

$$\begin{aligned}\bar{\alpha}_0\Omega(0) + \bar{\beta}_0\Omega'(0) = 0 \\ \bar{\alpha}_1\Omega(1) + \bar{\beta}_1\Omega'(1) = 0\end{aligned} \qquad (5.133)$$

To calculate the profile the Euler method can be used:

$$y(x, \alpha + k) = y(x, \alpha) + k\frac{\partial y(x, \alpha)}{\partial \alpha} = y(x, \alpha) + k\Omega(x, \alpha) \qquad (5.134)$$

Obviously, to determine a new profile $y(x, \alpha + k)$ it is necessary to integrate a single second-order linear differential equation for Ω [i.e., (5.132)], subjected to boundary conditions (5.133). The coefficients in this equation are calculated from the values of $y(x, \alpha)$. Of course, any arbitrary marching integration technique can be used to integrate (5.131) (e.g., the Runge–Kutta method). However, for each calculation of the right-hand sides in this type of method, the variational equation for Ω must be solved. Of course, the numerical values of the profiles $y(x, \alpha)$ can be stored only at discrete points.

It is obvious that this method, where the linear boundary value problems (5.132) and (5.133) are replaced by a three-point difference formula, is identical with the aforementioned method of lines, [see (5.129)]; moreover, using the Euler integration formula, both procedures (5.130) and (5.134) are identical to the finite-difference approximation (5.117).

The method using the variational variables can also be easily employed for systems of first-order differential equations:

$$y_i' = f_i(x, y_1, \ldots, y_n, \alpha) \qquad i = 1, 2, \ldots n \qquad (5.135)$$

subject to the boundary conditions

$$G_i[y(0), y(1), \alpha] = 0 \qquad i = 1, 2, \ldots, n \qquad (5.136)$$

On denoting

$$p_i(x) = \frac{\partial y_i(x)}{\partial \alpha} \qquad (5.137)$$

differentiation with respect to α yields

$$p_i' = \sum_{j=1}^{n} \frac{\partial f_i(x, y_1, \ldots, y_n, \alpha)}{\partial y_j} p_j + \frac{\partial f_i(x, y_1, \ldots, y_n, \alpha)}{\partial \alpha}$$
$$i = 1, 2, \ldots, n \qquad (5.138)$$

subject to the boundary conditions

$$\sum_{j=1}^{n} \left[\frac{\partial G_i}{\partial y_j(0)} p_j(0) + \frac{\partial G_i}{\partial y_j(1)} p_j(1) \right] + \frac{\partial G_i}{\partial \alpha} = 0 \qquad i = 1, 2, \ldots, n \qquad (5.139)$$

To calculate $y(x, \alpha + \Delta\alpha)$ the Euler method can be adopted:

$$y(x, \alpha + \Delta\alpha) = y(x, \alpha) + \Delta\alpha p(x, \alpha) \qquad (5.140)$$

The initial condition $y(x, \alpha_0)$ is to be calculated from Eq. (5.135); sometimes, however, these profiles can be evaluated easily without any calculation or may be determined readily from linear differential equations. Below three examples are presented to illustrate the methods discussed above. To integrate (5.137) or (5.131), multistep integration formulas can be adopted. Evidently, the computer-time expenditure is lower; however, in turn, a number of "old" profiles of $p_i(x)$ and $\Omega(x)$ must be stored. In addition, the problem of starting the integration procedure appears.

5.3.3 Examples

Example 5.11

The problem of calculating the Faulkner–Skan boundary layer similarity profiles is described by the following equations:[†]

$$f''' + ff'' + \beta(1 - f'^2) = 0 \qquad (5.141)$$

$$f(0) = f'(0) = 0, \qquad f'(\infty) = 1 \qquad (5.142)$$

We use β as the parameter for differentiation. Letting

$$\Omega(\xi) = \frac{\partial f}{\partial \beta} \qquad (5.143)$$

and differentiating (5.141) with respect to β, we obtain the following linear problems:

$$\Omega''' + f\Omega'' + \Omega f'' - 2\beta\Omega'f' = f'^2 - 1$$
$$\Omega(0) = \Omega'(0) = 0, \qquad \Omega'(\infty) = 0 \qquad (5.144)$$

where f and its derivatives appear as coefficients. A numerical integration of (5.143), for which the Runge–Kutta technique was used, was started at $\beta = 0$, corresponding to the Blasius flat-plate boundary layer (see Example 5.1).

The linear boundary value problem given by (5.144) has been solved by superposition; the relevant initial value problems have been integrated by the Runge–Kutta method with the step $h = 0.05$. The asymptotic boundary conditions have been replaced at the coordinate $\xi = 6.0$. In the interval $-0.1 \le \beta \le 1.0$ it was sufficient to integrate (5.143) with the step varying between 0.05 and 0.5. This step length can guarantee four significant digits. However, for $\beta < -0.1$ it was necessary to reduce the step length. The results of the computation are reported in Table 5-15.

TABLE 5-15
RESULTS FOR EXAMPLE 5.11

β	$f''(0)$	$f''(0)$ accurate
0.0	0.4696	0.4696
−0.1	0.3193	0.3193
−0.12	0.2818	.—
−0.14	0.2398	0.2397
−0.16	0.1908	0.1908
−0.18	0.1287	0.1286
−0.19	0.0859	0.0857
−0.194	0.0626	—
−0.195	0.0556	0.0552
−0.196	0.0479	—
−0.197	0.0384	—
−0.198	0.0264	—
−0.1985	0.0181	—
−0.19884	0.0093	—

Source: Taken from the paper by P. E. Rubbert and M. T. Landahl, *Phys. Fluids 10*, 831 (1967).

[†]P. E. Rubbert and M. T. Landahl, *Phys. Fluids 10*, 831 (1967).

Example 5.12

The boundary layer equations governing the flow of a compressible fluid become

$$f''' + ff'' + \beta(S - f'^2) = 0$$
$$S'' - fS' = 0 \tag{5.145}$$

The boundary conditions are

$$f(0) = f'(0) = 0, \qquad f'(\infty) = 1$$
$$S(0) = S_w, \qquad S(\infty) = 1 \tag{5.146}$$

Differentiation of (5.145) with respect to the parameter β gives

$$\Omega''' + f\Omega'' + f''\Omega + \beta(\psi - 2f'\Omega') + (S - f'^2) = 0$$
$$\psi'' + f\psi' + \Omega S' = 0 \tag{5.147}$$

subject to the boundary conditions

$$\Omega(0) = \Omega'(0) = \Omega'(\infty) = \psi(0) = \psi(\infty) = 0 \tag{5.148}$$

Here we have denoted

$$\frac{\partial f}{\partial \beta} = \Omega, \qquad \frac{\partial S}{\partial \beta} = \psi \tag{5.149}$$

The profiles of the variational variables Ω and ψ are governed, for known values of f and S, by linear differential equations. The new profiles $f(x, \beta + \Delta\beta)$ and $S(x, \beta + \Delta\beta)$ can be determined by integration of (5.149) (e.g., by the Runge–Kutta method). The initial profiles can take advantage of the solution of the Blasius problem (i.e., for $\beta = 0$) [the value $f''(0) = 0.4695999$ for $\beta = 0$ was used]. The results of the calculation are reported in Table 5-16.

TABLE 5-16
RESULTS FOR EXAMPLE 5.12

S_w	β	$f''(0)$	$S'(0)$
0.0	−0.3	0.318	0.426
	−0.2	0.387	0.448
	−0.15	0.412	0.455
	−0.1	0.434	0.460
	0.0	0.470	0.470
	0.5	0.581	0.494
	2.0	0.739	0.521
0.2	−0.125	0.395	0.362
	−0.1	0.412	0.365
	−0.075	0.428	0.368
	−0.050	0.442	0.371
	−0.025	0.456	0.373
	0.0	0.470	0.376
	0.5	0.655	0.404
	1.5	0.869	0.427
	2.0	0.948	0.433

Source: Taken from the paper by C. L. Narayana and P. Ramamoorthy, *AIAA J.* *10*, 1085 (1972).

Example 5.13

The steady-state flow of a viscous incompressible fluid between two coaxial rotating disks can be described by†

$$F'' = \sqrt{\text{Re}}\, HF' + \text{Re}(F^2 - G^2 + k) \qquad (5.150a)$$

$$G'' = 2\, \text{Re}\, FG + \sqrt{\text{Re}}\, G'H \qquad (5.150b)$$

$$H' = -2\sqrt{\text{Re}}\, F \qquad (5.150c)$$

Here Re denotes the Reynolds number and k is a parameter which has to be determined, since we have one surplus boundary condition.

$$H(0) = F(0) = 0, \qquad G(0) = 1, \qquad (5.151)$$

$$H(1) = F(1) = 0, \qquad G(1) = s \qquad (5.152)$$

Here s denotes the ratio of the angular speed of the upper disk to the angular speed of the lower disk.

The differential equations (5.150) can be differentiated with respect to the parameter s and the method of lines used to solve the resulting equations. Divide $0 \le x \le 1$ into n equal parts, each of length $h = 1/n$. We denote, for brevity, $F_i(s) = F(x_i, s) = F(ih, s)$, $i = 0, 1, \ldots, n$. Now primes will be used to denote the derivatives with respect to s. On replacing the "space derivatives" by finite differences, we have

$$2(F'_{i-1} - 2F'_i + F'_{i+1}) - h\sqrt{\text{Re}}\,[H'_i(F_{i+1} - F_{i-1}) \qquad (5.153a)$$
$$+ H_i(F'_{i+1} - F'_{i-1})]$$
$$- 2h^2\, \text{Re}\,(2F_iF'_i - 2G_iG'_i + k') = 0 \qquad i = 1, \ldots, n-1$$

$$2(G'_{i-1} - 2G'_i + G'_{i+1}) - 4h^2\, \text{Re}\,(F_iG'_i + F'_iG_i)$$
$$- h\sqrt{\text{Re}}\,[(G'_{i+1} - G'_{i-1})H_i \qquad (5.153b)$$
$$+ (G_{i+1} - G_{i-1})H'_i] = 0 \qquad i = 1, \ldots, n-1$$

$$H'_i - H'_{i-1} + h\sqrt{\text{Re}}\,(F'_i + F'_{i-1}) = 0 \qquad i = 1, 2, \ldots, n \qquad (5.153c)$$

Boundary conditions obtained by differentiating (5.151) and (5.152) with respect to s yield

$$G'_0 = H'_0 = F'_0 = H'_n = F'_n = 0 \qquad G'_n = 1 \qquad (5.154)$$

The initial conditions can be easily calculated analytically for $s = 1$; in a form suitable for differential equations (5.153a–c) we have

$$H_i(1) = 0, \qquad F_i(1) = 0, \qquad G_i(1) = 1, \qquad k(1) = 1 \qquad (5.155)$$

For given values of H_i, F_i, G_i, and k, the corresponding derivatives H'_i, F'_i, G'_i, and k' can be considered to be unknown. To calculate $3n + 4$ unknowns H'_i, F'_i, G'_i, and $k'(i = 0, 1, \ldots, n)$, $3n + 4$ linear algebraic equations (5.153) and (5.154) have to be solved. Let us group the variables according to

$$Y = (H_1, F_1, G_1, H_2, F_2, G_2, \ldots, H_{n-1}, F_{n-1}, G_{n-1}, k)^T$$

With regard to this grouping, (5.153) and (5.154) can be rewritten in a concise

†M. H. Rogers, and G. N. Lance, *J. Fluid. Mech.* **7**, 617 (1960).

matrix form

$$A(Y)Y' = B(Y) \qquad (5.156)$$

where the rows in $A(Y)$ are progressively ordered in the sequence (5.153c), (5.153a), (5.153b), and so on. The vector $B(Y)$ results after inserting (5.154) into (5.153).

It is obvious that for this grouping of variables the matrix A is almost seven diagonal. Gaussian elimination has been modified to solve matrices like this (see Chapter 2). On making use of this method, the vector of derivatives Y' results. To integrate the implicit differential equations, any reasonable adequate subroutine for the numerical solution of an initial value problem can be adopted (e.g., Runge–Kutta methods); however, the Euler formula also yields satisfactory results. All computations reported here were carried out using a simple Euler method

$$Y(s + \Delta s) = Y(s) + \Delta s \, Y'(s) \qquad (5.157)$$

or a modified Euler method

$$\begin{aligned} Y(s + \tfrac{1}{2}\Delta s) &= Y(s) + \tfrac{1}{2}\,\Delta s \, Y'(s) \\ Y(s + \Delta s) &= Y(s) + \Delta s \, Y'(s + \tfrac{1}{2}\,\Delta s) \end{aligned} \qquad (5.158)$$

All calculations have been performed using double-precision arithmetic on an IBM 360/40 computer. For high values of Reynolds number (Re = 625), single-precision arithmetic did not lead to satisfactory results. The results calculated have been compared with the reference ones obtained by quasi-linearization. Table 5-17 reveals the effect of the space increment h as well as the length of integration step Δs on the accuracy reached. For $n = 50$ and $n = 100$, the results are comparable with reference solutions. Referring to Table 5-17, the superiority of the modified Euler method is obvious.

The technique described also enables us to calculate the dependence $k(s)$. For Re = 100 the function $k(s)$ is shown in Fig. 5-15 ($n = 20$, $\Delta s = -0.02$, modified Euler method).

To check the accuracy of the method suggested, we can make use of the symmetry of F, G, and H profiles with respect to $x = 0.5$, for $s = -1$ (see Table 5-18).

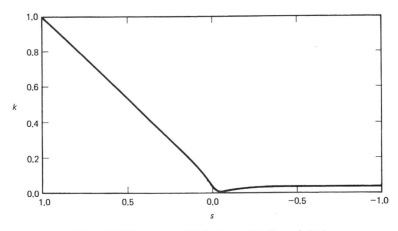

Figure 5-15 Dependence $k(s)$ for Re = 100: Example 5.13.

TABLE 5-17

EXAMPLE 5.13: RESULTS FOR $Re = 625$, $s = 0.8$
[EULER METHOD (5.157), MODIFIED EULER METHOD (5.158)]

Method	E	ME	E	ME	
n	20	20	100	100	
Δs	−0.01	−0.01	−0.005	−0.01	Reference Results[a]
$x = 0.1$	6.738E–3	6.795E–3	6.123E–3	6.156E–3	6.143E–3	
0.2	−4.381E–4	−4.390E–4	−6.975E–4	−6.986E–4	−7.120E–4	
0.3	−5.441E–6	−6.010E–6	3.991E–5	3.954E–5	4.219E–5	
0.4	2.597E–6	2.651E–6	−1.116E–7	−5.009E–8	−2.912E–7	
0.5	1.769E–7	1.915E–7	5.591E–7	5.958E–7	6.051E–7	$F(x)$
0.6	−4.714E–6	−4.978E–6	2.468E–7	9.549E–8	6.463E–7	
0.7	7.088E–6	8.065E–6	−7.815E–5	−7.887E–5	−8.466E–5	
0.8	5.690E–4	5.790E–4	1.038E–3	1.051E–3	1.078E–3	
0.9	−7.117E–3	−7.200E–3	−6.903E–3	−6.962E–3	−6.972E–3	
$x = 0.1$	0.8931	0.8929	0.8907	0.8906	0.8905	
0.2	0.8962	0.8960	0.8967	0.8966	0.8966	
0.5	0.8965	0.8963	0.8967	0.8966	0.8966	$G(x)$
0.8	0.8969	0.8967	0.8967	0.8966	0.8966	
0.9	0.9003	0.9001	0.9038	0.9037	0.9039	
1.0	0.8000	0.8000	0.8000	0.8000	0.8000	
$x = 0.1$	−0.0684	−0.0686	−0.1032	−0.1033	−0.1051	
0.2	−0.0766	−0.0770	−0.1046	−0.1048	−0.1062	
0.5	−0.0758	−0.0761	−0.1039	−0.1040	−0.1054	$H(x)$
0.8	−0.0769	−0.0773	−0.1050	−0.1051	−0.1065	
0.9	−0.0684	−0.0686	−0.1043	−0.1044	−0.1062	
	0.8032	0.8034	0.8038	0.8039	0.8040	k

[a]Reference results taken from M. Holodniok (Ph.D. diss., Institute of Chemical Technology, Prague, 1980).

TABLE 5-18
EXAMPLE 5.13: RESULTS FOR $Re = 100$, $s = −1.0$,
$n = 20$, $\Delta s = −0.02$, MODIFIED EULER METHOD
(5.158)

x	F	G	H
0.1	0.1604	0.5332	−0.2237
0.2	0.0614	0.2966	−0.4550
0.3	−0.0515	0.1756	−0.4613
0.4	−0.1242	0.0859	−0.2798
0.5	−0.1488	0.0002	−0.0007
0.6	−0.1244	−0.0856	0.2786
0.7	−0.0517	−0.1755	0.4606
0.8	0.0612	−0.2968	0.4548
0.9	0.1603	−0.5333	0.2236

5.4 Method of Differentiation with Respect to Boundary Conditions

This method can work successfully in such cases for which the algorithms presented in Section 5.3 fail (see Example 5.9).

We will assume, for simplicity, a single differential equation of second order

$$y'' - f(x, y, y', \alpha) = 0 \qquad (5.159)$$

where f is a nonlinear function. The boundary conditions are

$$\bar{\alpha}_0 y(0) + \bar{\beta}_0 y'(0) = \bar{\gamma}_0$$
$$\bar{\alpha}_1 y(1) + \bar{\beta}_1 y'(1) = \bar{\gamma}_1 \qquad (5.160)$$

Now supposing that $\bar{\beta}_0 \neq 0$, we will choose

$$y(0) = A \qquad (5.161)$$

Here A will be considered as a new parameter.

Let us note that in this case the new parameter is of an artificial nature, unlike the problems, which contain in the boundary conditions the parameter in question (see Example 5.13). Furthermore, we assume that the function f is differentiable with respect to all arguments; on differentiation we get for $y(x, A)$ a partial differential equation

$$\frac{\partial^3 y}{\partial x^2 \partial A} = \frac{\partial f}{\partial y} \frac{\partial y}{\partial A} + \frac{\partial f}{\partial y'} \frac{\partial^2 y}{\partial x \partial A} + \frac{\partial f}{\partial \alpha} \frac{d\alpha}{dA} \qquad (5.162)$$

For a given A the boundary conditions are

$$\bar{\alpha}_0 y(0, A) + \bar{\beta}_0 \frac{\partial y(0, A)}{\partial x} = \bar{\gamma}_0 \qquad (5.163)$$

$$\bar{\alpha}_1 y(1, A) + \bar{\beta}_1 \frac{\partial y(1, A)}{\partial x} = \bar{\gamma}_1 \qquad (5.164)$$

$$y(0, A) = A \qquad (5.165)$$

Thus we have three boundary conditions. It seems that the problem is overdetermined; however, in the partial differential equation (5.162) we have one unknown parameter,

$$C = \frac{d\alpha}{dA} \qquad (5.166)$$

We note that the problem is completely determined. Now on choosing

$$A = A_0: \qquad \alpha = \alpha_0, \qquad y(x, A_0) = y_0(x) \qquad (5.167)$$

we can integrate (5.162); here $y_0(x)$ is the solution of (5.159) with boundary conditions (5.160) and an additional condition $y_0(a) = A$. Frequently, the solution $y_0(x)$ may be readily calculated; sometimes for $\alpha = \alpha_0$ a trivial solution exists.

For the numerical solution of (5.162), a finite-difference method can be used. The replacement of the particular derivatives are analogous to approximations presented in Section 5.3 [see the development of (5.117)].

Let us denote $y_i^j \sim y(x_i, A^j) = y(ih, A^j)$. The nonlinear terms $\partial f/\partial y$, $\partial f/\partial y'$, and $\partial f/\partial \alpha$ have to be computed at the old profile j. It is obvious that an evaluation of the new profile demands that we solve a set of linear algebraic equations having a special, almost tridiagonal matrix. For the new profile we must calculate the values of both y_i^{j+1} ($i = 0, 1, \ldots, n$) and C. The resulting difference equations are in the form

$$s_{i-1}^j y_{i-1}^{j+1} + s_i^j y_i^{j+1} + s_{i+1}^j y_{i+1}^{j+1} + r_i^j C = q_i^j \qquad i = 0, 1, \ldots, n \qquad (5.168)$$

where the coefficients s_{i-1}^j, s_i^j, s_{i+1}^j, r_i^j, and q_i^j are dependent on the known values $\tilde{y} = y^j$ and $\tilde{\alpha} = \alpha^j$. For the approximation of boundary conditions (5.163)–(5.165), we have

$$\tilde{\alpha}_0 y_0^{j+1} + \frac{\tilde{\beta}_0}{2h}(y_1^{j+1} - y_{-1}^{j+1}) = \tilde{\gamma}_0 \qquad (5.169)$$

$$\tilde{\alpha}_1 y_n^{j+1} + \frac{\tilde{\beta}_1}{2h}(y_{n+1}^{j+1} - y_{n-1}^{j+1}) = \tilde{\gamma}_1 \qquad (5.170)$$

$$y_0^{j+1} = A^j + k \qquad (5.171)$$

Note that the profile j has been calculated for A^j. This procedure yields a set of $n + 4$ linear equations (5.168)–(5.171) for $n + 4$ variables y_{-1}^{j+1}, y_0^{j+1}, ..., y_{n+1}^{j+1}, C. The set of equations has a special form so that the following calculation procedure can be used:

1. Guess $C = C_1$. For $i = 0$, the values y_{-1}^{j+1}, y_0^{j+1}, and y_1^{j+1} can be calculated from (5.168), (5.169), and (5.171). The recurrence relation (5.168) yields $y_2^{j+1}, y_3^{j+1}, \ldots, y_{n+1}^{j+1}$. In general, the values y_{n-1}^{j+1}, y_n^{j+1}, and y_{n+1}^{j+1} do not meet (5.170), and a residue $R(C_1)$ depending on the guess of C, $C = C_1$, results.
2. Guess $C = C_2 \neq C_1$ and repeat step 1. Because the residue $R(C)$ is a linear function of C, the correct value of C can be calculated by linear interpolation.
3. For the correct value of C, a new profile $j + 1$ must be calculated or interpolated from two profiles evaluated for $C = C_1$ and $C = C_2$.
4. This new profile y_i^{j+1} may be employed to correct the coefficients $s_{i-1}^j, s_i^j, s_{i+1}^j, r_i^j$, and q_i^j; these will be calculated for the value y, defined as the arithmetical mean; that is,

$$\tilde{y}_i = \tfrac{1}{2}(y_i^j + y_i^{j+1}), \qquad \tilde{\alpha} = \alpha^j + \tfrac{1}{2}kC \qquad (5.172)$$

For the corrected coefficients, repeat steps 1 and 2. The recomputed profile is considered to be the new profile $j + 1$ (compare with the modified Euler method).

5. Calculate a new value A^{j+1}; $A^{j+1} = A^j + k$ and the corresponding value of

$$\alpha^{j+1} = \alpha^j + kC \qquad (5.173)$$

Thus the new profile is completely determined. The method described above will be illustrated by an example.

Example 5.14

Consider a nonlinear boundary value problem that arises in chemical reaction engineering:

$$\frac{dy^2}{dx^2} + \frac{a}{x}\frac{dy}{dx} = \delta y \exp\left[\frac{\gamma\beta(1-y)}{1+\beta(1-y)}\right] \tag{5.174}$$

$$x = 0: \quad \frac{dy}{dx} = 0$$
$$x = 1: \quad y = 1 \tag{5.175}$$

In accordance with (5.161) let us choose the boundary condition

$$y(0) = A$$

as a parameter.

Differentiation of (5.174) with respect to A yields the following third-order partial differential equation:

$$\frac{\partial^3 y}{\partial x^2 \partial A} + \frac{a}{x}\frac{\partial^2 y}{\partial x \partial A} - \delta\left[1 - \frac{\gamma\beta y}{(1+\beta(1-y))^2}\right]\exp\left[\frac{\gamma\beta(1-y)}{1+\beta(1-y)}\right]\frac{\partial y}{\partial A}$$
$$= y\exp\left[\frac{\gamma\beta(1-y)}{1+\beta(1-y)}\right]\frac{d\delta}{dA} \tag{5.176}$$

subject to the boundary conditions

$$\frac{\partial y(0, A)}{\partial x} = 0, \quad y(1, A) = 1, \quad y(0, A) = A \tag{5.177}$$

For numerical integration of this equation we must choose an appropriate initial condition; for $\delta = 0$, a trivial profile $y(x, 1) \equiv 1$ results.

A simple comparison of the numerical accuracy can be carried out for the isothermal case ($\beta = 0$) and plate geometry ($a = 0$), where the analytical solution of (5.174) and (5.175) is simple. The dependence between δ and $y(0)$ can be evaluated from the transcendental equation

$$\cosh\sqrt{\delta} = \frac{1}{y(0)}$$

The results calculated without correction (step 4 in algorithm) are presented in Table 5-19; the corrected results are in Table 5-20. It is obvious that the correction yields more accurate results.

TABLE 5-19

EFFECT OF GRID SPACING ON THE ACCURACY OF CALCULATED
PARAMETER δ: $y(0) = 0.5$, $a = 0$, $\beta = 0$[a]

			n		
k	5	10	20	40	80
0.05	1.62077	1.61490	1.61344	1.61307	1.61298
0.02	1.69264	1.68582	1.68411	1.68369	1.68358
0.01	1.71812	1.71095	1.70916	1.70871	1.70860
0.005	1.73117	1.72381	1.72198	1.72152	

[a]The case without correction. Exact value $\delta = 1.73438$.

TABLE 5-20

EFFECT OF GRID SPACING ON THE ACCURACY OF CALCULATED
PARAMETER δ: $y(0) = 0.5$, $a = 0$, $\beta = 0$[a]

	n				
k	5	10	20	40	80
0.05	1.74056	1.73310	1.73124	1.73077	1.73066
0.02	1.74378	1.73626	1.73438	1.73391	1.73379
0.01	1.74426	1.73673	1.73385	1.73438	1.73426
0.005	1.74439	1.73685	1.73496	1.73450	

[a]The case with correction. Exact value $\delta = 1.73438$.

TABLE 5-21

EFFECT OF GRID SPACING ON THE ACCURACY OF CALCULATED
PARAMETER δ: $y(0) = 0.5$, $a = 0$, $\gamma = 20$, $\beta = 0.05$[a]

		n			
Correction	k	5	10	20	40
No	0.05	1.12763	1.12659	1.12634	1.12628
	0.02	1.14437	1.14317	1.14288	1.14281
	0.01	1.15028	1.14903	1.14872	1.14864
	0.005	1.15331	1.15202	1.15170	1.15162
Yes	0.05	1.15542	1.15412	1.15380	1.15372
	0.02	1.15622	1.15490	1.15458	1.15450
	0.01	1.15633	1.15501	1.15469	1.15461
	0.005	1.15636	1.15504	1.15472	1.15464

[a]Exact value $\delta = 1.15462$.

TABLE 5-22

EFFECT OF GRID SPACING ON THE ACCURACY OF
CALCULATED PARAMETER δ: $y(0) = 0.5$, $a = 2$,
$\gamma = 40$, $\beta = 0.2$[a]

	n			
k	5	10	20	40
0.05	0.40696	0.41279	0.41414	0.41447
0.02	0.41242	0.41805	0.41934	0.41965
0.01	0.41311	0.41871	0.42000	0.42031
0.005	0.41328	0.41887	0.42015	0.42047

[a]The case with multiple solutions. Algorithm with correction.

For the nonisothermal case $\beta \neq 0$) the results are given in Table 5-21. The exact values are calculated by the GPM technique (see Section 5.5). Again, the correction essentially improves the accuracy attained. The results calculated for the event without correction and halved step $k/2$, which requires approximately the same computational effort as the algorithm with correction, are worse than those computed with correction. In practice, the relatively sparse grid ($N = 10$, $k = 0.02$) and the correction procedure yields satisfactory results. On using this method multiple solutions do not complicate the computation as far as $y(x, A)$ and $\alpha(A)$ are continuously differentiable. The case that has been solved by the differentiation with respect to an actual parameter with serious difficulties (see Example 5.19) can be calculated by this method in a straightforward way. The results for multiple solutions are reported in Table 5-22.

For a set of differential equations

$$
\begin{aligned}
y'' &= f(y, y', T, T', \alpha) \\
T'' &= g(y, y', T, T', \alpha)
\end{aligned}
\tag{5.178}
$$

with two-point boundary conditions, the preceding algorithm can be used in an analogous way. Here a seven-diagonal band matrix results.

The technique can be considered as an alternative to differentiation with respect to an actual parameter. For the case of multiple solutions, we can make use of this new method; on the other hand, for unique solutions the differentiation with respect to an actual parameter can be adopted. The results of calculation show that a relatively sparse grid of mesh points yields satisfactory accuracy. For example, branching points can be readily found. Here the parameter C is zero. As previously pointed out, the method can be extended to other classes of nonlinear boundary value problems.

Now consider a general set of first-order differential equations

$$
y_i' = f_i(x, y_1, \ldots, y_n, \alpha) \qquad i = 1, 2, \ldots, n
\tag{5.179}
$$

subject to nonlinear boundary conditions

$$
G_i[y_1(0), \ldots, y_n(0), y_1(1), \ldots, y_n(1), \alpha] = 0 \qquad i = 1, 2, \ldots, n
\tag{5.180}
$$

Let us select the variable $y_j(0)$ as a parameter (for given j)

$$
y_j(0) = A
\tag{5.181}
$$

After denoting

$$
p_i(x) = \frac{\partial y_i(x)}{\partial A}, \qquad \frac{d\alpha}{dA} = C
\tag{5.182}
$$

and on differentiation with respect to A, we have

$$
p_i' = \sum_{s=1}^{n} \frac{\partial f_i}{\partial y_s} p_s + \frac{\partial f_i}{\partial \alpha} C \qquad i = 1, \ldots, n
\tag{5.183}
$$

Differentation of boundary conditions (5.180) yields

$$
\sum_{s=1}^{n} \left[\frac{\partial G_i}{\partial y_s(0)} p_s(0) + \frac{\partial G_i}{\partial y_s(1)} p_s(1) \right] + \frac{\partial G_i}{\partial \alpha} C = 0 \qquad i = 1, 2, \ldots, n
\tag{5.184a}
$$

$$
p_j(0) = 1
\tag{5.184b}
$$

The variables $p_i(x)$ and C can be calculated from n linear differential equations (5.183) (with variable coefficients depending on y and α) which are subjected to $n + 1$ boundary conditions. Again the problem is not overdetermined because the variable C is not known. After solving the linear boundary value problem, (5.183) and (5.184), we get the profiles of the variables $p_i(x)$. The new profiles of the dependent variable y and the parameter α can be readily calculated by making use of the Euler integration method

$$y(x, A + \Delta A) = y(x, A) + \Delta A p(x, A)$$
$$\alpha(A + \Delta A) = \alpha(A) + \Delta A C \tag{5.185}$$

Let us note that we can differentiate with respect to any variable $y_j(\bar{x})$,

$$y_j(\bar{x}) = A \tag{5.186}$$

where $\bar{x} \in \langle 0, 1 \rangle$. In this case the governing linear differential equations (5.183) as well as the boundary conditions (5.184a) are identical, and the boundary condition (5.184b) becomes

$$p_j(\bar{x}) = 1 \tag{5.187}$$

It is evident that the linear boundary value problem given by (5.183), (5.184a), and (5.187) is a three-point problem. Analogously to the case of one second-order differential equation, the starting solution of (5.179) and (5.180) (for certain values of A_0 and α_0) must be found to be able to perform the integration described above.

5.5   General Parameter Mapping Technique

For problems where the pertinent initial value problems are stable at least in one direction, a general technique may be developed taking advantage of the implicit function theorem and the shooting method as well. This method will be referred to as the general parameter mapping technique (GPM method). This approach represents a whole family of methods that may be used for investigation of parametric dependence. The gist of this approach is illustrated for a single second-order differential equation.

5.5.1   GPM Algorithm for One Second-Order Equation

Consider the boundary value problem

$$y'' = f(x, y, y', \alpha) \tag{5.188}$$

subject to the linear boundary conditions

$$\alpha_0 y(a) + \beta_0 y'(a) = \gamma_0 \tag{5.189a}$$
$$\alpha_1 y(b) + \beta_1 y'(b) = \gamma_1 \tag{5.189b}$$

as a function of the parameter α. On choosing the initial condition

$$y(a) = A \tag{5.190}$$

and assuming that $\beta_0 \neq 0$, (5.189a) yields

$$y'(a) = \frac{1}{\beta_0}(\gamma_0 - \alpha_0 A) \tag{5.191}$$

If $\beta_0 = 0$, then the choice

$$y'(a) = B$$

results in a relation

$$y(a) = \frac{\gamma_0}{\alpha_0}$$

We consider only the first event. Assuming that the function f is continuously differentiable with respect to all independent variables, the following differential equation results:

$$p'' = \frac{\partial f}{\partial y}p + \frac{\partial f}{\partial y'}p' \tag{5.192}$$

with initial conditions

$$p(a) = 1 \tag{5.193}$$
$$p'(a) = -\frac{\alpha_0}{\beta_0}$$

Here we have denoted

$$p(x) = \frac{\partial y(x)}{\partial A} \tag{5.194}$$

In an analogous way, (5.195) may be developed:

$$q'' = \frac{\partial f}{\partial y}q + \frac{\partial f}{\partial y'}q' + \frac{\partial f}{\partial \alpha} \tag{5.195}$$

with initial conditions

$$q(a) = 0 \tag{5.196}$$
$$q'(a) = 0$$

where we have denoted

$$q(x) = \frac{\partial y(x)}{\partial \alpha} \tag{5.197}$$

The solution of (5.188) with initial conditions (5.190) and (5.191) depends on the value of the initial condition A as well as on the value of the parameter α:

$$y = y(x, A, \alpha) \tag{5.198}$$

However, for corresponding values of A and α, the boundary condition (5.189b) must be satisfied. Referring to (5.198), we have

$$\alpha_1 y(b, A, \alpha) + \beta_1 y'(b, A, \alpha) = \gamma_1 \tag{5.199}$$

Equation (5.199) may be rewritten in a more concise form:

$$F(A, \alpha) = 0 \tag{5.200}$$

At the curve $A(\alpha)$ obeying (5.200), we always have the following identity:

$$\frac{\partial F}{\partial A}\,dA + \frac{\partial F}{\partial \alpha}\,d\alpha = 0 \qquad (5.201)$$

Assuming that $(\partial F/\partial A) \neq 0$ or $(\partial F/\partial \alpha) \neq 0$, (5.201) may be rewritten in the form

$$\frac{dA}{d\alpha} = -\frac{\partial F/\partial \alpha}{\partial F/\partial A} = -\frac{\alpha_1 q(b) + \beta_1 q'(b)}{\alpha_1 p(b) + \beta_1 p'(b)} \qquad (5.202)$$

or, alternatively,

$$\frac{d\alpha}{dA} = -\frac{\partial F/\partial A}{\partial F/\partial \alpha} = -\frac{\alpha_1 p(b) + \beta_1 p'(b)}{\alpha_1 q(b) + \beta_1 q'(b)} \qquad (5.203)$$

Equations (5.202) and (5.203) may be considered as a new differential equation that must be satisfied by each smooth curve $A(\alpha)$ solving (5.200). Initial conditions for (5.202) or (5.203) must be chosen in accordance with relation (5.200); that is,

$$F(A_0, \alpha_0) = 0 \qquad (5.204)$$

which corresponds to

$$A(\alpha_0) = A_0 \quad \text{or} \quad \alpha(A_0) = \alpha_0 \qquad (5.205)$$

The integration of (5.202) or (5.203) may be readily performed numerically. The right-hand side in these equations can be enumerated as a result of an integration of the initial value problem given by (5.188), (5.190)–(5.193), (5.195), and (5.196). During the calculation the integration of (5.202) may be replaced by that of (5.203), and vice versa, to enhance the accuracy of the process. The accuracy of matching the boundary condition (5.189b) may be checked after each complete integration. Details of the computational aspects for a single equation are given in the following example.

Example 5.15

Heat and mass transfer and an exothermic chemical reaction occurring simultaneously in a porous catalyst involve a nonlinear two-point boundary value problem

$$\frac{d^2 y}{dx^2} + \frac{a}{x}\frac{dy}{dx} = \delta y \exp\left[\frac{\gamma\beta(1 - y)}{1 + \beta(1 - y)}\right] \qquad (5.188')$$

subject to the boundary conditions

$$\frac{dy(0)}{dx} = 0 \qquad (5.189a')$$

$$y(1) = 1 \qquad (5.189b')$$

Consider the missing condition

$$y(0) = y_0 \qquad (5.190')$$

Differentiation of (5.188') with respect to y_0 yields ($A = y_0$)

$$p'' + \frac{a}{x}p' = \delta \exp\left[\frac{\gamma\beta(1 - y)}{1 + \beta(1 - y)}\right]\left[1 - \frac{\gamma\beta y}{(1 + \beta(1 - y))^2}\right]p \qquad (5.192')$$

with the initial conditions

$$p(0) = 1, \qquad p'(0) = 0 \tag{5.193'}$$

For this particular problem, (5.195) may be rewritten ($\alpha = \delta$)

$$q'' + \frac{a}{x}q' = \delta \exp\left[\frac{\gamma\beta(1-y)}{1+\beta(1-y)}\right]\left[1 - \frac{\gamma\beta y}{(1+\beta(1-y))^2}\right]q$$
$$+ y \exp\left[\frac{\gamma\beta(1-y)}{1+\beta(1-y)}\right] \tag{5.195'}$$

The initial conditions associated with (5.195') are given by

$$q(0) = q'(0) = 0 \tag{5.196'}$$

Equation (5.199) is then

$$y(1, y_0, \delta) = 1 \tag{5.199'}$$

Thus (5.203) can be written

$$\frac{d\delta}{dy_0} = -\frac{p(1)}{q(1)} = -\frac{p(1, y_0, \delta)}{q(1, y_0, \delta)} \tag{5.203'}$$

The selection of an initial condition is easy; for $\delta = 0$ we have the solution of (5.188') and (5.189') in the form $y(x) \equiv 1$; that is, $(y_0)_0 = 1$. We aim at getting the whole dependence $\delta = \delta(y_0)$ for $y_0 \in (0.1)$, which obey the differential equation (5.203') and the initial condition

$$\delta(1) = 0 \tag{5.205'}$$

The simultaneous integration of the associated initial value problems (5.188'), (5.189a'), (5.190'), (5.192'), (5.193'), (5.195'), and (5.196') has been performed by Merson's modification of the Runge–Kutta method. For practical calculation it appears satisfactory to apply the step-size control procedure (maximum permissible error in one step $\sim 10^{-3}$) only to (5.188'). For integration of the differential equation (5.203') with (5.205'), the standard fourth-order Runge–Kutta procedure has been adopted. The enumeration of the right-hand side requires a simultaneous integration of six first-order differential equations given by (5.188'), (5.192'), and (5.195'). Thus for N steps in the Runge–Kutta procedure, $4N$ initial value problems for six first-order equations have to be solved. The results of an integration for two different steps in the Runge–Kutta procedure are reported in Table 5-23. The

TABLE 5-23
INTEGRATION OF (5.203'): $a = 0$, $\gamma = 20$, $\beta = 0.4$

$y(0)$	Integration step $h = -0.1$		Integration step $h = -0.05$	
	δ	$y(1)$	δ	$y(1)$
1.0	0.00000	1.00000	0.00000	1.00000
0.9	0.11429	0.99987	0.11444	0.99999
0.8	0.13734	0.99982	0.13748	0.99999
0.7	0.12975	0.99980	0.12985	0.99999
0.6	0.11425	0.99980	0.11432	0.99999
0.5	0.09913	0.99980	0.09918	0.99999
0.4	0.08726	0.99980	0.08730	0.99999
0.3	0.07980	0.99981	0.07983	0.99999
0.2	0.07824	0.99983	0.07826	0.99999
0.1	0.08854	1.00017	0.08853	0.99998

accuracy of the procedure may be extracted from column $y(1)$. In the first case, three decimal points are guaranteed, whereas in the second case, five digits are assured. It is evident that this procedure yields both the dependence $\delta = \delta(y_0)$ and the corresponding set of profiles solving the boundary value problem for a particular value of the parameter δ.

5.5.2 GPM Algorithm for Two Second-Order Equations

In this section we develop the algorithm for a set of two differential equations; the general algorithm is presented below.

Consider a set of two differential equations

$$
\begin{aligned}
y_1'' &= f_1(x, y_1, y_2, y_1', y_2', \alpha) \\
y_2'' &= f_2(x, y_1, y_2, y_1', y_2', \alpha)
\end{aligned}
\tag{5.206}
$$

subject to linear boundary conditions

$$\alpha_0^1 y_1(a) + \beta_0^1 y_1'(a) = \gamma_0^1 \tag{5.207a}$$

$$\alpha_0^2 y_2(a) + \beta_0^2 y_2'(a) = \gamma_0^2 \tag{5.207b}$$

$$\alpha_1^1 y_1(b) + \beta_1^1 y_1'(b) = \gamma_1^1 \tag{5.207c}$$

$$\alpha_1^2 y_2(b) + \beta_1^2 y_2'(b) = \gamma_1^2 \tag{5.207d}$$

Let us select the two initial conditions

$$y_1(a) = A_1, \qquad y_2(a) = A_2 \tag{5.208}$$

and assuming, without loss of generality, that $\beta_0^1 \neq 0$ and $\beta_0^2 \neq 0$, we arrive at

$$
\begin{aligned}
y_1'(a) &= \frac{1}{\beta_0^1}(\gamma_0^1 - \alpha_0^1 A_1) \\
y_2'(a) &= \frac{1}{\beta_0^2}(\gamma_0^2 - \alpha_0^2 A_2)
\end{aligned}
\tag{5.209}
$$

After denoting

$$p_{ij}(x) = \frac{\partial y_i(x)}{\partial A_j} \qquad i, j = 1, 2 \tag{5.210}$$

a set of associated equations results:

$$p_{ij}'' = \frac{\partial f_i}{\partial y_1} p_{1j} + \frac{\partial f_i}{\partial y_2} p_{2j} + \frac{\partial f_i}{\partial y_1'} p_{1j}' + \frac{\partial f_i}{\partial y_2'} p_{2j}' \qquad i, j = 1, 2 \tag{5.211}$$

assuming that both functions f_1 and f_2 are continuously differentiable with respect to all variables. The initial conditions are in the form

$$p_{ij}(a) = \begin{cases} 0, & i \neq j \\ 1, & i = j \end{cases} \qquad p_{ij}'(a) = \begin{cases} 0, & i \neq j \\ -\dfrac{\alpha_0^i}{\beta_0^i}, & i = j \end{cases} \tag{5.212}$$

According to previous considerations and on denoting

$$q_i(x) = \frac{\partial y_i(x)}{\partial \alpha} \tag{5.213}$$

one can readily derive

$$q_i'' = \frac{\partial f_i}{\partial y_1} q_1 + \frac{\partial f_i}{\partial y_2} q_2 + \frac{\partial f_i}{\partial y_1'} q_1' + \frac{\partial f_i}{\partial y_2'} q_2' + \frac{\partial f_i}{\partial \alpha} \qquad i = 1, 2 \qquad (5.214)$$

with the initial conditions

$$q_i(a) = 0, \qquad q_i'(a) = 0, \qquad i = 1, 2 \qquad (5.215)$$

Thus both solutions $y_1(x)$ and $y_2(x)$ depend on the selected initial conditions A_1 and A_2 as well as on the value of the parameter α:

$$y_1 = y_1(x, A_1, A_2, \alpha), \qquad y_2 = y_2(x, A_1, A_2, \alpha) \qquad (5.216)$$

After inserting both relations (5.216) into the boundary conditions (5.207c) and (5.207d), the following relations have to be satisfied:

$$\alpha_1^k y_k(b, A_1, A_2, \alpha) + \beta_1^k y_k'(b, A_1, A_2, \alpha) = \gamma_1^k \qquad k = 1, 2 \qquad (5.217a)$$

or in a more concise notation,

$$F_k(A_1, A_2, \alpha) = 0 \qquad k = 1, 2 \qquad (5.217b)$$

These two relations may be considered as two nonlinear equations for three variables A_1, A_2, and α. At the curves $A_1(\alpha)$ and $A_2(\alpha)$ obeying (5.217), the following relations are permanently valid:

$$\frac{\partial F_1}{\partial A_1} dA_1 + \frac{\partial F_1}{\partial A_2} dA_2 + \frac{\partial F_1}{\partial \alpha} d\alpha = 0$$
$$\frac{\partial F_2}{\partial A_1} dA_1 + \frac{\partial F_2}{\partial A_2} dA_2 + \frac{\partial F_2}{\partial \alpha} d\alpha = 0 \qquad (5.218)$$

According to the implicit function theorem and supposing that the matrix Γ_{A_1, A_2}

$$\Gamma_{A_1, A_2}(A_1, A_2, \alpha) = \begin{pmatrix} \dfrac{\partial F_1}{\partial A_1} & \dfrac{\partial F_1}{\partial A_2} \\ \dfrac{\partial F_2}{\partial A_1} & \dfrac{\partial F_2}{\partial A_2} \end{pmatrix} \qquad (5.219)$$

is regular, we get

$$\frac{dA}{d\alpha} = -\Gamma_{A_1, A_2}^{-1}(A_1, A_2, \alpha) \frac{\partial F}{\partial \alpha} \qquad (5.220)$$

where we have denoted

$$\frac{dA}{d\alpha} = \begin{pmatrix} \dfrac{dA_1}{d\alpha} \\ \dfrac{dA_2}{d\alpha} \end{pmatrix}, \qquad \frac{\partial F}{\partial \alpha} = \begin{pmatrix} \dfrac{\partial F_1}{\partial \alpha} \\ \dfrac{\partial F_2}{\partial \alpha} \end{pmatrix}$$

Similarly, for a regular matrix $\Gamma_{A_2, \alpha}$,

$$\Gamma_{A_2, \alpha}(A_1, A_2, \alpha) = \begin{pmatrix} \dfrac{\partial F_1}{\partial A_2} & \dfrac{\partial F_1}{\partial \alpha} \\ \dfrac{\partial F_2}{\partial A_2} & \dfrac{\partial F_2}{\partial \alpha} \end{pmatrix} \qquad (5.221)$$

equations (5.218) can be rearranged

$$\begin{pmatrix} \dfrac{dA_2}{dA_1} \\ \dfrac{d\alpha}{dA_1} \end{pmatrix} = -\Gamma_{A_2,\alpha}^{-1}(A_1, A_2, \alpha) \begin{pmatrix} \dfrac{\partial F_1}{\partial A_1} \\ \dfrac{\partial F_2}{\partial A_1} \end{pmatrix} \tag{5.222}$$

The partial derivatives occurring in the matrices Γ and also in (5.220) and (5.222) may be readily calculated with regard to the definition of F [see (5.217)]. Hence the following relations can be developed:

$$\frac{\partial F_k}{\partial A_j} = \alpha_1^k p_{kj}(b) + \beta_1^k p'_{kj}(b) \qquad j, k = 1, 2 \tag{5.223}$$

$$\frac{\partial F_k}{\partial \alpha} = \alpha_1^k q_k(b) + \beta_1^k q'_k(b) \qquad k = 1, 2 \tag{5.224}$$

Consider that for

$$A_1 = A_{10}, \qquad A_2 = A_{20}, \qquad \alpha = \alpha_0 \tag{5.225}$$

relations (5.226) are satisfied:

$$\begin{aligned} F_1(A_{10}, A_{20}, \alpha_0) &= 0 \\ F_2(A_{10}, A_{20}, \alpha_0) &= 0 \end{aligned} \tag{5.226}$$

These relations are permanently satisfied at the curves, being the solution of (5.220) or (5.222) with initial conditions corresponding to (5.225); thus

$$A_1(\alpha_0) = A_{10}, \qquad A_2(\alpha_0) = A_{20} \tag{5.227a}$$

or, alternatively,

$$A_2(A_{10}) = A_{20}, \qquad \alpha(A_{10}) = \alpha_0 \tag{5.227b}$$

The integration of differential equations (5.220) or (5.222) can be performed numerically; for enumeration of coefficients in these equations, an initial value problem, given by (5.206), (5.211), and (5.214), with initial conditions (5.208), (5.209), (5.212), and (5.215), must be solved.

Example 5.16

Steady-state axial heat and mass transfer with an exothermic first-order reaction may be described by two nonlinear ordinary differential equations of boundary value nature:

$$\frac{1}{\text{Pe}_M} y'' - y' + \text{Da}(1 - y) \exp\left(\frac{\theta}{1 + \epsilon\theta}\right) = 0$$

$$\frac{1}{\text{Pe}_H} \theta'' - \theta' - \beta(\theta - \theta_c) + B\,\text{Da}(1 - y) \exp\left(\frac{\theta}{1 + \epsilon\theta}\right) = 0 \tag{5.206'}$$

The boundary conditions are

$$\begin{aligned} y'(1) &= 0, \qquad \theta'(1) = 0 \\ \text{Pe}_M y(0) - y'(0) &= 0, \qquad \text{Pe}_H \theta(0) - \theta'(0) = 0 \end{aligned} \tag{5.207'}$$

Consider the initial conditions

$$y(1) = A_1, \qquad \theta(1) = A_2 \tag{5.208'}$$

For instance, the Damköhler number Da will be regarded as the variable parameter α. Of course, any other parameter may be chosen. After rewriting dimensionless mass and heat balances in accordance with (5.206), one gets

$$f_1 = \mathrm{Pe}_M y'_1 - \alpha(1 - y_1) \exp\left(\frac{y_2}{1 + \epsilon y_2}\right)$$

$$f_2 = \mathrm{Pe}_H y'_2 + \beta(y_2 - \theta_c) - B\alpha(1 - y_1) \exp\left(\frac{y_2}{1 + \epsilon y_2}\right)$$

Referring to (5.211) the auxiliary equations are, for example,

$$p''_{11} = \alpha \exp\left(\frac{y_2}{1 + \epsilon y_2}\right)p_{11} - \frac{\alpha(1 - y_1)}{(1 + \epsilon y_2)^2} \exp\left(\frac{y_2}{1 + \epsilon y_2}\right)p_{21} + \mathrm{Pe}_M p'_{11} \quad (5.211')$$

and so on, according to (5.214)

$$q''_2 = B\alpha \exp\left(\frac{y_2}{1 + \epsilon y_2}\right)q_1 + \left[\beta - \frac{B\alpha(1 - y_1)}{(1 + \epsilon y_2)^2} \exp\left(\frac{y_2}{1 + \epsilon y_2}\right)\right]q_2$$

$$+ \mathrm{Pe}_H q'_2 - B(1 - y_1) \exp\left(\frac{y_2}{1 + \epsilon y_2}\right) \quad (5.214')$$

and so on. Since the matrix (5.221) is regular, the differential equation of the type (5.222) will be integrated. The initial conditions (5.225) can be readily established for $\alpha_0 = \mathrm{Da} = 0$ because the homogeneous linear boundary value problem with constant coefficients

$$\frac{1}{\mathrm{Pe}_M}y'' - y' = 0, \qquad \frac{1}{\mathrm{Pe}_H}\theta'' - \theta' - \beta(\theta - \theta_c) = 0 \quad (5.228)$$

subject to boundary conditions (5.207'), may be integrated analytically.

The results of calculation for parameters $\mathrm{Pe}_M = \mathrm{Pe}_H = 2$, $\beta = 2$, $\theta_c = \epsilon = 0$, $B = 12$ are summarized in Table 5-24. The initial value problem given by (5.206'),

TABLE 5-24
COURSE OF INTEGRATION OF (5.222)

$A_1 = y(1)$	$A_2 = \theta(1)$	$\alpha = \mathrm{Da}$	R_1	R_2	$\dfrac{dA_2}{dA_1}$	$\dfrac{d\alpha}{dA_1}$
0.00	0.00000	0.00000	0.00000	0.00000	4.5087	1.0000
0.05	0.22692	0.04316	−0.00000	−0.00002	4.5693	0.7372
0.20	0.92879	0.11256	−0.00000	−0.00005	4.8033	0.2498
0.35	1.67274	0.13130	−0.00001	−0.00006	5.1329	0.0306
0.40	1.93261	0.13179	−0.00001	−0.00006	5.2635	−0.0090
0.60	3.03663	0.12140	−0.00001	−0.00006	5.7184	−0.0740
0.78	4.00517	0.11150	−0.00001	−0.00007	4.2912	−0.0061
0.80	4.08483	0.11160	−0.00001	−0.00007	3.6377	0.0167
0.86	4.19987	0.11531	−0.00001	−0.00007	−0.5697	0.1114
0.91	3.97953	0.12113	−0.00000	−0.00001	−8.8540	0.0578
0.92	3.88320	0.12137	−0.00000	−0.00001	−10.3255	−0.0120
0.965	3.39723	0.11568	0.00002	0.00017	−9.7498	−0.0829
0.97	3.34932	0.11548	0.00002	0.00022	−9.4206	0.0083
0.99	3.16442	0.12638	0.00005	0.00068	−9.8894	1.8042
0.996	3.09767	0.14757	0.00011	0.00144	−13.3197	6.93

(5.211), and (5.214) has been integrated by Merson's method with specified accuracy $\sim 10^{-3}$. The step-size control mechanism has been applied to (5.206') only. The initial value problem (5.222) has been integrated by the standard fourth-order Runge–Kutta method. The step size has been externally changed according to the value of $y(1)$ (see Table 5-25).

TABLE 5-25
INTEGRATION STEP FOR
RUNGE–KUTTA METHOD

Region of $A_1 \equiv y(1)$	Integration step
0–0.7	0.05
0.7–0.9	0.02
0.9–0.95	0.01
0.95–0.99	0.005
0.99–0.998	0.002

To check the accuracy of this technique, the matching of boundary conditions for $x = 0$,

$$R_1 = y'(0) - \text{Pe}_M y(0)$$

$$R_2 = \theta'(0) - \text{Pe}_H \theta(0)$$

is also reported in Table 5-24.

The overall computer-time expenditure has been approximately 8 minutes on an IBM 7040. The dependences $A_2(A_1)$ and $\alpha(A_1)$ (i.e., $\theta(1)[y(1)]$ and $\text{Da}[y(1)]$) are shown in Figs. 5-16 and 5-17. This method yields results practically identical with

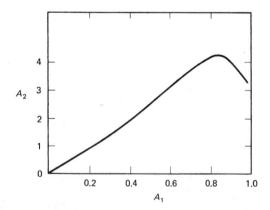

Figure 5-16 Dependence $A_2 = \theta(1)$ on $A_1 = y(1)$. Example 5.16.

those obtained by the shooting procedure for solving boundary value problems for particular values of parameters (e.g., of Da).† However, the algorithm suggested does not require human–machine interaction, which is time consuming. For calculation of the entire dependence by the Newton–Fox shooting procedure (see Section 4.6), approximately 70 guesses must be considered, each requiring on the average eight iterations. This amounts to approximately 500 initial value problems for 12 first-order ordinary differential equations. On the other hand, the method described allows us to get the same result after 160 integrations of 16 first-order ordinary differential equations. Thus the computer-time expenditure is reduced by a factor of 3. However, the new method enables one to perform calculations in a straightforward way.

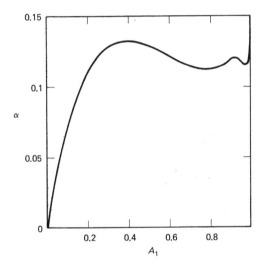

Figure 5-17 Dependence $\alpha = $ Da on $A_1 = y(1)$: occurrence of multiplicity. Example 5.16.

5.5.3 Predictor–Corrector GPM Algorithm for Two Second-Order Equations

The development of the procedure is presented for a set of two second-order equations; a general algorithm is presented below.

For a given parameter α the calculated values of initial conditions A_1 and A_2 can be improved by making use of the Newton shooting method (see Section 4.6); a sequence of values A_1^k and A_2^k, converging to the solution of (5.217), A_1^* and A_2^*, will be constructed:

$$A_1^{k+1} = A_1^k + \Delta A_1^k, \qquad A_2^{k+1} = A_2^k + \Delta A_2^k \tag{5.229}$$

†V. Hlaváček, H. Hofmann, and M. Kubíček, *Chem. Eng. Sci.* 26, 1629 (1971).

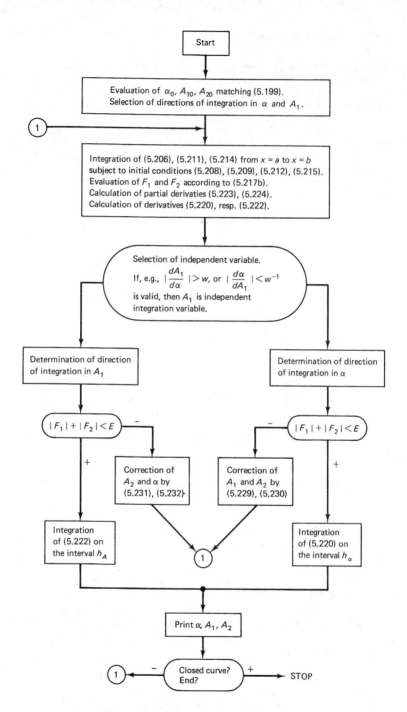

Figure 5-18 Flowchart of the GPM method.

Here the increment ΔA becomes

$$\Delta A^k = -\Gamma_{A_1, A_2}^{-1}(A_1^k, A_2^k, \alpha)F(A_1^k, A_2^k, \alpha) \qquad (5.230)$$

We have denoted

$$\Delta A^k = (\Delta A_1^k, \Delta A_2^k)^T, \qquad F = (F_1, F_2)^T$$

Analogously, a modification of this method can be derived; that is, for a given A_1 a sequence of values A_2^k and α^k will be constructed converging to the solution of (5.217), A_2^* and α^*. It can be written

$$A_2^{k+1} = A_2^k + \Delta A_2^k, \qquad \alpha^{k+1} = \alpha^k + \Delta \alpha^k \qquad (5.231)$$

where the increments ΔA_2^k and $\Delta \alpha^k$ may be calculated from (5.232):

$$\begin{pmatrix} \Delta A_2^k \\ \Delta \alpha^k \end{pmatrix} = -\Gamma_{A_2, \alpha}^{-1}(A_1, A_2^k, \alpha^k)F(A_1, A_2^k, \alpha^k) \qquad (5.232)$$

Now we can check at any point of the dependence $A(\alpha)$, computed by integration of (5.220) or (5.222), whether the relation (5.217) is matched within the predetermined tolerance. If not, we can reevaluate either A_1 and A_2 from (5.229) and (5.230) or A_2 and α from (5.231) and (5.232). By means of this correction, which can be controlled by the accuracy attained during the calculation process, the algorithm reported in Section (5.5.2) may be improved. To obtain higher accuracy of results, it is convenient to interchange the integration variables (α and A_1) according to the magnitude of the derivatives $dA_1/d\alpha$ or $d\alpha/dA_1$. The flowchart (see Fig. 5-18) for set of two second-order equations is presented. However, this rather cumbersome algorithm can be substantially simplified. We shall construct a simple algorithm of predictor–corrector type (see the flowchart in Fig. 5-19), which forms an internal part of the general algorithm in Fig. 5-18. For each step of the algorithm (5.206), (5.211), and (5.214) must be integrated ($N_{corr} + 1$) times. Figure 5-20 presents a modified procedure requiring only one integration per step. By employing this method the relation among α, A_1, and A_2 can be calculated; an evaluation of corresponding profiles $y_1(x)$ and $y_2(x)$ demands an additional calculation. Clearly, the profiles are calculated for the original, that is, uncorrected values of the parameters α, A_1, and A_2.

Compared with the sequential use of the standard shooting procedure, the GPM algorithm leads to substantial improvements, (i.e., there are no difficulties in crossing the branching point). In this event the simple shooting procedure fails; however, the GPM algorithm is successful after the integration variables have been interchanged. Both refined algorithms described in the flowcharts in Figs. 5-19 and 5-20 are illustrated by the problem presented in Example 5.16. We can thus compare the efficiency of all three versions.

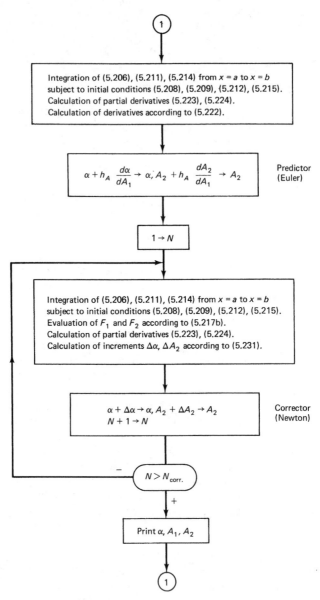

Figure 5-19 Flowchart of the predictor–corrector GPM method. Only the left branch of the flowchart from Fig. 5-18 is presented.

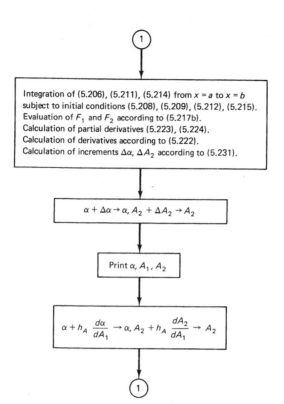

Figure 5-20 Flowchart of the modified predictor–corrector GPM method. Only the left branch of the flowchart from Fig. 5-18 is presented.

Example 5.17

Consider a nonlinear boundary value problem describing axial dispersion of heat and mass in a tubular reactor. The governing equations (5.206′) and (5.207′) are presented in Example 5.16. The initial conditions will be denoted by A_1 and A_2 according to (5.208′). As the mapping parameter α the Damköhler number Da will be taken. The initial conditions will be chosen for $\alpha_0 = 0$; the differential equation for y yields $A_{10} = 0$; the initial condition for A_{20} [$\sim \theta(1)$] will be calculated from linear BVP (5.228). The results for $\text{Pe}_M = \text{Pe}_H = 2$, $\beta = 2$, $\theta_c = \epsilon = 0$, and $B = 12$, calculated according to the flowchart drawn in Fig. 5-19, are presented for different values of N_{corr} and h_A in Tables 5-26 through 5-29. In the last column the values of the function

$$R_2 = \theta'(0) - \text{Pe}_H \theta(0)$$

in semilogarithmic for are calculated.

For instance, after comparison of the results for $A_1 = 0.6$ from Table 5-26 with those from Table 5-24, we notice that the accuracy is approximately the same. How-

293

TABLE 5-26
DEPENDENCES $\alpha(A_1)$ AND $A_2(A_1)$
CALCULATED BY PREDICTOR–CORRECTOR
GPM METHOD: $N_{\mathrm{corr}} = 1$, $h_A = 0.05$

A_1	α	A_2	R_2
0.1	0.07460	0.45696	$-3\mathrm{E}{-3}$
0.2	0.11253	0.92875	$-1\mathrm{E}{-3}$
0.3	0.12849	1.41916	$-4\mathrm{E}{-4}$
0.4	0.13179	1.93262	$-2\mathrm{E}{-4}$
0.5	0.12819	2.47254	$-4\mathrm{E}{-5}$
0.6	0.12140	3.03660	$-5\mathrm{E}{-6}$
0.7	0.11444	3.60324	$-2\mathrm{E}{-5}$
0.8	0.11160	4.08474	$-2\mathrm{E}{-4}$
0.9	0.12075	4.04934	$-4\mathrm{E}{-3}$

TABLE 5-27
DEPENDENCES $\alpha(A_1)$ AND $A_2(A_1)$
CALCULATED BY PREDICTOR–CORRECTOR
GPM METHOD: $N_{\mathrm{corr}} = 1$, $h_A = 0.1$

A_1	α	A_2	R_2
0.1	0.07292	0.45491	$-6\mathrm{E}{-2}$
0.2	0.11202	0.92812	$-2\mathrm{E}{-2}$
0.3	0.12830	0.41897	$-9\mathrm{E}{-3}$
0.4	0.13172	1.93263	$-3\mathrm{E}{-3}$
0.5	0.12817	2.47258	$-1\mathrm{E}{-3}$
0.6	0.12140	3.03664	$-2\mathrm{E}{-4}$
0.7	0.11444	3.60320	$-5\mathrm{E}{-5}$
0.8	0.11159	4.08420	$-2\mathrm{E}{-3}$
0.9	0.12227	4.01369	$-2\mathrm{E}{-2}$

TABLE 5-28
DEPENDENCES $\alpha(A_1)$ AND $A_2(A_1)$
CALCULATED BY PREDICTOR–CORRECTOR
GPM METHOD: $N_{\mathrm{corr}} = 1$, $h_A = 0.2$

A_1	α	A_2	R_2
0.2	0.09682	0.90880	$-7\mathrm{E}{-1}$
0.4	0.13074	1.93357	$-5\mathrm{E}{-2}$
0.6	0.12119	3.03711	$-1\mathrm{E}{-2}$
0.8	0.11152	4.08362	$-7\mathrm{E}{-3}$

TABLE 5-29
DEPENDENCES $\alpha(A_1)$ AND $A_2(A_1)$
CALCULATED BY PREDICTOR–CORRECTOR
GPM METHOD: $N_{corr} = 2$, $h_A = 0.2$

A_1	α	A_2	R_2
0.2	0.11204	0.92718	−2E–2
0.4	0.13178	1.93259	−6E–4
0.6	0.12140	3.03664	−9E–6
0.8	0.11160	4.08486	−4E–6

ever, for calculation of these values only 24 integrations of the initial value problem (5.206), (5.211), and (5.214) have been necessary. On the other hand, the uncorrected GPM method requires 48 integrations (see Table 5-24). In both cases the Merson modification of the Runge–Kutta method with a tolerance $\sim 10^{-3}$ has been used. It can be inferred from Tables 5-26 through 5-29 that the accuracy of results depends on the integration step h_A, as well as on N_{corr}. On using the procedure according to the flowchart in Fig. 5-20 the results reported in Table 5-30 are obtain-

TABLE 5-30
DEPENDENCES $\alpha(A_1)$ CALCULATED
BY MODIFIED PREDICTOR–CORRECTOR
GPM METHOD

h_A	0.05	0.1	0.2
A_1	α	α	α
0.1	0.07461	0.07292	
0.2	0.11253	0.11214	0.09682
0.3	0.12849	0.12835	
0.4	0.13179	0.13174	0.13182
0.5	0.12819	0.12817	
0.6	0.12140	0.12140	0.12124
0.7	0.11444	0.11444	
0.8	0.11160	0.11159	0.11147
0.9	0.12077	0.12230	

ed. The accuracy of these results, computed by the modified procedure, are approximately the same as those in Tables 5-26 through 5-28. However, (5.206), (5.211), and (5.214) are integrated for $h_A = 0.05$ and $A_1 = 0.6$ only 12 times (i.e., the number of integrations is further reduced by a factor of 2). For this case the method does not yield the profiles $y(x)$ and $\theta(x)$ for a particular A_1; thus the value of R_2 is also lacking. Hence, evidently, the last event is appropriate for the evaluation of the dependences $\alpha(A_1)$ and $A_2(A_1)$, that is, for the relations Da $\sim y(1)$ and $\theta(1) \sim y(1)$.

5.5.4 GPM Algorithm for a System of First-Order Differential Equations

Consider a set of ordinary differential equations

$$\frac{dy_i}{dx} = f_i(x, y_1, y_2, \ldots, y_n, \alpha) \qquad i = 1, 2, \ldots, n \qquad (5.233)$$

subject to the linear two-point boundary conditions

$$\sum_{j=1}^{n} a_{ij} y_j(a) = c_i \qquad i = 1, 2, \ldots, r \qquad (5.234)$$

$$\sum_{j=1}^{n} b_{ij} y_j(b) = d_i \qquad i = 1, 2, \ldots, n - r \qquad (5.235)$$

It is the purpose of this section to derive the governing equations describing the dependence of the solution on the considered parameter α. It will be convenient to denote the unknown initial conditions

$$y_1(a) = \eta_1, y_2(a) = \eta_2, \ldots, y_{n-r}(a) = \eta_{n-r} \qquad (5.236)$$

and the matrices

$$A_2 = \begin{pmatrix} a_{1,n-r+1} & a_{1,n-r+2} & \cdots & a_{1,n} \\ a_{2,n-r+1} & & \cdots & a_{2,n} \\ \cdot & & & \\ \cdot & & & \\ \cdot & & & \\ a_{r,n-r+1} & & & a_{r,n} \end{pmatrix} \qquad (5.237)$$

$$A_1 = \begin{pmatrix} a_{1,1} & a_{1,2} & \cdots & a_{1,n-r} \\ a_{2,1} & & \cdots & a_{2,n-r} \\ \cdot & & & \\ \cdot & & & \\ \cdot & & & \\ a_{r,1} & a_{r,2} & \cdots & a_{r,n-r} \end{pmatrix}$$

For a regular matrix A_2, the boundary conditions (5.234) may be written in a more concise form:

$$\begin{pmatrix} y_{n-r+1}(a) \\ y_{n-r+2}(a) \\ \cdot \\ \cdot \\ \cdot \\ y_n(a) \end{pmatrix} = A_2^{-1}c - A_2^{-1}A_1\eta \qquad (5.238)$$

where the notation $\eta = (\eta_1, \eta_2, \ldots, \eta_{n-r})^T$ and $c = (c_1, c_2, \ldots, c_r)^T$ has been used. If the matrix A_2 is singular, then for a correctly formulated problem it is always possible to reorder the unknown values in such a way that a regular matrix A_2 results. By means of (5.236) and (5.238) all initial values $y_1(a)$, $y_2(a)$,

$\ldots, y_n(a)$ are thus in hand. In addition, the right-hand sides of differential equations are considered to be continuous. The continuous derivatives with respect to both y and α are supposed to exist in a sufficiently large domain of these arguments. Hence two sets of differential equations can be derived:

$$\frac{dp_{ij}(x)}{dx} = \sum_{k=1}^{n} \frac{\partial f_i}{\partial y_k} p_{kj} \qquad \begin{matrix} i = 1, 2, \ldots, n \\ j = 1, 2, \ldots, n - r \end{matrix} \tag{5.239}$$

$$\frac{dq_i(x)}{dx} = \sum_{k=1}^{n} \frac{\partial f_i}{\partial y_k} q_k + \frac{\partial f_i}{\partial \alpha} \qquad i = 1, 2, \ldots, n \tag{5.240}$$

where we have denoted

$$p_{kj}(x) = \frac{\partial y_k(x)}{\partial \eta_j}, \qquad q_k(x) = \frac{\partial y_k(x)}{\partial \alpha} \tag{5.241}$$

The initial conditions are, for (5.239),

$$p_{kj}(a) = \begin{cases} 0, & k \neq j \\ 1, & k = j; \quad k, j = 1, 2, \ldots, n - r \end{cases} \tag{5.242a}$$

The missing initial conditions may be evaluated after differentiation of (5.238), which after some manipulations leads to

$$P = -A_2^{-1} A_1 \tag{5.242b}$$

where

$$P = \begin{pmatrix} p_{n-r+1, 1}(a) & p_{n-r+1, 2}(a) & \cdots & p_{n-r+1, n-r}(a) \\ p_{n-r+2, 1}(a) & & \cdots & p_{n-r+2, n-r}(a) \\ \cdots\cdots\cdots\cdots\cdots\cdots\cdots\cdots\cdots\cdots\cdots\cdots\cdots \\ p_{n, 1}(a) & p_{n, 2}(a) & \cdots & p_{n, n-r}(a) \end{pmatrix} \tag{5.242c}$$

Finally, the initial conditions for (5.240) become

$$q_i(a) = 0 \qquad i = 1, 2, \ldots, n \tag{5.243}$$

The solution $y(x)$ of initial value problem (5.233) subject to initial conditions (5.236) and (5.238) depends on the selection of initial values η chosen as well as on the value of α; that is,

$$y = y(x, \eta_1, \eta_2, \ldots, \eta_{n-r}, \alpha) = y(x, \eta, \alpha) \tag{5.244}$$

The solution (5.244) obeys the original boundary value problem (5.233)–(5.235) if boundary conditions (5.235) are satisfied:

$$\sum_{j=1}^{n} b_{ij} y_j(b, \eta, \alpha) = d_i \qquad i = 1, 2, \ldots, n - r \tag{5.245}$$

Equation (5.245) is a set of $n - r$ equations for $n - r + 1$ unknowns $\eta_1, \eta_2, \ldots, \eta_{n-r}, \alpha$:

$$F_i(\eta, \alpha) = 0 \qquad i = 1, 2, \ldots, n - r \tag{5.246}$$

The Jacobi matrix $\{\partial F_i/\partial \eta_j\}$ may be written

$$\Gamma_\eta(\eta, \alpha) = \begin{pmatrix} \dfrac{\partial F_1}{\partial \eta_1} & \dfrac{\partial F_1}{\partial \eta_2} & \cdots & \dfrac{\partial F_1}{\partial \eta_{n-r}} \\[2mm] \dfrac{\partial F_2}{\partial \eta_1} & & \cdots & \dfrac{\partial F_2}{\partial \eta_{n-r}} \\ \cdot & & & \\ \cdot & & & \\ \cdot & & & \\ \dfrac{\partial F_{n-r}}{\partial \eta_1} & & \cdots & \dfrac{\partial F_{n-r}}{\partial \eta_{n-r}} \end{pmatrix} \qquad (5.247)$$

where

$$\frac{\partial F_i}{\partial \eta_j} = \sum_{k=1}^{n} b_{ik} p_{kj}(b) \qquad (5.248)$$

It can be written for a regular matrix (5.247) using the implicit function theorem

$$\frac{d\eta}{d\alpha} = -\Gamma_\eta^{-1}(\eta, \alpha) \frac{\partial F}{\partial \alpha} \qquad (5.249)$$

where

$$\frac{d\eta}{d\alpha} = \left(\frac{d\eta_1}{d\alpha}, \frac{d\eta_2}{d\alpha}, \ldots, \frac{d\eta_{n-r}}{d\alpha}\right)^T$$

and

$$\frac{dF}{d\alpha} = \left(\frac{\partial F_1}{\partial \alpha}, \frac{\partial F_2}{\partial \alpha}, \ldots, \frac{\partial F_{n-r}}{\partial \alpha}\right)^T$$

We note that

$$\frac{\partial F_i}{\partial \alpha} = \sum_{j=1}^{n} b_{ij} q_j(b) \qquad i = 1, 2, \ldots, n - r \qquad (5.250)$$

Consider a modified regular Jacobi matrix $\Gamma_\alpha(\eta, \alpha)$:

$$\Gamma_\alpha(\eta, \alpha) = \begin{pmatrix} \dfrac{\partial F_1}{\partial \alpha} & \dfrac{\partial F_1}{\partial \eta_2} & \cdots & \dfrac{\partial F_1}{\partial \eta_{r-r}} \\[2mm] \dfrac{\partial F_2}{\partial \alpha} & & \cdots & \dfrac{\partial F_2}{\partial \eta_{n-r}} \\ \cdot & & & \\ \cdot & & & \\ \cdot & & & \\ \dfrac{\partial F_{n-r}}{\partial \alpha} & \dfrac{\partial F_{n-r}}{\partial \eta_2} & \cdots & \dfrac{\partial F_{n-r}}{\partial \eta_{n-r}} \end{pmatrix} \qquad (5.251)$$

where the particular partial derivatives are determined by (5.248) and (5.250). We obtain in an analogous way

$$\begin{pmatrix} \dfrac{d\alpha}{d\eta_1} \\[2mm] \dfrac{d\eta_2}{d\eta_1} \\ \cdot \\ \cdot \\ \cdot \\ \dfrac{d\eta_{n-r}}{d\eta_1} \end{pmatrix} = -\Gamma_\alpha^{-1}(\eta, \alpha) \frac{\partial F}{\partial \eta_1} \qquad (5.252)$$

where

$$\frac{\partial F}{\partial \eta_1} = \left(\frac{\partial F_1}{\partial \eta_1}, \frac{\partial F_2}{\partial \eta_1}, \ldots, \frac{\partial F_{n-r}}{\partial \eta_1}\right)^T$$

Clearly, we can differentiate with respect to any arbitrary η_j $(j = 1, 2, \ldots, n - r)$. The solution of these differential equations satisfies continuously the original boundary problem (5.233)–(5.235) assuming that the initial conditions (5.253) and (5.254)

$$\eta(\alpha_0) = \eta_0 \tag{5.253}$$

$$\alpha(\eta_{1,0}) = \alpha_0, \qquad \eta_i(\eta_{1,0}) = \eta_{i\,0} \qquad i = 2, 3, \ldots, n - r \tag{5.254}$$

may be found for (5.249) and (5.252), respectively. Of course, the conditions (5.253) and (5.254) must obey (5.246); that is,

$$F_i(\eta_0, \alpha_0) = 0 \qquad i = 1, 2, \ldots, n - r \tag{5.255}$$

The initial conditions (5.253) and (5.254) can be calculated from (5.233)–(5.235) for $\alpha = \alpha_0$. Sometimes the initial condition may be established analytically.

It should be noted that this approach is fairly general; that is, after minor modifications it can be adopted to such events as nonlinear boundary conditions containing the parameter α, mixed boundary conditions, more parameters considered, and so on.

Consider, for instance, the boundary conditions (5.235) in the form

$$g_i[y_1(b), \ldots, y_n(b), \alpha] = 0 \qquad i = 1, 2, \ldots, n - r \tag{5.256}$$

where g_i is an arbitrary nonlinear function. Now (5.248) and (5.250) become

$$\frac{\partial F_i}{\partial \eta_j} = \sum_{k=1}^{n} \frac{\partial g_i}{\partial y_k(b)} p_{kj}(b) \tag{5.248'}$$

and

$$\frac{\partial F_i}{\partial \alpha} = \sum_{j=1}^{n} \frac{\partial g_i}{\partial y_j(b)} q_j(b) + \frac{\partial g_i}{\partial \alpha} \tag{5.250'}$$

Using these expressions, the algorithm may also be adopted for nonlinear boundary conditions. Moreover, also having the nonlinear conditions for $x = a$,

$$G_i[y_1(a), \ldots, y_n(a), \alpha] = 0 \qquad i = 1, 2, \ldots, r \tag{5.257}$$

the procedure for determination of the unknown initial conditions (5.238) is of the trial-and-error type. The initial conditions (5.242) and (5.243) for auxiliary equations can be developed by differentiation of (5.257) with respect to η and α. However, another way to handle nonlinear boundary conditions (5.256) and (5.257) is possible. All initial conditions at $x = a$ are guessed:

$$y_i(a) = \eta_i \qquad i = 1, 2, \ldots, n \tag{5.258}$$

and n nonlinear equations (5.256) and (5.257) for $n + 1$ variables $\eta_1, \ldots, \eta_n$ and α must be satisfied if the initial guess is correct. Of course, in an analogous way the governing equations can also be developed for mixed boundary conditions (see Example 5.18).

The analytical integration of equations presented here is obviously rather limited; however, numerical integration may be readily performed. Next, some aspects of the construction of the numerical algorithm are dealt with briefly.

It is necessary to integrate many times the initial value problem (5.233), (5.239), and (5.240) subject to initial conditions (5.236), (5.238), (5.242), and (5.243); that is, the set of $n(n - r + 2)$ ordinary differential equations of first order must be integrated. If the dimensionality of the problem in question is high, the matrix of auxiliary equations is huge and hence the method of integration ought to be deliberately chosen. The Runge–Kutta methods, especially those with automatic step-size control (e.g., Merson's modification), appear to work effectively. Step-size control is adopted for checking the y value only. To construct the dependence $\eta(\alpha)$ and $\alpha(\eta_1)$, it is necessary to integrate the differential equations (5.249) and (5.252) subject to initial conditions (5.253) and (5.254), respectively. The right-hand sides in these equations may be established after solving a set of linear algebraic equations (or after inversion of Γ), where the elements of the matrix Γ as well as the right-hand sides of these linear equations can be calculated from (5.248) and (5.250) using the values $y_i(b)$, $p_{ij}(b)$, and $q_i(b)$. The Runge–Kutta–Merson method again provides a powerful alternative to integration of either (5.249) or (5.252). It is sufficient for the event of the unique solution $\eta(\alpha)$ to integrate relations (5.249). However, for branching points at the curve $\eta(\alpha)$, the Jacobian matrix (5.248) becomes singular and (5.249) cannot be integrated. Assuming that $\eta(\alpha)$ is a smooth curve, integration of relations (5.252) can be adopted. Because of this interchange in both variables, multistep integration techniques (e.g., the Adams methods) are not recommended. This approach results in the dependence $\eta(\alpha)$, and the solution $y(x)$ of the original problem (5.233) for mesh values η and α can also be obtained.

The accuracy of the solution obtained may be readily improved further taking into consideration the Newton method

$$\eta_{\text{corr}} = \eta + \Delta\eta \tag{5.259}$$

The increment $\Delta\eta$ obeys

$$\Gamma_\eta(\eta, \alpha)\, \Delta\eta = -F(\eta, \alpha) \tag{5.260}$$

where

$$F = (F_1, F_2, \ldots, F_{n-r})^T$$

The development, of course, can be reversed and (5.233)–(5.235) may be handled as being an initial value problem at $x = b$. Generally, it seems convenient to convert the problem (5.233)–(5.235) into an associated initial value problem at $x = a$ if $r < n/2$. However, a contingent inherent instability may impose severe restrictions on the direction of integration.

For problems where branching points occur, the differential equations (5.249) must be integrated. However, in the vicinity of the particular branching point the algorithm is switched over to integration of (5.252), or vice versa.

Another possibility for overcoming this difficulty is to introduce a new independent variable—the arc length of the solution locus t.

Differentiation of (5.246) with respect to t gives

$$\frac{dF_i}{dt} = \sum_{j=1}^{n} \frac{\partial F_i}{\partial \eta_j} \frac{d\eta_j}{dt} + \frac{\partial F_i}{\partial \alpha} \frac{d\alpha}{dt} = 0 \quad i = 1, 2, \dots, n - r \qquad (5.261)$$

An additional equation

$$\sum_{i=1}^{n-r} \left(\frac{d\eta_i}{dt}\right)^2 + \left(\frac{d\alpha}{dt}\right)^2 = 1 \qquad (5.262)$$

determines the parameter t as the arc length of the curve $\eta(\alpha)$ in the space $\eta - \alpha$. Equation (5.261) forms a system of $n - r$ linear equations for $n - r + 1$ unknowns $d\eta_i/dt$, $i = 1, \dots, n - r$, $d\alpha/dt$. Suppose that the $(n - r) \times (n - r)$ matrix Γ_k constructed from the Jacobian matrix $\{\partial F/\partial \eta_1, \dots, \partial F/\partial \eta_{n-r}, \partial F/\partial \alpha\}$ by eliminating the kth column is regular (for certain t and k, $1 \leq k \leq n - r + 1$). Equation (5.261) can be then presolved in the form

$$\frac{d\eta_i}{dt} = \beta_i \frac{d\eta_k}{dt} \quad i = 1, 2, \dots, k - 1, k + 1, \dots, n - r \qquad (5.263a)$$

$$\frac{d\alpha}{dt} = \beta_{n-r+1} \frac{d\eta_k}{dt} \qquad (5.263b)$$

On substituting (5.263) into (5.262), we obtain

$$\left(\frac{d\eta_k}{dt}\right)^2 = \left(1 + \sum_{\substack{i=1 \\ i \neq k}}^{n-r} \beta_i^2 + \beta_{n-r+1}^2\right)^{-1} \qquad (5.264)$$

The sign of $d\eta_k/dt$ is given by the orientation of the parameter t along the curve. All other derivatives $d\eta_i/dt$ and $d\alpha/dt$ are then determined by (5.263).

Next, and example with mixed boundary conditions is presented.

Example 5.18

A simple recycle stream in chemical tubular reactors may be described by

$$\frac{dy_1}{dx} = \alpha(1 - y_1) \exp\left(\frac{y_2}{1 + \epsilon y_2}\right) \qquad (5.265)$$

$$\frac{dy_2}{dx} = \alpha H(1 - y_1) \exp\left(\frac{y_2}{1 + \epsilon y_2}\right) - \beta(y_2 - c) \qquad (5.266)$$

subject to the boundary conditions

$$-y_1(0) + \lambda y_1(1) = 0 \qquad (5.267)$$

$$-y_2(0) + \mu y_2(1) = 0 \qquad (5.268)$$

On choosing

$$y_1(0) = \eta_1, \qquad y_2(0) = \eta_2 \qquad (5.269)$$

the initial value problem (5.265), (5.266), (5.239), and (5.240) must be integrated with initial conditions (5.269), (5.242), and (5.243). Equations (5.246) are now

$$F_1(\eta, \alpha) = \lambda y_1(1, \eta, \alpha) - \eta_1 = 0 \qquad (5.270a)$$

$$F_2(\eta, \alpha) = \mu y_2(1, \eta, \alpha) - \eta_2 = 0 \qquad (5.270b)$$

The Runge–Kutta–Merson technique has been used for integration of (5.265), (5.266), (5.239), and (5.240). In addition, the same procedure has been employed for integration of (5.249) or (5.252). The results of integration are reported in the Table 5-31. It is obvious that the results depend weakly on the accuracy of the inner-loop integration method. However, a strong dependence on the accuracy of the outer-loop integration method may be noted. For $\beta = 0$ the relation $\eta_2 = H\eta_1$ should be valid.

TABLE 5-31

RESULTS FOR EXAMPLE 5.18: $H = 6$, $\epsilon = 0.05$, $\beta = 0$, $c = 0$, $\lambda = \mu = 0.5^a$

	IM1 accuracy $\sim 10^{-4}$				IM1 accuracy $\sim 10^{-6}$			
	IM2 accuracy				IM2 accuracy			
	$\sim 10^{-4}$		$\sim 10^{-6}$		$\sim 10^{-4}$		$\sim 10^{-6}$	
η_1	α	η_2	α	η_2	α	η_2	α	η_2
0.0	0.00000	0.00000	0.00000	0.00000	0.00000	0.00000	0.00000	0.00000
0.1	0.05009	0.60000	0.05009	0.60000	0.05020	0.60000	0.05020	0.60000
0.2	0.05938	1.19999	0.05638	1.19999	0.05646	1.19999	0.05646	1.19999
0.3	0.05320	1.79999	0.05320	1.79999	0.05325	1.79999	0.05325	1.79999
0.35	0.05164	2.09999	0.05164	2.09999	0.05168	2.09999	0.05168	2.09999
0.4	0.05126	2.39999	0.05126	2.39999	0.05131	2.39999	0.05131	2.39998
0.45	0.05391	2.69999	0.05391	2.69998	0.05397	2.69998	0.05397	2.69998
0.46	0.05530	2.75999	0.05530	2.75998	0.05536	2.75998	0.05537	2.75998
0.47	0.05733	2.81998	0.05733	2.81998	0.05471	2.81998	0.05741	2.81998
0.48	0.06052	2.87998	0.06052	2.87998	0.06063	2.87998	0.06063	2.87997

^{a}IM1, method of integration of (5.252): outer loop; IM2, method of integration of (5.265), (5.266), (5.239), and (5.240): inner loop.

5.6 Evaluation of Branching Points in Nonlinear Boundary Value Problems

Nonlinear boundary value problems often exhibit more than one solution for a given set of parameters. The kinds of problems we treat arise in a wide variety of applications. In particular, some problems in chemical engineering, such as axial mixing in tubular reactors, heat and mass transfer within and outside a porous catalyst, boundary layer problems for exothermic reaction, and equilibrium of neighboring drops at different potentials, provide an excellent motivation for our study. Of course, it is a difficult task to establish all possible solutions if more than one is expected. If the solutions are very close to each other, the majority of straightforward methods fail to calculate the relevant profiles because of oscillation of a particular iteration procedure between both solutions. If both neighboring profiles coincide, a branching profile results. Because in a functional space this profile corresponds to a point, the term "branching point" is used below. A great savings of effort can usually be made if it is possible to evaluate all branching points before the calculation of solutions for particular values of governing parameters. It is the purpose of this section to present a straightforward method for an evaluation of the branching points appearing in a two-point nonlinear boundary value problem. The method of solution will be outlined and employed to solve three practical problems.

5.6.1 Branching Points for One Second-Order Equation

Consider the boundary value problem

$$y'' = f(x, y, y', \alpha) \tag{5.271}$$

subject to the linear boundary conditions

$$\alpha_0 y(a) + \beta_0 y'(a) = \gamma_0 \tag{5.272a}$$

$$\alpha_1 y(b) + \beta_1 y'(b) = \gamma_1 \tag{5.272b}$$

as a function of the parameter α. On choosing the initial condition

$$y(a) = A \tag{5.273}$$

and assuming that $\beta_0 \neq 0$, (5.272a) yields

$$y'(a) = \frac{1}{\beta_0}(\gamma_0 - \alpha_0 A) \tag{5.274}$$

In accordance with the preceding section a set of differential equations for the auxiliary variables

$$p(x) = \frac{\partial y(x)}{\partial A}, \qquad q(x) = \frac{\partial y(x)}{\partial \alpha} \tag{5.275}$$

can be developed:

$$p'' = \frac{\partial f}{\partial y}p + \frac{\partial f}{\partial y'}p', \qquad p(a) = 1, \quad p'(a) = -\frac{\alpha_0}{\beta_0} \tag{5.276}$$

$$q'' = \frac{\partial f}{\partial y}q + \frac{\partial f}{\partial y'}q' + \frac{\partial f}{\partial \alpha}, \qquad q(a) = q'(a) = 0 \tag{5.277}$$

After rewriting (5.272b), we have

$$F_1(A, \alpha) = \alpha_1 y(b, A, \alpha) + \beta_1 y'(b, A, \alpha) - \gamma_1 = 0 \tag{5.278}$$

A necessary condition for a branching point† yields

$$\frac{d\alpha}{dA} = -\frac{\partial F_1/\partial A}{\partial F_1/\partial \alpha} = 0$$

This equation gives, assuming that $\partial F_1/\partial \alpha \neq 0$,

$$F_2(A, \alpha) = \frac{\partial F_1}{\partial A} = \alpha_1 p(b, A, \alpha) + \beta_1 p'(b, A, \alpha) = 0 \tag{5.279}$$

---

†Here limit point is considered, i.e. a point where two branches of the solution coincide and then disappear.

Equations (5.278) and (5.279) are sufficient to establish the unknown coordinates $(\bar{A}, \bar{\alpha})$ of a particular branching point. For instance, using the Newton method to solve (5.278) and (5.279), the Jacobian matrix

$$\Gamma = \begin{pmatrix} \dfrac{\partial F_1}{\partial A} & \dfrac{\partial F_1}{\partial \alpha} \\[2mm] \dfrac{\partial F_2}{\partial A} & \dfrac{\partial F_2}{\partial \alpha} \end{pmatrix} \tag{5.280}$$

must be determined.

The values of the first derivatives of F_1 can be evaluated because

$$\frac{\partial F_1}{\partial A} = \alpha_1 p(b, A, \alpha) + \beta_1 p'(b, A, \alpha)$$

$$\frac{\partial F_1}{\partial \alpha} = \alpha_1 q(b, A, \alpha) + \beta_1 q'(b, A, \alpha) \tag{5.281}$$

To calculate the first derivatives of F_2, two additional auxiliary equations for the variables

$$r = \frac{\partial^2 y}{\partial A^2}, \qquad s = \frac{\partial^2 y}{\partial A \, \partial \alpha} \tag{5.282}$$

must be developed.

These equations result after differentiation of (5.276) with respect to A and α as well:

$$r'' = \frac{\partial^2 f}{\partial y^2} p^2 + 2 \frac{\partial^2 f}{\partial y \, \partial y'} p p' + \frac{\partial^2 f}{\partial (y')^2} (p')^2 + \frac{\partial f}{\partial y} r + \frac{\partial f}{\partial y'} r' \tag{5.283}$$

$$s'' = \frac{\partial^2 f}{\partial y^2} pq + \frac{\partial^2 f}{\partial y \, \partial y'} (pq' + p'q) + \frac{\partial^2 f}{\partial y \, \partial \alpha} p$$

$$+ \frac{\partial^2 f}{\partial (y')^2} p'q' + \frac{\partial^2 f}{\partial y' \, \partial \alpha} p' + \frac{\partial f}{\partial y} s + \frac{\partial f}{\partial y'} s' \tag{5.284}$$

Evidently, the initial conditions are

$$r(a) = r'(a) = s(a) = s'(a) = 0 \tag{5.285}$$

Now the first derivatives of F_2 can be evaluated easily:

$$\frac{\partial F_2}{\partial A} = \alpha_1 r(b, A, \alpha) + \beta_1 r'(b, A, \alpha)$$

$$\frac{\partial F_2}{\partial \alpha} = \alpha_1 s(b, A, \alpha) + \beta_1 s'(b, A, \alpha) \tag{5.286}$$

A general algorithm for the calculation of $\bar{\alpha}$ and $\bar{A}$ is the following:

1. Initial guess of A and α.
2. Integration of five second-order differential equations (5.271),(5.276), (5.277), (5.283), and (5.284) with the initial conditions (5.273), (5.274), (5.276), (5.277), and (5.285) from $x = a$ to $x = b$.
3. Enumeration of F_1 and F_2 according to (5.278) and (5.279).
4. Enumeration of the elements of the Jacobian matrix (5.280) according to (5.281) and (5.286).

5. Evaluation of a new approximation of A and α using the Newton method:

$$\begin{pmatrix} A \\ \alpha \end{pmatrix}^{\text{new}} = \begin{pmatrix} A \\ \alpha \end{pmatrix}^{\text{old}} - \lambda \Gamma^{-1} \begin{pmatrix} F_1 \\ F_2 \end{pmatrix}^{\text{old}} \qquad (5.287)$$

where $\lambda \in (0, 1>$ is a damping factor.

6. The calculation is stopped for $\| (A, \alpha)^{\text{new}} - (A, \alpha)^{\text{old}} \| < \epsilon$ (ϵ is the prescribed tolerance); if not, go to step 2.

The method developed is illustrated now for a calculated example.

Example 5.19

Heat and mass transfer in a porous catalyst of the plate shape is described by

$$\frac{d^2 y}{dx^2} = \delta y \exp\left[\frac{\gamma \beta (1 - y)}{1 + \beta (1 - y)}\right]$$

$$\frac{dy(0)}{dx} = 0, \qquad y(1) = 1$$

For $y(0) = A$ and setting $\alpha = \delta$, we can readily develop a set of differential equations (5.276), (5.277), (5.283), and (5.284); for example, for r, we have

$$r'' = \delta \left[\frac{\gamma \beta}{(1 + \beta(1-y))^2} \left(\frac{\gamma \beta y}{(1 + \beta(1-y))^2} - 1 \right) - \frac{\gamma \beta (1 + \beta(1-y)) + 2\gamma \beta^2 y}{(1 + \beta(1-y))^3} \right]$$
$$\times \exp\left[\frac{\gamma \beta (1-y)}{1 + \beta(1-y)} \right] p^2 + \delta \left[1 - \frac{\gamma \beta y}{(1 + \beta(1-y))^2} \right]$$
$$\times \exp\left[\frac{\gamma \beta (1-y)}{1 + \beta(1-y)} \right] r$$

The relevant initial value problems have been integrated by virtue of Merson's method. Some results of our computations are shown in Table 5-32. The convergence properties of this procedure are very good.

TABLE 5-32
COURSE OF ITERATION FOR A SINGLE SECOND-ORDER EQUATION:
$\gamma = 20$, $\beta = 0.4$, $\lambda = 1$

Iteration	A	$\alpha = \delta$	F_1	F_2
0	0.8000	0.1000	−0.04772	0.23304
1	0.7954	0.1367	−0.00104	0.01286
2	0.7929	0.1376	0.00002	−0.00003
3	0.7928	0.1376		
0	0.2000	0.1000	0.15636	−0.02447
1	0.2218	0.0771	−0.00660	0.04470
2	0.2272	0.0779	−0.00016	0.00103
3	0.2273	0.0779		
0	0.6000	0.1000	−0.03936	−0.31023
1	0.8832	0.1451	0.01903	0.24825
2	0.8024	0.1463	0.01065	−0.01402
3	0.7928	0.1372	−0.00040	0.00165
4	0.7928	0.1376		

5.6.2 Branching Points for Two Second-Order Equations

In this section we develop an algorithm for a set of two second-order differential equations; the general algorithm is presented below.

Consider a set of two differential equations

$$y_1'' = f_1(x, y_1, y_2, y_1', y_2', \alpha)$$
$$y_2'' = f_2(x, y_1, y_2, y_1', y_2', \alpha)$$

(5.288)

subject to the linear boundary conditions

$$\alpha_0^i y_i(a) + \beta_0^i y_i'(a) = \gamma_0^i \qquad i = 1, 2 \tag{5.289a}$$

$$\alpha_1^i y_i(b) + \beta_1^i y_i'(b) = \gamma_1^i \qquad i = 1, 2 \tag{5.289b}$$

Let us select the following two initial conditions:

$$y_1(a) = A_1, \qquad y_2(a) = A_2 \tag{5.290}$$

and supposing that $\beta_0^i \neq 0$, $i = 1, 2$, we arrive at

$$y_i'(a) = \frac{1}{\beta_0^i}(\gamma_0^i - \alpha_0^i A_i) \qquad i = 1, 2 \tag{5.291}$$

In the same way as in the one-dimensional case, a set of differential equations for auxiliary variables

$$p_{ij} = \frac{\partial y_i(x)}{\partial A_j}, \qquad q_i(x) = \frac{\partial y_i(x)}{\partial \alpha}$$

$$p_{ijk} = \frac{\partial^2 y_i(x)}{\partial A_j \partial A_k}, \qquad q_{ij}(x) = \frac{\partial^2 y_i(x)}{\partial A_j \partial \alpha} \qquad i, j, k = 1, 2; \quad k \geq j \tag{5.292}$$

can be developed with the initial conditions

$$p_{ij}(a) = p_{ij}'(a) = 0 \qquad i \neq j$$

$$p_{ii}(a) = 1$$

$$p_{ii}'(a) = -\frac{\alpha_0^i}{\beta_0^i}$$

$$q_i(a) = q_i'(a) = p_{ijk}(a) = p_{ijk}'(a)$$
$$= q_{ij}(a) = q_{ij}'(a) = 0 \qquad i, j, k = 1, 2; \quad k \geq j \tag{5.293}$$

For example, for q_{ij} we have

$$q_{ij}'' = \sum_{m=1}^{2}\left[\sum_{s=1}^{2}\left(\frac{\partial^2 f_i}{\partial y_m \partial y_s}q_m p_{sj} + \frac{\partial^2 f_i}{\partial y_m \partial y_s'}q_m p_{sj}'\right.\right.$$
$$+ \frac{\partial^2 f_i}{\partial y_m' \partial y_s}q_m' p_{sj} + \frac{\partial^2 f_i}{\partial y_m' \partial y_s'}q_m' p_{sj}'\right)$$
$$\left. + \frac{\partial f_i}{\partial y_m}q_{mj} + \frac{\partial f_i}{\partial y_m'}q_{mj}'\right] + \sum_{s=1}^{2}\left(\frac{\partial^2 f_i}{\partial y_s \partial \alpha}p_{sj}\right.$$
$$\left. + \frac{\partial^2 f_i}{\partial y_s' \partial \alpha}p_{sj}'\right) \qquad i, j = 1, 2.$$

After inserting the values of y and y' [which are the results of integration of (5.288), (5.290), and (5.291) into (5.289b)], two nonlinear equations can be written

$$F_k(A_1, A_2, \alpha) = \alpha_1^k y_k(b, A_1, A_2, \alpha) + \beta_1^k y'_k(b, A_1, A_2, \alpha) - \gamma_1^k = 0$$
$$k = 1, 2 \tag{5.294}$$

The necessary condition for the branching point is

$$F_3(A_1, A_2, \alpha) = \det \Gamma(A_1, A_2, \alpha) = 0 \tag{5.295}$$

Here Γ is the Jacobian matrix

$$\Gamma = \begin{pmatrix} \dfrac{\partial F_1}{\partial A_1} & \dfrac{\partial F_1}{A_2} \\ \dfrac{\partial F_2}{\partial A_1} & \dfrac{\partial F_2}{\partial A_2} \end{pmatrix} \tag{5.296}$$

This condition results from the implicit function theorem.

Clearly, the branching point can be evaluated as a result of solving three nonlinear equations (5.294) and (5.295), which yield the unknowns A_1, A_2, and α. If the Newton method is used to solve these nonlinear equations, the terms $\partial F_i/\partial A_j$, $\partial F_i/\partial \alpha$, $i = 1, 2, 3; j = 1, 2$, must be evaluated. To evaluate all these terms, the following relations can be developed:

$$\frac{\partial F_i}{\partial A_j} = \alpha_1^i p_{ij}(b) + \beta_1^i p'_{ij}(b) \qquad i, j = 1, 2 \tag{5.297a}$$

$$\frac{\partial F_i}{\partial \alpha} = \alpha_1^i q_i(b) + \beta_1^i q'_i(b) \qquad i = 1, 2 \tag{5.297b}$$

$$\frac{\partial F_3}{\partial A_j} = \frac{\partial F_1}{\partial A_1} \frac{\partial^2 F_2}{\partial A_2 \, \partial A_j} + \frac{\partial F_2}{\partial A_2} \frac{\partial^2 F_1}{\partial A_1 \, \partial A_j} - \frac{\partial F_1}{\partial A_2} \frac{\partial^2 F_2}{\partial A_1 \, \partial A_j}$$
$$- \frac{\partial F_2}{\partial A_1} \frac{\partial^2 F_1}{\partial A_2 \, \partial A_j} \qquad j = 1, 2 \tag{5.298a}$$

Here, for example,

$$\frac{\partial^2 F_2}{\partial A_2 \, \partial A_j} = \alpha_1^2 p_{22j}(b) + \beta_1^2 p'_{22j}(b)$$

and

$$\frac{\partial F_3}{\partial \alpha} = \frac{\partial F_1}{\partial A_1} \frac{\partial^2 F_2}{\partial A_2 \, \partial \alpha} + \frac{\partial F_2}{\partial A_2} \frac{\partial^2 F_1}{\partial A_1 \, \partial \alpha} - \frac{\partial F_1}{\partial A_2} \frac{\partial^2 F_2}{\partial A_1 \, \partial \alpha} - \frac{\partial F_2}{\partial A_1} \frac{\partial^2 F_1}{\partial A_2 \, \partial \alpha} \tag{5.298b}$$

Here, for example, we have

$$\frac{\partial^2 F_2}{\partial A_2 \, \partial \alpha} = \alpha_1^2 q_{22}(b) + \beta_1^2 q'_{22}(b)$$

To illustrate the method for a set of two second-order differential equations, an example is presented.

Example 5.20

Axial heat and mass transfer in tubular reactors is described by two second-order differential equations

$$y'' - \text{Pe}_M y' + \text{Pe}_M \text{Da} (1 - y) \exp\left(\frac{\theta}{1 + \epsilon\theta}\right) = 0$$

$$\theta'' - \text{Pe}_H \theta' - \text{Pe}_H \beta(\theta - \theta_c) + \text{Pe}_H \text{Da} B(1 - y) \exp\left(\frac{\theta}{1 + \epsilon\theta}\right) = 0$$

subject to the boundary conditions

$$y'(1) = \theta'(1) = 0$$

$$\text{Pe}_M y(0) - y'(0) = 0, \qquad \text{Pe}_H \theta(0) - \theta'(0) = 0$$

We denote

$$y(1) = A_1, \qquad \theta(1) = A_2, \qquad \alpha = \text{Da}$$

To find the branching points it is necessary to integrate 18 differential equations of second order (initial value problem). The course of the iteration process is reported in Table 5-33. The particular branching points are displayed in Fig. 5-21, where the

TABLE 5-33
COURSE OF ITERATION FOR THE VALUES OF PARAMETERS
$B = 12$, $\beta = 2$, $\epsilon = \theta_c = 0$, $\text{Pe}_H = \text{Pe}_M = 2$, $\lambda = 1$

Iteration	$A_1 = y(1)$	$A_2 = \theta(1)$	$\alpha = \text{Da}$	F_1	F_2	F_3
0	0.2000	1.0000	0.1000	0.00469	0.36868	−0.99197
1	0.3759	1.8061	0.1330	0.00003	0.00138	−0.00444
2	0.3871	1.8652	0.1319	0.00000	0.00001	−0.00000
3	0.3871	1.8652	0.1319			
0	0.7500	3.8000	0.1150	−0.01787	−0.02230	0.45113
1	0.7933	4.0761	0.1108	0.00424	0.00815	−0.11603
2	0.7860	4.0313	0.1115	0.00018	0.00056	−0.00357
3	0.7858	4.0296	0.1115			
0	0.9000	4.0000	0.1400	−0.03548	0.50492	−1.89113
1	0.9713	3.6593	0.1338	0.23797	2.44774	8.58409
2	0.9533	3.6495	0.1206	0.06945	0.61371	3.94400
3	0.9309	3.8366	0.1230	0.02078	0.22125	0.80386
4	0.9207	3.8911	0.1216	0.00393	0.03630	0.11828
5	0.9185	3.8990	0.1214	0.00016	0.00145	0.00421
6	0.9184	3.8993	0.1214			
0	0.9800	3.8000	0.1150	0.47920	3.97732	20.8162
1	0.9677	3.3720	0.1175	0.00328	0.09019	1.31692
2	0.9701	3.3476	0.1154	−0.00113	−0.01242	−0.40002
3	0.9696	3.3527	0.1155	−0.00003	−0.00038	−0.01564
4	0.9696	3.3529	0.1155			

dependences $A_1(\alpha)$ and $A_2(\alpha)$ are drawn. It will be noted that the rate of convergence of the Newton iteration process is high.

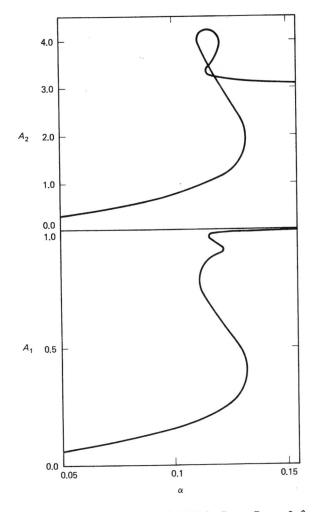

Figure 5-21 Dependences $A_1(\alpha)$ and $A_2(\alpha)$ for $\mathrm{Pe}_H = \mathrm{Pe}_M = 2$, $\beta = 2$, $\theta_c = \epsilon = 0$, $B = 12$. Example 5.20.

5.6.3 Branching Points for a System of Nonlinear Algebraic Equations

Consider a system of nonlinear algebraic equations depending on an actual parameter α:

$$F_i(x_1, \ldots, x_N, \alpha) = 0 \qquad i = 1, 2, \ldots, N \qquad (5.299)$$

Let us study a dependence of the solution $x(\alpha)$ of (5.299) on the parameter α with regard to the determination of branching points. The necessary condition to establish the branching points follows from the implicit function theorem:

$$F_{N+1}(x_1, \ldots, x_N, \alpha) = \det G(x_1, \ldots, x_N, \alpha) = 0 \qquad (5.300)$$

Here G is the Jacobian matrix with the elements

$$g_{ij} = \frac{\partial F_i(x_1, \ldots, x_N, \alpha)}{\partial x_j} \qquad i, j = 1, 2, \ldots, N \qquad (5.301)$$

It is evident that (5.300) and (5.299) are capable of determining the coordinates $(x_1^+, \ldots, x_N^+, \alpha^+)$ of branching points in question. Equations (5.300) and (5.299) constitute a system of $N + 1$ nonlinear algebraic equations for $N + 1$ unknowns; to solve them we can make use of the Newton method:

$$X^{k+1} = X^k - \lambda \Gamma_F^{-1}(X^k) F(X^k) \qquad (5.302)$$

Here we have denoted $X = (x_1, \ldots, x_N, \alpha)^T$, $F = (F_1, \ldots, F_{N+1})^T$, $\lambda \in (0, 1 >$. The elements of the matrix Γ_F, γ_{ij} are

$$\gamma_{ij} = \frac{\partial F_i}{\partial x_j} \qquad i = 1, 2, \ldots, N + 1 \quad j = 1, 2, \ldots, N$$

$$\gamma_{i, N+1} = \frac{\partial F_i}{\partial \alpha} \qquad i = 1, 2, \ldots, N + 1 \qquad (5.303)$$

To avoid the divergence of the iteration process, the condition

$$\| F(X^{k+1}) \| < \| F(X^k) \|$$

is to be tested. Each iteration is started with $\lambda = 1$. If the above presented inequality is not fulfilled, the value of λ is, for example, halved. The expressions for γ_{ij}, $i \neq N + 1$, can be developed easily. From the definition of the determinant, we obtain

$$\frac{\partial \det G}{\partial g_{ij}} = (-1)^{i+j} G_{ij} \qquad (5.304)$$

where G_{ij} is a minor, that is, a determinant of the matrix $(N - 1) \times (N - 1)$ which arises from G by eliminating the row i and the column j. For partial derivatives of (5.300), we have

$$\gamma_{N+1, m} = \frac{\partial F_{N+1}}{\partial x_m} = \sum_{i, j=1}^{N} (-1)^{i+j} G_{ij}(X) \frac{\partial^2 F_i}{\partial x_j \, \partial x_m} \qquad (5.304b)$$

$$m = 1, 2, \ldots, N$$

and

$$\gamma_{N+1, N+1} = \frac{\partial F_{N+1}}{\partial \alpha} = \sum_{i, j=1}^{N} (-1)^{i+j} G_{ij}(X) \frac{\partial^2 F_i}{\partial x_j \, \partial \alpha} \qquad (5.304c)$$

Numerical calculation for a large set of equations based on (5.300) may be associated with computational difficulties. Recently, Seydel† proposed an alter-

---

†R. Seydel, *Numer. Math. 33*, 339 (1979).

native approach in which condition (5.300) is replaced by

$$\sum_{j=1}^{N} g_{ij}\phi_j = 0 \qquad i = 1, 2, \ldots, N \tag{5.305a}$$

$$\phi_k = 1 \qquad k \text{ fixed, } 1 \leq k \leq N \tag{5.305b}$$

To determine the branching point it is necessary to solve $2N + 1$ equations (5.299), (5.305a), and (5.305b) for $2N + 1$ variables $\alpha, x_1, \ldots, x_N, \phi_1, \ldots, \phi_N$. This set of equations is larger, but the numerical realization is simpler.

5.6.4 Branching Points for a System of First-Order Differential Equations

Consider a set of ordinary differential equations

$$\frac{dy_i}{dx} = f_i(x, y_1, \ldots, y_n, \alpha) \qquad i = 1, 2, \ldots, n \tag{5.306}$$

subject to the linear two-point boundary conditions

$$\sum_{j=1}^{n} a_{ij} y_j(a) = c_i \qquad i = 1, 2, \ldots, r \tag{5.307a}$$

$$\sum_{i=1}^{n} b_{ij} y_j(b) = d_i \qquad i = 1, 2, \ldots, n - r \tag{5.307b}$$

On choosing

$$y_i(a) = \eta_i \qquad i = 1, 2, \ldots, n - r \tag{5.308}$$

and after solving (5.307a) with respect to $y_i(a)$, $i = n - r + 1, \ldots, n$, we can obtain the relations

$$y_i(a) = h_i + \sum_{j=1}^{n-r} h_{ij} \eta_j \qquad i = n - r + 1, \ldots, n \tag{5.309}$$

Here h_i and h_{ij} are the calculated coefficients, described in detail in Section 5.5. Differentiation of (5.306) with respect to η_j gives rise to the auxiliary differential equations

$$p'_{ij} = \sum_{m=1}^{n} \frac{\partial f_i}{\partial y_m} p_{mj} \qquad i = 1, \ldots, n; \quad j = 1, \ldots, n - r \tag{5.310}$$

subject to the initial conditions

$$p_{ii}(a) = 1, \qquad p_{ij}(a) = 0 \qquad i, j = 1, 2, \ldots, n - r \tag{5.311a}$$

$$p_{ij}(a) = h_{ij} \qquad \begin{array}{l} i = n - r + 1, \ldots, n; \\ j = 1, 2, \ldots, n - r \end{array} \tag{5.311b}$$

Here we have denoted

$$p_{kj}(x) = \frac{\delta y_k(x)}{\delta \eta_j} \tag{5.312}$$

The system of equations, which must be satisfied at a branching point, is

$$F_i(\eta_1, \ldots, \eta_{n-r}, \alpha) = \sum_{j=1}^{n} b_{ij} y_j(b, \eta, \alpha) - d_i = 0 \qquad i = 1, 2, \ldots, n - r$$

(5.313a)

and

$$F_{n-r+1}(\eta_1, \ldots, \eta_{n-r}, \alpha) = \det G(\eta_1, \ldots, \eta_{n-r}, \alpha) = 0 \qquad (5.313b)$$

where the elements $g_{im} = \partial F_i / \partial \eta_m$ of the Jacobian matrix G are

$$g_{im} = \sum_{j=1}^{n} b_{ij} p_{jm}(b, \eta, \alpha) \qquad i, m = 1, 2, \ldots, n - r \qquad (5.314)$$

To adopt the Newton method it is necessary to evaluate the second derivatives of the functions F_i (see Section 5.6.3). To develop the equations for the second derivatives, we differentiate the set of differential equations (5.306) and (5.310) with respect to α and η_k;

$$q_i' = \sum_{j=1}^{n} \frac{\delta f_i}{\partial y_j} q_j + \frac{\delta f_i}{\partial \alpha} \qquad i = 1, \ldots, n \qquad (5.315a)$$

$$p_{ijk}' = \sum_{m=1}^{n} \left[\frac{\partial f_i}{\partial y_m} p_{mjk} + \sum_{s=1}^{n} \frac{\partial^2 f_i}{\partial y_m \partial y_s} p_{sk} p_{mj} \right]$$

$$i = 1, 2, \ldots, n \quad j, k = 1, 2, \ldots, n - r, \quad j \leq k; \quad p_{ijk} = p_{ikj}$$

(5.315b)

$$q_{ik}' = \sum_{j=1}^{n} \left[\sum_{s=1}^{n} \frac{\partial^2 f_i}{\partial y_j \partial y_s} q_s p_{jk} + \frac{\partial f_i}{\partial y_j} q_{jk} + \frac{\partial^2 f_i}{\partial y_j \partial \alpha} p_{jk} \right]$$

$$i = 1, \ldots, n; \quad k = 1, \ldots, n - r$$

(5.315c)

The initial conditions are

$$q_i(a) = 0, \qquad p_{ijk}(a) = 0, \qquad q_{ik}(a) = 0 \qquad (5.316)$$

Here we have denoted

$$q_i(x) = \frac{\partial y_i(x)}{\partial \alpha}, \qquad p_{ijk}(x) = \frac{\partial^2 y_i(x)}{\partial \eta_j \partial \eta_k}, \qquad q_{ik}(x) = \frac{\partial^2 y_i(x)}{\partial \eta_k \partial \alpha} \qquad (5.317)$$

Clearly, the second and first derivatives are

$$\frac{\partial^2 F_i}{\partial \eta_m \partial \eta_k} = \sum_{j=1}^{n} b_{ij} p_{jmk}(b, \eta, \alpha) \qquad i, m, k = 1, 2, \ldots, n - r \qquad (5.318a)$$

$$\frac{\partial^2 F_i}{\partial \eta_m \partial \alpha} = \sum_{j=1}^{n} b_{ij} q_{jm}(b, \eta, \alpha) \qquad i, m = 1, 2, \ldots, n - r \qquad (5.318b)$$

$$\frac{\partial F_i}{\partial \eta_m} = \sum_{j=1}^{n} b_{ij} p_{jm}(b, \eta, \alpha) \qquad (5.319a)$$

$$\frac{\partial F_i}{\partial \alpha} = \sum_{j=1}^{n} b_{ij} q_j(b, \eta, \alpha) \qquad (5.319b)$$

We have all variables necessary for making use of the Newton method (5.302) where $N = n - r$, $x_i = \eta_i$, see Section 5.6.3. To recapitulate the algorithm, the following steps are to be performed:

1. Initial guess of $\eta_1^0, \ldots, \eta_{n-r}^0, \alpha^0$: set $k = 0$.
2. Integrate equations (5.306), (5.310), and (5.315) with initial conditions (5.308), (5.309), (5.311), and (5.316) from $x = a$ to $x = b$. The overall number of differential equations to be integrated is $n(n - r + 1)(2 + \dfrac{n - r}{2})$.
3. Calculate the values F_i, $i = 1, 2, , \ldots, n - r + 1$, from (5.313).
4. Determine the partial derivatives defined by (5.318), (5.319), and (5.305).
5. Perform one step of the Newton method with an appropriately chosen λ.
6. If $\| \eta^{k+1} - \eta^k \| + |\alpha^{k+1} - \alpha^k| > \epsilon$, set $k = k + 1$ and go to step 2. ϵ is the prescribed tolerance.

The method can be extended easily to problems with nonlinear boundary conditions containing at $x = b$ the parameter α, that is, instead of (5.307b), the conditions

$$\varphi_i(y_1(b), \ldots, y_n(b), \alpha) = 0 \qquad i = 1, 2, \ldots, n - r \qquad (5.320)$$

are considered. Now (5.313a), (5.314), (5.318a), (5.318b), (5.319a), and (5.319b) have to be rewritten on considering (5.320):

$$g_{im} = \frac{\partial F_i}{\partial \eta_m} = \sum_{j=1}^{n} \frac{\partial \varphi_i}{\partial y_j(b)} p_{jm}(b) \qquad (5.314'; \; 5.319a')$$

$$\frac{\partial^2 F_i}{\partial \eta_m \, \partial \eta_k} = \sum_{j=1}^{n} \left[\left(\sum_{s=1}^{n} \frac{\partial^2 \varphi_i}{\partial y_j(b) \, \partial y_s(b)} p_{sk}(b) \right) p_{jm}(b) + \frac{\partial \varphi_i}{\partial y_j} p_{jmk}(b) \right] \qquad (5.318a')$$

$$\frac{\partial^2 F_i}{\partial \eta_m \, \partial \alpha} = \sum_{j=1}^{n} \left[\left(\sum_{s=1}^{n} \frac{\partial^2 \varphi_i}{\partial y_j(b) \, \partial y_s(b)} q_s(b) + \frac{\partial^2 \varphi_i}{\partial y_j(b) \, \partial \alpha} \right) p_{jm}(b) \right.$$
$$\left. + \frac{\partial \varphi_i}{\partial y_j(b)} q_{jm}(b) \right] \qquad (5.318b')$$

$$\frac{\partial F_i}{\partial \alpha} = \sum_{j=1}^{n} \frac{\partial \varphi_i}{\partial y_j(b)} q_j(b) + \frac{\partial \varphi_i}{\partial \alpha} \qquad (5.319b')$$

For mixed two-point boundary conditions,

$$\varphi_i(y_1(a), \ldots, y_n(a), y_1(b), \ldots, y_n(b), \alpha) = 0 \qquad i = 1, 2, \ldots, n \qquad (5.321)$$

it seems convenient to choose all initial conditions

$$y_i(a) = \eta_i \qquad i = 1, 2, \ldots, n \qquad (5.322)$$

and the system of n nonlinear equations

$$F_i(\eta_1, \ldots, \eta_n, \alpha) = \varphi_i(\eta_1, \ldots, \eta_n, y_1(b, \eta, \alpha),$$
$$\ldots, y_n(b, \eta, \alpha), \alpha) = 0 \qquad i = 1, 2, \ldots, n \qquad (5.323)$$

can be handled in a similar way. The method is again both straightforward and simple; however, the dimensionality may be a limiting factor for large systems of equations. The procedure described is illustrated next for an example arising in the theory of explosion of solid materials.

Example 5.21

The explosion of solid explosives is described by

$$\frac{d^2\theta}{dx^2} + \frac{a}{x}\frac{d\theta}{dx} + \alpha \exp\left[\frac{\theta}{1 + \theta/\gamma}\right] = 0 \tag{5.324}$$

$$\frac{d\theta(0)}{dx} = 0, \qquad \text{Nu } \theta(1) + \frac{d\theta(1)}{dx} = 0 \tag{5.325}$$

Here γ, Nu, and $a \, (= 0,1,2)$ are physical parameters and θ is a dimensionless temperature. To solve (5.324) and (5.325), we guess the missing initial condition at $x = 0$:

$$\theta(0) = \eta \tag{5.326}$$

The auxiliary equations (5.276), (5.277), (5.283), and (5.284) are solved together with (5.324) to evaluate the values of branching points. The resulting α^+ and η^+ for a sequence of values of the parameter Nu are shown in Table 5-34. Two branches of the solution of the boundary value problem (5.324) and (5.325) coincide for $\alpha = \alpha^+$ and disappear for $\alpha > \alpha^+$.

TABLE 5-34
BRANCHING POINTS FOR
EXAMPLE 5.21: $\gamma = 20$, $a = 2$

Nu	α^+	η^+
100	3.457	1.820
50	3.389	1.820
20	3.196	1.815
10	2.906	1.801
5	2.430	1.753
2	1.564	1.577
1	0.951	1.400
0.5	0.526	1.271

PROBLEMS

1. An explosion of solid explosives requires us to evaluate the branching point of the nonlinear boundary value problem:

$$\theta'' + (a/x)\theta' = -\delta \exp\left(\frac{\theta}{1 + \theta/\gamma}\right)$$

$$x = 0: \qquad \theta' = 0$$

$$x = 1: \qquad \nu\theta + \theta' = 0$$

Here a, δ, γ, and ν are parameters. Use the GPM technique for evaluation of the branching point.

Reference: KUBÍČEK, M., AND HLAVÁČEK, V.: *J. Comp. Phys.* 17, 79 (1975).

Results ($\gamma = 40$):

	$a = 0$		$a = 1$		$a = 2$	
ν	δ	$\theta(0)$	δ	$\theta(0)$	δ	$\theta(0)$
100	0.885	1.253	2.016	1.467	3.352	1.706
20	0.819	1.252	1.866	1.465	3.100	1.702
5	0.636	1.240	1.433	1.435	2.359	1.646
0.5	0.161	1.123	0.333	1.169	0.512	1.201

2. Approximate the differential operator in Problem 1 by finite differences calculated from three mesh points at $x = 0$, $x = \bar{x}$, and $x = 1$. Take $\bar{x}$ as a first root of Jacobi polynomial ($\bar{x} = 0.4472$, 0.5773, or 0.6547 for $a = 0$, 1 or 2, respectively). For this simple approximation, calculate the branching points and compare these approximate results with the results of Problem 1.

3. The differential equations governing the finite bending of circular tubes can be written as

$$\frac{d^2\beta}{d\xi^2} = \alpha^2 f \sin(\xi + \beta)$$

$$\frac{d^2 f}{d\xi^2} = \cos(\xi + \beta)$$

subject to the boundary conditions

$$\frac{df(0)}{d\xi} = \beta(0) = f\left(\frac{\pi}{2}\right) = \beta\left(\frac{\pi}{2}\right) = 0$$

The results are usually expressed in terms of a dimensionless moment m:

$$m = -\frac{4\alpha}{\pi} \int_0^{\pi/2} f \cos(\xi + \beta)\, d\xi$$

Using differentiation with respect to an actual parameter, calculate the dependence of the dimensionless moment m on the value α.

Reference: NA, T. Y., AND TURSKI, C. E.: *Aeronaut. Q.* (February 1974), p. 14.

4. The Falkner-Skan boundary layer similarity profiles are governed by the equations

$$f''' + ff'' + \beta(1 - f'^2) = 0$$
$$f(0) = f'(0) = 0, \qquad f'(\infty) = 1$$

Calculate the dependence of $f''(0)$ on the value of the parameter β.

Reference: RUBBERT, P. E., AND LANDAHL, M. T.: *Phys. Fluids.* 10, 831 (1967).

Results:

β	$f''(0)$
-0.10	0.3193
-0.14	0.2398
-0.16	0.1908
-0.18	0.1287
-0.195	0.0556

5. The Hartmann problem for the viscous laminar flow of an electrically conducting liquid between parallel walls with a transverse magnetic field is described by

$$\frac{d\theta}{d\xi} = \psi$$

$$\frac{dQ}{d\xi} = \Phi$$

$$\frac{d\psi}{d\xi} = Nv\Phi - M^2v\left[P^2 + \frac{M^2(P-J)^2 \cosh 2M}{\sinh^2 M} - \frac{2MP(P-J)\cosh M}{\sinh M}\right]$$

$$\frac{d\Phi}{d\xi} = 3\omega^2 Q + 16\omega\theta^3\psi$$

subject to the boundary conditions

$$\xi = -1: \quad \theta = 1, \quad \Phi = \omega\left(\frac{4}{\epsilon_1} - 2\right)Q$$

$$\xi = 1: \quad \theta = \vartheta, \quad \Phi = -\omega\left(\frac{4}{\epsilon_2} - 2\right)Q$$

Here N, v, M, P, J, ω, ϵ_1, and ϵ_2 are parameters. Calculate the dependence of $\theta(0)$ and $Q(-1)$ on M for $N = 0.01$, $v = 5$, $J = 1$, $\omega = 0.1$, and $\epsilon_1 = \epsilon_2 = 1$. P is given by

$$P = J + \frac{1}{M \coth M - 1}$$

Reference: HELLIWELL, J. B.: *J. Eng. Math.* 7, 347 (1973).

6. Diffusion with an autocatalytic reaction (Brusselator model) may be described by

$$\frac{D_x}{L^2}x'' + f(x, y) = 0$$

$$\frac{D_y}{L^2}y'' + g(x, y) = 0$$

with respect to the boundary conditions

$$\text{BC1:} \quad x(0) = x(1) = \bar{x}, \quad y(0) = y(1) = \bar{y}$$

or

$$\text{BC2:} \quad x'(0) = x'(1) = 0, \quad y'(0) = y'(1) = 0$$

Here

$$f(x, y) = A + x^2y - (B + 1)x, \qquad \bar{x} = A$$

$$g(x, y) = Bx - x^2y, \qquad\qquad \bar{y} = \frac{B}{A}$$

$$A = 2, \qquad D_x = 0.0016, \qquad D_y = 0.008$$

(a) Show that for $B = 3.7$, BC2, and $L = 0.2$, a unique solution exists.
(b) Show that for $B = 3.7$, BC2, and $L = 0.25$, three solutions exist.
(c) Show that for $B = 4.6$, BC2, and $L = 0.1$, three solutions exist, whereas for $L = 0.2$ five solutions are possible.
(d) Show that for $B = 4.6$, BC1 and $L = 0.1$, three solutions occur; for $L = 0.2$, seven solutions are possible; and for $L = 0.3$, nine solutions may be found. Use the GPM method.
(e) Calculate the branching points ($B = 4.6$):

BC2 and BC1: $x(z) \equiv \bar{x}, \qquad y(z) \equiv \bar{y}, \qquad L = 0.079844, 0.221112$

BC1: $x'(0) = 6.274, \qquad y'(0) = -2.984, \qquad L = 0.1698$

References: KUBÍČEK, M., RÝZLER, V., AND MAREK, M.: *Biophys. Chem.* 8, 235 (1978)
KUBÍČEK, M., AND MAREK, M.: *Appl. Math. Comput.* 5, 253 (1979)

7. Finite deflections of a nonlinearly elastic bar are described by the following system of differential equations:

$$\frac{du}{dz} = \sin \theta$$

$$\frac{d\theta}{dz} = \psi$$

$$\frac{d\lambda}{dz} = 0$$

$$\frac{d\psi}{dz} = -\frac{p^2(1 - \epsilon)(\pi^2/4)\lambda \sin \theta}{p^2(1 - \epsilon) - (\pi^2\lambda u/4 + \epsilon\psi)^2}$$

subject to the boundary conditions

$$u(0) = 0, \qquad \psi(0) = 0, \qquad \theta(0) = \alpha, \qquad \theta(1) = 0$$

Here ϵ, α, and p are physical constants. Calculate the dependence of the load parameter λ on p. Consider $\alpha = 40°$ and $\epsilon = 0.25$. It is known that for $p = 0.1789$, $\lambda = 0.6380$.
Results:

p	0.1789	0.1951	0.2260	0.2772	0.5180
λ	0.6380	0.6469	0.6634	0.6900	0.8012

Reference: NA, T. Y., AND KURAJIAN, G. M.: *AIAA J. 13*, 220 (1975).

BIBLIOGRAPHY

Differentiation with respect to actual parameter or boundary condition is described by:

RUBBERT, P. E., AND LANDAHL, M. T.: Solution of nonlinear flow problems through parametric differentiation. *Phys. Fluids 10*, 831 (1967).

NA, T. Y., AND TURSKI, C. E.: Solution of the nonlinear differential equations for finite bending of a thin-walled tube by parameter differentiation. *Aeronaut. Q.*, February 1974, p. 14.

TAN, C. W., AND DIBANO, R.: A study of the Falker—Skan problem with mass transfer. *AIAA J. 10*, 923 (1972).

NARAYANA, C. C., AND RAMAMOORTHY, P.: Compressible boundary layer equations solved by the method of parameter differentiation. *AIAA J. 10*, 1085 (1972).

NA, T. Y., AND KURAJIAN, G. M.: Solution of eigenvalue problem in solid mechanics employing parameter differentiation. *AIAA J. 13*, 220 (1975).

NATH, G.: Solution of nonlinear problems in magnetofluiddynamics and non-Newtonian fluid mechanics through parameter differentiation. *AIAA J. 11*, 1429 (1973).

JISCHKE, M. C., AND BARON, J. R.: Application of the method of parametric differentiation to radiative gasdynamics. *AIAA J. 7*, 1326 (1969).

RUBBERT, P. E., AND LANDAHL, M. T.: *AIAA J. 5*, 470 (1967).

GOLDBERG, J. E., AND RICHARDS, R. H.: Analysis of nonlinear structures: *J. Struct. Division, Proc. ASCE 91*, 57 (1965).

KUBÍČEK, M., AND HLAVÁČEK, V.: Solution of nonlinear boundary value problems: III. A novel method: differentiation with respect to an actual parameter. *Chem. Eng. Sci. 26*, 705 (1971).

KUBÍČEK, M., HOLODNIOK, M. AND HLAVÁČEK, V.: Calculation of flow between two coaxial rotating disks by differentiation with respect to an actual parameter. *Comput. Fluids 4*, 51 (1976).

KUBÍČEK, M., AND HLAVÁČEK, V.: Solution of nonlinear boundary value problems: VII. A novel method: differentiation with respect to boundary condition. *Chem. Eng. Sci. 28*, 1049 (1973).

For calculation of parametric dependence via the GPM concept, see:

KUBÍČEK, M., AND HLAVÁČEK, V.: Solution of nonlinear boundary value problems: Va. A novel method: general parameter mapping (GPM). *Chem. Eng. Sci. 27*, 743 (1972).

KUBÍČEK, M., AND HLAVÁČEK, V.: Solution of nonlinear boundary value problems: Vb. Predictor–corrector GPM method. *Chem. Eng. Sci. 27*, 2095 (1972).

KUBÍČEK, M., AND HLAVÁČEK, V.: General parameter mapping technique—a procedure for solution of nonlinear boundary value problems depending on an actual parameter. *J. Inst. Math. Anal. 12*, 287 (1973).

HLAVÁČEK, V., AND VAN ROMPAY, P.: Calculation of parametric dependence and finite-difference methods, *AICHE Journal*, (in press).

KELLER, H. B.: Continuation methods in computational fluid dynamics, in *Numerical and Physical Aspects of Aerodynamic Flow* (Ed. T. Cebeci), Springer-Verlag, New York, 1981.

For transformation of boundary value problems to initial value problems using the parameter mapping technique, see:

KLAMKIN, M. S.: On the transformation of a class of boundary value problems into initial value problems for ordinary differential equations. *SIAM Rev. 4*, 43 (1962).

KLAMKIN, M. S. Transformation of boundary value problems into initial value problems. *J. Math. Anal. Appl. 32*, 308 (1970).

NA, T. Y.: Further extension on transforming from boundary value to initial value problems. *SIAM Rev. 10*, 83 (1968).

LIN, S. H., AND FAN, L. T.: Examples of the use of the initial value method to solve nonlinear boundary value problems. *AICHE J. 18*, 654 (1972).

KUBÍČEK, M. AND HLAVÁČEK, V.: Solution of Troesch's two-point boundary value problem by shooting technique. *J. Comp. Phys. 17*, 95 (1975).

For evaluation of branching points, see:

KELLER, H. B.: Numerical solution of bifurcation and nonlinear eigenvalue problems, in *Application of Bifurcation Theory* (Ed. P.H. Rabinowicz), Academic Press, New York, 1977.

MITTELMANN, H. D., AND WEBER, H.: Numerical methods for bifurcation problems— A survey and classification, in *Bifurcation and Their Numerical Solution* (Eds. H.D. Mittelmann, and H. Weber), Birkhauser-Verlag, Basel (1980).

SIMPSON, R. B.: A method for the numerical determination of bifurcation states of nonlinear systems of equations. *SIAM J. Numer. Anal. 12*, 439–451 (1975).

STAKGOLD, I.: Branching of solutions of nonlinear equations. *SIAM Rev. 13*, 283–332 (1971).

BAUER, L., KELLER, H. B., AND REISS, E. L.: Multiple eigenvalues lead to secondary bifurcation. *SIAM Rev. 17*, 101–122 (1975).

KUBÍČEK, M., AND HLAVÁČEK, V.: Direct evaluation of branching points for equations arising in the theory of explosions of solid explosives. *J. Comp. Phys. 17*, 79 (1975).

KUBÍČEK, M., AND HLAVÁČEK, V.: Solution of nonlinear boundary value problems: VIII. Evaluation of branching points based on shooting method and GPM technique. *Chem. Eng. Sci. 29*, 1695 (1974).

KUBÍČEK, M., AND HLAVÁČEK, V.: Solution of nonlinear boundary value problems: IX. Evaluation of branching points based on differentation with respect to boundary condition. *Chem. Eng. Sci. 30*, 1439 (1975).

KUBÍČEK, M.: Evaluation of branching points for nonlinear boundary value problems based on the GPM technique. *Appl. Math. Comput. 1*, 341 (1975).

KUBÍČEK, M., AND MAREK, M.: Evaluation of limit and bifurcation points for algebraic and nonlinear boundary value problems. *Appl. Math. Comput. 5*, 253 (1979).

SEYDEL, R.: Numerical computation of branch points in ordinary differential equations, *Numer. Math. 33*, 51 (1979).

A survey of numerical techniques used in the bifurcation theory is:

KUBÍČEK, M., AND MAREK, M.: *Computational Methods in Bifurcation Theory and Dissipative Structures*, Lecture Notes in Computational Physics, Springer-Verlag, Berlin, to be published.

Index